REASON AND
WONDER

Figure 1: Gaseous Pillars in M16: from the Hubble Space Telescope. (NASA, ESA, STScI, J. Hester and P. Scowen, Arizona State University.)

REASON AND WONDER

*A Copernican Revolution
in Science and Spirit*

DAVE PRUETT

ALBA ENTERPRISES LLC
Harrisonburg, Virginia

Reason and Wonder

Publisher Cataloging-in-Publication Data

Pruett, Dave (Charles David), 1948 –
Reason and wonder: a Copernican revolution in science and spirit / Dave Pruett.
Includes bibliographical references and index.
Paperback with new Introduction, by permission of the hardback publisher.
ISBN: 978-0-692-56874-3 (Paperback)
1. Religion and science. 2. Cosmology. 3. Metaphysics. 4. Mind, Body, Spirit.

Typeset by the author using the MEMOIR *class of* LATEX.
Front cover design by Silverander Communications.
Spine and back cover design by the author.
Cover photo: NASA, Hubble Space Telescope image of "Mystic Mountain."

Hardback and ebook editions published in May 2012 by Praeger, an imprint of
ABC-CLIO LLC, 130 Cremona Drive, P.O. Box 1911, Santa Barbara, CA 93116.
ISBN: 978-0-313-39919-0 (Hardback)
EISBN: 978-0-313-39920-6 (Ebook)
Visit www.abc-clio.com for details regarding purchase of the ebook.

Printed in the United States of America by IngramSpark.

To Connie Kay Nelson Harkey (1948-1991),
for unwavering faith in her friends,
especially this one.

Contents

List of Figures

Preface

Awe is the beginning of wisdom.

– Rabbi Heschel

Enchantment is the natural state of children. It is common among the aboriginal peoples of the world (Fig. 2), and also among religious mystics. Alas, among Western adults, enchantment is all too rare. Insulated from nature, bombarded by technology, compulsively busy, and accustomed to noise, we leave precious little space for wonder. If there is a universal criticism of the Western world, it is that "we have lost our awe."[1]

Years ago, I met a "wisdomkeeper."[2] Thereafter, I developed the habit of dropping in on my Native American friend and his wife whenever my

Figure 2: The enchantment of aboriginal humans is displayed in their rock art. (James Q. Jacobs, Archaeo Art Media, jqjacobs.net.)

travels took me near their small Indian reservation in Tidewater Virginia. Arranged visits seldom seemed to pan out, so I took to dropping in unannounced. Serendipitous visits, by contrast, often turned magical.

Despite the cramped quarters of a small, artifact-cluttered house, Sun Eagle and Gentle Wind were gracious and playful hosts, who welcomed and celebrated guests, customarily by drumming, singing, and storytelling. Blessing songs were especially powerful. Sun Eagle would first purify the room with sacred smoke from burning sage and sweet grass. Next followed the "heartbeat" of a deerskin drum he had crafted and Gentle Wind had graced with symbol-rich art. And then came the voices: haunting chants in harmony. Individualized for each guest – male or female, youth or elder, parent or child – blessing songs could evoke tears and usually did.

I once took four university students to the reservation, on the return from a conference in Hampton, Virginia. Time constraints permitted only an hour's detour. But even an hour in the presence of this deeply spiritual couple left its mark. Back on the road, many minutes elapsed before anyone dared to speak, each afraid to break the spell that had been cast upon us. When the silence finally lifted, a student confessed, "Never in my life have I felt so completely welcomed."[3] Such moments pierce a veil of which we are only dimly aware.

By then I had unexpectedly uncovered the source of the sudden emotion that occasionally overpowers Westerners in the presence of wisdom-keepers. The epiphany came upon reading a magazine article by cultural historian Richard Tarnas. In "The Great Initiation," Tarnas outlines the dilemma of the West. On the one hand, we live in a scientific age that promises the continual betterment of humankind. On the other hand, the materialism of the scientific paradigm has largely stripped our lives of meaning by breaking our connection to the anima mundi, the world soul that pervades the mythologies of indigenous peoples. Tarnas invokes the apt phrase "disenchantment of the universe" to name the subliminal loss of myth that afflicts the psyche of most Westerners. Our technological progress has been purchased at great price, for which we are only beginning to account.

The most endearing quality of Sun Eagle's stories was their message of respect: respect for mother earth, respect for her manifold creatures, respect even for the stones. My favorite of Sun Eagle's stories was about fishing. The details are lost to memory, but it contained the oft-repeated refrain, "So I thanked the worm for helping me to catch the fish." The simple refrain captures two defining qualities of Sun Eagle, Gentle Wind, and most Native Americans I have encountered: respect for all things, living and non-living, no matter how seemingly lowly, and gratitude to the Great Spirit of the universe.

Somehow, I always thought I'd be able to drive to the Rez for an infusion of enchantment when supplies were running low. It was not to be. Sun Eagle passed to the spirit world in November 2004 at the age of sixty-one. All those who felt the embrace of his bear hug or his huge spirit miss him sorely. Sun Eagle taught me many things. Most of all, he taught me that inside each disenchanted Westerner lies an inner child, eager to crawl into the lap of mother earth, filled with reverent awe for the Great Holy Mystery, and longing for stories that will reconnect him or her to the world soul.

C. DAVID PRUETT
Harrisonburg, Virginia
October 2011

Acknowledgments

Acknowledgments for the Hardback Edition (2012)

Writing a book is as close as a man gets to giving birth. I would like to first thank the midwives for their dexterous blend of encouragement and forceps: Margot and Vitae Bergman, Bob Bersson, Rusty Carlock, Katherine Kross, Ed Parker, Karen Risch, Fritz Rosebrook, and Louise Temple-Rosebrook.

To each of the following a hearty thanks for sundry and timely gifts – references, insights, suggestions, corrections, inspiration, or encouragement: Bobbi Sue Bales, David Brakke, Jorge Bruno, Andy Currant, Taz Daughtrey, Eileen Dight, Lois Carter Fay, Fred Fiederlein, Charlie Finn, Joanne Gabbin, Peyman Givi, John Goerdt, Jane Hagaman, Tom Hawkins, Gregg Henriques, Emily Iekel, Bill Ingham, Steve Keffer, Ursula King, Peter Kohn, Metro Lazorack, Terry Una Lee, Noel Lyons, Judy McConnell, Allison Pesce Monroe, Jon Monroe, Katie Pesce Monroe, Jill Moses, Dave Moyer, Susan Myers, Andrea Pesce, Maria Prytula, Henry and Janis Reed, Uli Rist, Ken Rutherford, Sue Savage-Rumbaugh, Jim Sochacki, Kate Monger Stevens, Sharon Rae Sun-Eagle, Michelle Surat, Cheryl Talley, Bill Voige, and Penelope Yungblut.

To psychologist Lawrence LeShan and mathematical physicist Freeman Dyson – whose thoughts grace many of these pages and whose worlds are not that far apart – go my deepest esteem for their rare combination of shining intellect and courageous heart.

To the honors students of James Madison University (JMU) who have accompanied me along segments of this journey go my admiration and appreciation for all you have taught me. Thanks also to the Templeton Foundation and the Center for Theology and the Natural Sciences for stimulating dialogue through their Science-Religion Course Awards and Science and Transcendence Advanced Research Series (STARS) conferences.

To JMU's class of 1958 goes my lasting gratitude for their generous Mengebier Endowed Professorship, a precious gift: time for writing. To the good people of Praeger/ABC-CLIO, I am deeply indebted: to Michael Wilt for having faith in this project, to Beth Ptalis and Erin Ryan for shepherding it to publication, to Jennifer Green for carefully tending the marketing aspects, and to Suba Ramya Nambiaruran for deft and gracious handling of the final typesetting. And to my lucky stars go my awe and appreciation for directing my path to two amazing editors: Aimee Brasseur and Alan Krome.

To my brother William Pruett and sister Barbara Green go an elder brother's gratefulness, respect, and warm affection. To Danny and Edna Pruett, my parents, I owe tender thanks for setting me, accidentally, upon this adventure of a lifetime. Love you, Mom (1923-2011).

Finally, my deepest gratitude goes to two extraordinary spiritual guides, both now departed – John Sun Eagle and John Yungblut – as it does also to my wife, Suzanne, for her constant support in this and all endeavors, and to our daughter, Elena, the light of our lives.

Acknowledgments for the Paperback Edition (2015)

Try as one might to produce an error-free manuscript, "to err is human."

For this second chance, I'm grateful, especially to Margot Bergman, Suzanne Fiederlein (my wife), James Leary, and Walter Shropshire, who collectively caught numerous errors in the hardback edition. Walter, in particular, performed yeoman's service by providing not only a review of *Reason and Wonder* for the online science-and-religion site WesleyNexus but also two full pages of errata. Moreover, Walter introduced me to Lawrence Fagg, sections of whose book *The Becoming of Time* (2003) share uncanny resemblances to parts of *Reason and Wonder*. Strikingly, in his first chapter, Larry outlines, as do I, the advancement of science through three successive revolutions: the first due to Copernicus, the second to Darwin, and the third – which may bring science and religion back into harmony – in ferment. After meeting, Larry and I exchanged copies of our books, each inscribed serendipitously to a "kindred spirit." Fittingly, *The Becoming of Time* has been added to the bibliography of *Reason and Wonder*, even though I remained unaware of Larry's insightful book when *Reason and Wonder* was published in 2012.

Recently, memory has served up a glaring omission in the original acknowledgments: world-renowned religion scholar Huston Smith, whom I met briefly at a conference in 2000. Smith's sage advice about writing struck home, launching this would-be author.

For the second time, I'm indebted to Michael Wilt, who gave *Reason and Wonder* its first life in hardback and proofread its reincarnation in paperback. Thanks also to Anthony Chiffolo, Editorial Director, Print, Praeger/ABC-CLIO, for generous permissions to revise the Introduction and to reuse the hardback's cover design for the paperback.

Finally, between the publication dates of the hardback and paperback, my father, Danny Pruett (1923-2014) – small-town physician, World War II infantryman, and a liberator of the Nazi concentration camp at Dachau – passed away. Miss you, Pop. You were my North Star.

Introduction

> We are in trouble just now because we do not have a good story. We
> are between stories.
>
> **– Father Thomas Berry**

Our species – *Homo sapiens sapiens* – is endangered.

Never before has humankind faced simultaneous crises on so many
fronts: over-population; a rapidly deteriorating biosphere; competition
for oil, water, and arable land; wars and rumors of wars; failed nation
states; unprecedented extremes of wealth and poverty; terrorism on a
global scale; and a climate on the brink of spinning out of control.

It seems utterly naive to suggest, as does eco-theologian Thomas Berry
(1914-2009), that our woes stem from an inadequate "story." By "story"
Berry means "mythology": a grand meta-narrative that guides our relationships with the creator, with one another, and with all creation.

On second thought, what could be more vital than getting *this* story
right?

In the West especially, our prevailing mythology is woefully flawed.
It's flawed because it's incomplete. And it's incomplete because, as Berry
recognizes, "we are between stories." We have yet to integrate into a seamless narrative the scientific and religious stories of our origins.

From time immemorial, religion informed our guiding myths. "Be
fruitful, multiply, fill the earth, and subdue it!" commands God in Genesis
1:28. "Be masters over the fish in the ocean, the birds that fly, and every
living thing that crawls on the earth!" Mainstream religious traditions (in
contrast to indigenous ones) have long taught that humans arose by divine
fiat, were made in the image of God, occupy the center of a static cosmos,
and exist on an earth created expressly to satisfy our needs.

The religious worldview went largely unquestioned until 1543, when
Nicolaus Copernicus (1473-1543) – a Polish astronomer, mathematician,
and cleric – upset the mythological apple cart. By shifting from a geocentric cosmological perspective to a heliocentric one, Copernicus literally
made the earth move. All hell broke loose. "What were we doing," Nietzsche would later lament in *The Parable of the Madman* (1882), "when we
unchained this earth from its sun?"

"Of all discoveries and opinions," observed Goethe, "none may have exerted a greater effect on the human spirit than the doctrine of Copernicus."[1] The deathbed publication of Copernicus' masterpiece, *On the Revolutions of the Heavenly Spheres*, launched a revolution in cosmology that dislodged humans from the center of the cosmos, exploded the size of the known universe, turned a static cosmos into a dynamic one, and called into question the religious story. Big Bang cosmology and the enthralling deep-space images of the Hubble Space Telescope are but twentieth- and twenty-first-century reverberations of the seismic paradigm shift Copernicus initiated nearly five centuries ago. Moreover, Copernicanism – in the skilled hands of Galileo, Kepler, and Newton – gave rise to modern science. The Age of Reason, the industrial age, the nuclear age, the space age, and the information age all followed in rapid succession.

Today, in deference to Copernicus, "Copernican" revolutions refer to paradigm shifts, originating in science, with profound implications for human self-perceptions. Copernicus was the first scientific revolutionary but hardly the last.

While the Copernican revolution was mid-stride, the publication in 1859 of another scientific work rocked Victorian England, sending intellectual shock waves around the world. Prior to Darwin's *Origin of Species*, humans sat atop the pinnacle of divine creation. *Origin* and its sequel, the *Descent of Man* (1871), displaced humans from that seat of high honor and relegated us to one branch of a "tree of life," no different in kind from other living organisms, only different in degree. Adding insult to injury, Darwin's theory hinted that *Homo sapiens* arose by fortuitous happenstance among random events rather than by inspired design.

One hundred fifty years hence, humanity still hasn't made peace with *Origin*, as attested by continuing controversies over genetic engineering, embryonic stem-cell research, "intelligent design," and the teaching of evolution in public schools.

The back-to-back punches thrown by Copernicus and Darwin disfigured the human face in the mirror of self-perception. Are we not what we thought we were: the central focus of the physical and biological universes? The message from science is strangely dissonant to that of religion, which yet proclaims our divine origins and exalted status. It's like having two parents, one who underscores our uniqueness and the other our commonness. Which are we to believe?

Following Galileo, science and religion separated. After Darwin, they divorced.[2] Science has thrived, but humans, the children of that divorce, have suffered what cultural historians term "disenchantment of the universe."

Whether at the hand of science or of religion, nature has been diminished from "thou" to "it." On the one hand, science's materialism has destroyed "the ancient alliance" with nature; "man knows at last that he is alone in the universe's indifferent immensity out of which he emerged only by chance."[3] By demoting nature to an "it" to be conquered rather than a "thou" to be revered, science has desacralized the world.

On the other hand, by exalting the human, to whom all creatures and all of creation are deemed subservient, traditional Judaic-Christian-Islamic religion has largely absolved man from his responsibility as creation's caretaker and has unraveled the web of life. Nature is to be exploited for man's benefit and his alone. And by its obsession with the soul of man, traditional religion has discounted as pagan the *anima mundi*, the world soul.

And so, the Western psyche faces a stark choice between competing mythologies, neither of which satisfies. The scientific story speaks to our rationalism, but is devoid of meaning. The religious story speaks to our intuition, but denies the facts. "Do we really have to make the tragic choice between an antiscientific philosophy and an alienating science?"[4] pleads Nobel laureate Ilya Prigogine.

The tragedy of the choice is compounded by the current myopia of conventional science and traditional religion, each believing itself to have cornered the market on ultimate truth. "The true disease of the age is . . . literalism," observes mythologist Michael Meade:

> Literalism has two factions that often oppose each other while secretly conspiring to reduce the mystery of the living world. One side champions positivism and a tyranny of scientism that obsesses over facts and figures and relies solely upon a statistical worldview. The opposite extreme insists upon fundamental religious beliefs that reject facts or alter them to conform to literalized stories. Each side gains some surety at the cost of a tragic loss of imagination and a dramatic reduction in the sense of wonder of the immediate world.[5]

"Without wonder the world becomes a marketplace" and everything just a commodity,[6] observes theologian Matthew Fox.

In isolation, science and religion each tend toward myopia, hubris, and dogma. Dialogue is the antidote to such innate – and destructive – tendencies. "Science without religion is lame; religion without science is blind,"[7] Einstein wisely noted. But there's a more basic reason for dialogue: science and religion are kinfolk. Both are grounded in awe.[8]

The separation of these kin into distinct domains of philosophical inquiry has historical roots. Countermanding Plato and Aristotle, the French philosopher-mathematician-scientist René Descartes (1596-1650) divorced matter from mind. Natural philosophy – modern-day science – claimed the former as its domain. Philosophy and religion claimed the latter. An uneasy truce prevailed, seeding the mistrust between science and religion that lingers into the present.

But early in the twentieth century, the Cartesian partition suddenly collapsed, the casualty of yet another scientific revolution: quantum mechanics. At the most fundamental levels, quantum mechanics challenges long-standing assumptions of classical physics, among them that the world is deterministic, mechanistic, and materialistic. The demise of the Cartesian partition has opened to science a frontier that was once taboo: mind. "Understanding the human mind . . . has emerged as the central challenge for science in the twenty-first century."[9] And with this opening comes opportunity that just might save us from ourselves.

While rank-and-file scientists remain largely in the thrall of the scientific materialism of classical mechanics, there are many indications that modern science, of necessity, is slowly softening its stance and opening to dialogue with other modalities of knowing:

- In 1962, Rachel Carson's *Silent Spring* – a classic in which science merges with spirituality – launches the modern environmental movement, raising consciousness worldwide regarding the fragility of planet Earth and the necessity of responsible stewardship of the earth's air, water, soil, and biodiversity.

- Classics such as *Black Elk Speaks* (Neihardt, 1932), the life story of a Lakota shaman, enjoy decades of popularity as Westerners, drawn in ever-increasing numbers to Native American spirituality and ecology, seek to reclaim lost wisdom. Symbolic of the complementary natures of ancient spiritual wisdom and modern scientific knowledge is the juxtaposition on the Washington Mall of the Smithsonian's stunning new National Museum of the American Indian and its ever popular Air and Space Museum.

- Modern medicine cautiously considers alternative and holistic modalities. Unlike conventional allopathic medicine, with its mechanistic focus on disease, holistic medicine acknowledges the synergy of mind, body, and spirit in both the prevention and manifestation of *dis*-ease. In recognition of the influence of mind on body, major medical centers now incorporate mindfulness practices for patients and caregivers.

- New global organizations are springing up to promote "the constructive engagement of science and religion"[10] and to explore multiple (intuitive and rational) modes of knowing. Among these are the Institute of Noetic Sciences, Metanexus Institute, The Center for Theology and the Natural Sciences, and Evolutionary Christianity.

- Perennial conferences – such as the Tucson Consciousness Conference and the SAND (Science and Non-Duality) Conferences – now bring together world-class scientists, spiritual seekers, and interested laypersons to investigate all aspects of science's most perplexing question: What is the nature of consciousness?

- Increasing numbers of scientists call for a more open science. Specifically, the September-October 2014 issue of *Explore – The*

Journal of Science and Healing – contains a bombshell of an essay titled "Manifesto for a Post-Materialist Science." To science perhaps what Luther's *Ninety-Five Theses* were to religion, the article calls for a scientific paradigm shift that "may be even more pivotal than the transition from geocentrism to heliocentrism." The essay follows closely on the heels of a like-minded open letter of January 27, 2014, in *Human Neuroscience*, signed by nearly 100 scientists and neuroscientists and calling for "more open, informed study of all aspects of consciousness," including those that fall into the categorization of "paranormal."

As science reluctantly awakens to the numinous qualities of the universe,[11] official pronouncements from the Church herself herald a new era of dialogue.

At the writing of this revised introduction to *Reason and Wonder*, Pope Francis' remarkable encyclical *Laudato Si'* (August 2015) has just created a global stir. In paragraph 200 of a section titled "Religions in Dialogue with Science," Francis acknowledges that religion at times has jumped the rails of its own principles, and yet he reasons:

Any technical solution [that] science claims to offer will be powerless to solve the serious problems of our world if humanity loses its compass, if we lose sight of the great motivations [that] make it possible for us to live in harmony, to make sacrifices and to treat others well.

Francis is not free-lancing in calling for dialogue. He is building upon the legacies of predecessors.

Three hundred fifty years following the "Galileo Affair" – the incident of 1616-1633 that forced the schism between science and religion that persists to this day – Pope John Paul II commissioned a thorough review of Galileo's encounter with the Inquisition. In 1983, to dress still festering wounds, the pope set a new tone:

The Church is convinced that there can be no real contradiction between science and faith. It is certain that science and faith represent two different orders of knowledge, autonomous in their processes, but finally converging upon the discovery of reality in all its aspects.[12]

The full epistle of John Paul II astounds for many reasons. First, it stresses the complementarity – rather than the antagonism – of science and religion. Second, it grants autonomy to both revelatory processes, implying that neither should seek to manipulate or triumph over the other. Third, it suggests that ultimate truth arises only from the concerted efforts of internal and external explorations. And finally, it outlines the expansion of human understanding by successive stages, informed by both religion and science.

In the same epistle, the former pope makes explicit those broad facets of human understanding deserving of the most intense scrutiny: "[The Church] is always interested in research concerning the knowledge of the universe, whether physical, biological, or psychological."[13]

Analogously, Sigmund Freud observed, "[Darwinism] was the second, the biological, blow to human narcissism,"[14] encapsulating in just ten words nearly five *centuries* of Western intellectual history. Freud's terse summary refers obliquely to the sixteenth century's Copernican revolution, credits Darwin for a second "Copernican" revolution, and shrewdly sets the stage for his own claim to a third.

Taken together, the observations of Freud and Pope John Paul II provide an intriguing outline of how human self-awareness has unfolded during the past half-millennium. An unanticipated revolution in science challenges our naive – even narcissistic – human self-perceptions. Centuries may be required to fully process and integrate these new insights. Then, before we become complacent with new self-awareness, along comes another scientific blow. Like growth rings in the tree of knowledge, each stage pushes our comprehension into wider and wider compass: first to an understanding of our physical place in the cosmos, then to our biological place, and now to our psychic/spiritual place.

In the chapters to follow, we will relive seminal events in Western intellectual and scientific history that have brought us to the dawn of yet newer awareness. The tale unfolds through the stories of three successive "Copernican" revolutions, in chronological order of their respective inceptions.

- Part I – The Copernican Revolution: Cosmology – explores the monumental repercussions of the Copernican shift from geocentric to heliocentric cosmology. Part I culminates in Big Bang cosmology, its vast expansion of human perceptions of the spatial limits of the universe, and its ultimate redefinition of the physical place of humans in the cosmos.

- Part II – The Second "Copernican" Revolution: Biology – explores the life, theory, and legacy of Charles Darwin. Darwin's radical theory rests upon the notion of deep time, thereby exploding human perceptions of the temporal extent of the cosmos as Copernicanism exploded its spatial extent. Part II culminates in the successful search for the genetic code and evolution's ultimate redefinition of the biological place of humans in the cosmos.

- Interlude – Lifting the Veil – explores the advent of quantum mechanics, including the "damned facts" that failed to fit the mold of classical physics, the dual nature of light, the philosophical conundrums that provoked many of quantum's founders to despise the theory, the ultimate collapse of the Cartesian partition, and the notion of entanglement.

- Part III – The Third "Copernican" Revolution: Psychology & Spirituality – projects into the future based upon present scientific and spiritual trends. Like the German titan of philosophy and amateur astronomer Immanuel Kant – who speculated in 1750 that nebulae were in fact "island universes" like our Milky Way, but who lacked the tools to settle the issue – we can only dimly resolve the future. Still, the signs are heartening and point to the reconciliation of reason and wonder, to the ultimate integration of the scientific and religious stories, and to a transformational awareness our psychic/spiritual place within the cosmos.

Within each sub-tale, linear chronological ordering of events is abandoned to favor a more coherent narrative. The integrated story shies from "bleeding edge" science to incorporate mostly what has withstood the test of time. Quantum theory and Big Bang cosmology, for example, are well vetted. String theory or dark energy's role in cosmic inflation are not.

The story would be incomplete without tributes to those rare individuals who stand like mileposts, marking our progress, those who have struggled first to grasp and then to articulate the new paradigms of successive revolutions: Copernicus, Galileo, Kepler, Newton, Kant, Darwin, Einstein, Bohr, Heisenberg, and Schrödinger, among many others. The unique role of each is elucidated.

Since *Reason and Wonder*'s inception in 1999 (as an honors course at James Madison University), several sparkling biographies of key players have surfaced. Among these are *Galileo's Daughter* by Dava Sobel (1999); *Darwin, His Daughter, and Human Evolution* by Randall Keynes (2002); *Newton* by James Gleick (2003); *Kepler's Witch* by James Conner (2004); *Einstein* by Walter Isaacson (2007); and *Max Perutz and the Secret of Life* by Georgina Ferry (2007). If I, a non-historian, have borrowed too heavily from these gifted chroniclers of science, I beg their forgiveness. May the vignettes presented here entice readers to fuller explorations of science's most fascinating – and all too human – characters.

To summarize, perhaps no statement presages the present revelatory epoch better than that uttered by Julian Huxley, grandson of evolution's ardent defender Thomas Huxley:

Man discovers that he is nothing else than evolution become conscious of itself.[15]

In the human, evolution has reached a tipping point. We are presumably the first species – on earth at least – endowed with the ability to piece together the story of our own origins. And what a story it is: a tale of *cosmogenesis*. Cosmogenesis bespeaks a universe that dynamically unfolds for an apparent purpose: the emergence of higher consciousness.

Creation is not a done deal; it's ongoing, and we are participants. What lies around the next bend? Some see transformation. Others see only Armageddon. Both are possible given the universe's predilection for chaos and creativity.

Among the seers of the third Copernican revolution are the French paleontologist-priest Teilhard de Chardin (1881-1955) – who coined the term "cosmogenesis" and about whom much more will be said in later chapters – and the American mythologist Joseph Campbell (1904-1987). From a lifetime of studying the mythologies of the world's cultures, Campbell caught a glimpse of the future that is both sobering and hopeful at this moment of unprecedented peril:

> The world, as we know it, is coming to an end. The world as the center of the universe, the world divided from the heavens, the world bound by horizons in which love is reserved for the members of the in group: that is the world that is passing away. Apocalypse does not point to a fiery Armageddon but to the fact that our ignorance and our complacency are coming to an end.[16]

Above all, *Reason and Wonder* is the story of how we – *Homo sapiens sapiens* – arrived at this critical turning point.

PART I

The Copernican Revolution: Cosmology

Figure 3: Hubble Space Telescope image of Kant's "island universes." (NASA, ESA, J. Blakeslee and H. Ford, Johns Hopkins University.)

One

The Clockmaker and the Clockwork

> I do not know what I may appear to the world; but to myself I seem to
> have been only like a boy, playing on the seashore, and diverting my-
> self in now and then finding a smoother pebble or a prettier shell than
> ordinary, whilst the great ocean of truth lay all undiscovered before
> me.
>
> – Isaac Newton

By any measure, NASA's Apollo program was gargantuan. In scale, the Saturn V launch vehicle dwarfed the Statue of Liberty. To assemble and house the behemoth, NASA constructed the world's largest building. The nation sunk $20 billion into Apollo, a staggering sum in the 1960s, worth twenty times that in today's dollars. And at its zenith, Apollo consumed the expertise of 400,000 engineers, scientists, and mathematicians.

All these efforts focused to a single point: landing a man on the moon. Articulated to a joint session of Congress by President John F. Kennedy in 1961 and transformed into a national obsession, the goal of reaching the moon teetered on the edge of reality by 1969. If successful, Apollo 11 would climax the most daring voyage of discovery since Magellan. It would culminate a four-day journey of a quarter-million miles, eight years of national preoccupation, four and a quarter centuries of scientific progress since Copernicus, and, in a larger sense, 3.5 billion years of evolution on Earth, the unfathomable depth of time needed to produce beings capable of such a feat.

Spaceship Earth

It all came together at 3:17:39.9 p.m. Central Standard Time[1] on July 20, 1969, when the spidery lunar module (LM) of Apollo 11 touched down gently on the moon's barren surface, its thrusters kicking up a powdery dust. The final minute of flight was harrowing in the extreme. Those sixty seconds seemed an eternity to mission controllers, as mission commander Neil Armstrong, perilously low on fuel, deftly guided the LM, code-named *Eagle*, over a boulder-strewn field at the landing site on the moon's Sea of Tranquility.

The waiting world sighed its collective relief as Armstrong uttered the reassuring words, "Houston, Tranquility Base here. The *Eagle* has landed."

A few hours later, abandoning the flight plan's call for sleep, Armstrong and fellow astronaut Edwin "Buzz" Aldrin disembarked the LM. First to step off the vessel's ladder and to place a human footprint in untouched lunar soil, Armstrong offered simple words, now indelibly etched in the human record: "That's one small step for a man, one giant leap for mankind."

And what a leap it was. Merely 360 years from the moment Galileo had raised his telescope in awe at the rough and irregular surface of our nearest celestial neighbor, humans stood in the Sea of Tranquility. More remarkably, less than nine years had elapsed since the historic speech of May 1961 in which a visionary president had committed his country to "achieving the goal, before this decade is out, of placing a man on the moon and returning him safely to Earth."

"We came in peace for all mankind." Apollo 11's LM bore a plaque with these noble words, signed by President Richard Nixon and the three Apollo 11 astronauts. Yet the words could not mask that the moon landing had been born of fierce competitions: the Cold War and the space race. Whoever controlled space in a nuclear age, it was feared, controlled the world. Twice Russia had humiliated the United States: by placing in orbit first an artificial satellite, Sputnik, in 1957, and then a human being, cosmonaut Yuri Gagarin, in 1961. Kennedy's goal of a moon landing "before the decade is out" was cleverly chosen: distant enough to buy time to close the technology gap but complex enough that Russian success was by no means guaranteed. With luck – lots of it – the United States might just leapfrog the Russians.

On that July afternoon, however, nationalism and competition were set aside as 600 million people gathered around the world's televisions to partake in the fulfillment of a dream larger than individuals or nations. For one shining moment the aspirations of astronauts, Americans, and the denizens of the earth merged. In a phone call to the moon, President Nixon summed up the emotions of many: "For one priceless moment in the history of man, all the people of the earth are truly one." Ecstasy and awe, however fleeting, united the human race.

It was basic science that lassoed the moon for humankind. The quintessential achievement of classical physics, Apollo needed none of modern physics: relativity and quantum mechanics. The seventeenth-century legacy of Sir Isaac Newton, classical physics, painted nature as mechanistic and deterministic. Steam engines, the industrial age, the airplane, and rocketry followed. In 1901, 214 years after the publication of Newton's *Principia Mathematica*, which laid the foundations for the scientific era, German physicist Ernst Mach cast Newton's accomplishments in perspective: "All that has been accomplished in mechanics since his day has been a deductive, formal and mathematical development of mechanics on the basis of Newton's laws."[2] In other words, for more than two centuries, the world's best scientific minds had been merely mopping up after the scientific tidal waves Newton unleashed.

For all its grandeur and scale, the Apollo program required little new science beyond what Newton had bequeathed. Apollo's most significant challenges were those of manufacturing, organization, and reliability. Apollo necessitated the assembly of immensely complex machinery – the Saturn V booster, the command and service modules, and the LM – from hundreds of thousands of parts supplied by hundreds of contractors. Yet the end product had to perform its intended function with at least a reasonable chance of success. No one in the know expected success without paying a heavy price somewhere along the line. Apollo 8 astronaut Bill Anders once assessed his chances of complete success, of partial success with survival, and of perishing. With a fighter pilot's no-nonsense savvy, Anders evaluated the probability of each outcome at one in three.[3]

The program was promoted as scientific exploration and sold to the public by touting its spin-offs: everything from improved computer technology to medical advances to photovoltaic cells. In truth, but for the space race, Apollo would likely not have been attempted on its intrinsic merits alone. Still, no one anticipated the program's greatest legacy – a shift in consciousness.

The moment of awakening occurred during Apollo 8, a dress rehearsal for Apollo 11. Its mission plan called for the entire command module with attached LM, carrying astronauts Frank Borman, Jim Lovell, and Bill Anders, to enter lunar orbit without attempting to land. On the first pass around the moon, Anders set to filming the moon's "dark" side, never before witnessed by human eyes. Looking up halfway through a lunar orbit, Anders was stunned to see the earth appear above the lunar landscape (Fig. 4). The contrast was breathtaking: against the grey, cratered, and lifeless moon, the earth seemed to Anders a delicate and exquisitely beautiful marble suspended in the deep blackness of space.[4]

Chosen like Anders for their "right stuff," many of America's Apollo astronauts nevertheless experienced near-mystical states upon viewing their fragile home from the sublime detachment of deep space. With time, Anders' historic photo of earthrise has transformed how humans see their planet. A plaque in front of the Apollo 10 capsule at the British Science Museum in London acknowledges this unanticipated gift to the world:

> [Apollo's] most potent and enduring legacy perhaps lies in the views of earth that the astronauts captured. These "earthrise" photographs resonated with a developing environmental consciousness and helped awaken us to the fragility of our planet.

"The true voyage of discovery," insisted French novelist Marcel Proust, "consists not in seeking new landscapes but in seeing with new eyes."[5]

Figure 4: Earthrise from Apollo 8. (NASA.)

Lunar-Orbit Rendezvous

The Apollo program's primary technological hurdle was orbital mechanics: rendezvous in orbit was key. Initially two approaches were considered. The first mode – direct ascent – called for a brute-force, direct shot to the moon by a rocket of immense size and weight. Few thought it feasible. The second, earth-orbit rendezvous, or EOR for short, tried to finesse the size problem by sending into earth orbit two moderately sized rockets, one containing the astronauts and the other the fuel they would need for the push to the moon. But size was the Achilles heel for this plan as well. The moon-landing phase required a vehicle the size of the eight-story Atlas rocket that had launched American astronaut John Glenn into earth orbit in 1962. With no stable launch pad on the moon, this seemed risky at best.

Enter John Houbolt, a lone-wolf NASA engineer, with a novel but radical idea. The moon landing should be accomplished by a relatively small lunar excursion module, which would then rendezvous in lunar orbit with the remainder of the spacecraft for the return to Earth.[6] Initially ridiculed, the lunar-orbit rendezvous (LOR) surprised planners by working well in simulations and was eventually adopted, thereby greatly reducing the Apollo's takeoff weight.

Although untested, rendezvous in space involved no new science or mechanics. In fact, every phase of the Apollo flight plan, every sensitive orbital correction, from liftoff to midcourse corrections to lunar-orbit rendezvous to reentry, was guided precisely by principles Newton had laid down 300 years previously. Copernicus had set the stage for Galileo. Galileo had placed the study of motion on firm scientific footing, and his German contemporary Johannes Kepler had stated precise mathematical laws to describe orbital motions. For all their genius, however, Copernicus, Galileo, and Kepler went only so far as to provide elegant descriptions of *how* matter behaved when in motion. It took Newton to explain *why*.

The Wager Heard Round the World

Isaac Newton was born fatherless and premature in 1642, the year of Galileo's death. When Newton was three years old, his widowed mother remarried and placed him in the care of a grandmother. Late in life he confessed to having felt abandoned by his mother, which perhaps accounts for the solitariness and aloofness that characterized him. Of Newton, mathematician and chronicler of science Jacob Bronowski wrote: "All his life, he makes the impression of an unloved man. He never married. He never seems to have been able to flow out in that warmth that makes achievement a natural outcome of thought honed in the company of people."[7] It is interesting to speculate how different our world might be had Newton felt loved, for "as a man he was a failure; as a monster he was superb."[8]

In 1661, at the age of eighteen, having by his absent-mindedness convinced the family that he would never turn a profit at sheep farming,[9] Newton left his home in Woolsthorpe, England, to attend Trinity College at Cambridge University intending to become a cleric. At Trinity, he fell under the influence of professor Isaac Barrow, who occupied the prestigious Lucasian Chair of Mathematics. Recognizing Newton's extraordinary gifts, Barrow quickly turned the young man from theology toward mathematics and optics.

In early 1665, when the plague struck Europe, closing universities and sending students home, Newton returned to the rustic manor house where he had been born. There, in leisure and quiet, he studied alone for two years. At the age of seventy-three, Newton reminisced wistfully upon those golden years: "All of this was in the two plague years of 1665 and 1666, for in those days I was in the prime of my age for invention, and minded mathematics and philosophy [science] more than at any time since."[10] He speaks, of course, of the creative flurry during which he produced – from the ages of twenty-two to twenty-four – the calculus, which he termed "Fluxions"; the inverse-square law of gravitational attraction, which he used to calculate the period of the moon at $27\frac{1}{4}$ days; and a particle theory of light, which he developed through experiments with prisms.

Newton returned to Cambridge in 1667. In 1669, in a gesture of unheard-of magnanimity, Isaac Barrow resigned the Lucasian Chair, which passed to his student, Isaac Newton, whom Barrow recognized as an unequalled intellectual giant.

Despite spectacular achievements, Newton was little inclined to publish. He seems to have viewed the calculus as being of little intrinsic value, merely a secret tool with which to crack the hard nuts of physical phenomena. However, on the subject of light, he betrayed uncharacteristic enthusiasm, and he published his first paper, a work on optics, in 1672. Secretive by nature and abhorrent of controversy, Newton was stung by criticism of his theory by eminent scientist Robert Hooke. He confessed in a letter to his friend Wilhelm Leibniz, a powerful mathematician in his own right, "I was so persecuted with discussions arising from the publication of my theory of light, that I blamed my own imprudence for parting with so substantial a blessing as my quiet to run after a shadow."[11] Newton did not publish again on the subject until 1703, after Hooke's death.

Ironically, however, it was Hooke who instigated Newton's publication of *Principia Mathematica*, commonly regarded as the greatest contribution to science by an individual. Hooke, a towering genius, and Royal Astronomer Edmund Halley had independently concluded that Kepler's first law of planetary motion – that the orbits of planets are elliptical – could be explained if the gravitational attraction between the sun and a planet varied inversely with the square of the distance between them. Halley admitted he couldn't prove this conjecture, but Hooke boasted he could. It was their habit to meet regularly at a London inn with the famous architect Christopher Wren. During one such meeting, the freewheeling conversation turned to orbital mechanics. "To encourage the inquiry,"[12] and apparently to call Hooke's bluff, Wren offered the sum of forty shillings for a successful proof that inverse-square gravitation resulted in elliptical orbits. Neither Hooke nor Halley delivered on the wager within the two months allotted.

Some months later, however, on a visit to Cambridge, Halley paid Newton a call and revived the question that had been smoldering: What would be the curve described by the planets on the supposition that gravity diminishes as the square of distance? Without hesitation, Newton responded, "An ellipse." Astonished, Halley probed as to how Newton knew this to be true. "Why, I have calculated it."[13] Not surprisingly, Halley wished to see Newton's analysis. Newton, however, had misplaced his papers! Halley departed on Newton's promise to supply the proofs, and Newton kept his word. In the process, Newton rekindled his interest in the subject matter and undertook to systematically set down previously unpublished work. Eighteen feverish months later, *Principia Mathematica* emerged, paving the way for the Age of Reason.

Principia contained nearly all the mathematical machinery needed three centuries later by Apollo: the laws of motion, differential and integral calculus, the theory of universal gravitation, and the fundamentals of fluid mechanics. Most foundational were the laws of motion:

1. Objects at rest remain at rest, and objects in motion remain in uniform rectilinear motion unless acted upon by an unbalanced force.

2. Force equals the time rate of change of momentum.

3. For every action, there is an equal and opposite reaction.

Essentially a careful restatement of Galileo's first law of motion, Newton's first law implies that objects at rest or in motion possess an innate quality called *inertia*, the modern term for Galileo's *impeto*. Inertia is the property of resistance to change in motion. An object at rest "resists" being accelerated. An object in motion "resists" changes in either its speed or direction. Objects inherit inertia from their mass.

Closely related to inertia is Newton's "quantity of motion," *momentum* in modern parlance.[14] Momentum, the focus of Newton's second law, is the product of an object's mass and its velocity. Like velocity, momentum is a vector quantity, meaning that it possesses both magnitude and direction. Moving objects may exhibit great momentum because of great mass, like a train, or high velocity, like a bullet.

Collectively, the first two laws provided a natural "autopilot" for Apollo missions. Far from Earth, with no drag force in the deep vacuum of space, a coasting Apollo spacecraft would simply follow a preordained trajectory. Mission controllers needn't bother about constant course corrections. Changes to velocity were necessary only at critical junctures: upon leaving earth orbit, near the distant Lagrangian point where the gravitational tugs of the earth and moon are in delicate balance, and during lunar-orbit insertion.

Newton's third law remains a point of eternal embarrassment for NASA's managers, who in a rare moment of collective amnesia completely overlooked it. The omission occurred during the Gemini program, which formed a crucial bridge between the Mercury program, which placed astronauts in earth orbit, and the Apollo program, whose prize was the moon. As the name implied, Gemini's ultimate goal was the perfection of rendezvous techniques for twin orbiting spacecraft. Gemini spanned the years 1965 and 1966, exactly 300 years after Newton's *anni mirabili*, when he both surmised and validated the inverse-square law of gravitation.

> I deduced that the forces which keep the planets in their orbs must be reciprocally as the squares of their distances from the centres about which they revolve; and thereby compared the force requisite to keep the moon in her orb with the force of gravity at the surface of the earth; and found them answer pretty nearly.[15]

Newton arrived at an understanding of the problem of celestial mechanics by a bootstrap approach. He thought first of tossing an object, a ball perhaps, high into the air. Its parabolic trajectory, described mathematically by Galileo, was clearly understood by Newton. But what about orbital motion? What keeps the moon in its orbit? The moon, Newton reasoned, is like a ball thrown very fast. It constantly falls toward earth, but the earth's curved surface recedes beneath it so that it continually misses the earth. Orbital motion, therefore, is a kind of perpetual free fall.

Newton's predictions rang true for Mercury, Gemini, and Apollo. Because of free fall, the occupants of orbiting spacecraft are weightless. Moreover, the techniques of orbital rendezvous by two spacecraft each in free fall are counterintuitive. The slower craft, by virtue of its lower orbit, overtakes the faster. The first test of rendezvous in orbit took place in December 1965. The rendezvous and docking of the two Gemini spacecraft went without a hitch, precisely according to Newtonian mechanics. It was the piloting skills that needed testing, not the laws of physics. A year later, however, NASA neglected one of Newton's laws, with near catastrophic consequences.

In June 1966, Thomas Stafford, who had flown an earlier mission, flew again aboard Gemini IX-A, accompanied by rookie astronaut Eugene "Gene" Cernan. Their goals were to gain additional practice with rendezvous and to experiment with extravehicular activity (EVA). The first "space walk" by an American had been accomplished a year earlier by Ed White on Gemini IV.

Unlike White's exhilarating twenty-two-minute float in space, Cernan was expected to do real work. He quickly got into trouble. Mission planners assumed that weightlessness made movement easier. Not so. NASA had failed to provide handholds and footholds by which Cernan could anchor himself. In accordance with Newton's third law, Cernan's every slight movement produced a countermovement that was impossible to control. Within moments, his heart rate soared to 180 beats per minute. For more than two hours, Cernan struggled to control his movements. Only with Herculean effort was he able to wedge himself back into the tiny capsule, narrowly avoiding America's first death in space.[16]

Following the close call, NASA's EVA review panel found the major error to be a failure to account for Newton's third law of motion. On subsequent flights, anchor points – handholds and footholds – were provided; astronauts were trained for EVA in the simulated weightlessness of an underwater environment; and EVA became more-or-less routine.

The extraordinary success of the Gemini program lent confidence to Apollo's program directors. In particular, the perfection of orbital rendezvous techniques boosted stock in Houbolt's LOR approach. It began to look as if the United States might win the race to the moon. Then, on the heels of Gemini's triumph, disaster struck.

Everyone inside NASA expected the ultimate price to be paid at some point. No one expected it to happen on the ground. On January 27, 1967, three astronauts perished instantly in a flash fire during a ground test of the Apollo command module. An ugly combination of design flaws was uncovered in the aftermath. The pure-oxygen atmosphere contributed to a fire-friendly environment, the command module's electrical wiring was woefully substandard, and the escape hatch opened inward rather than outward. That perfect storm of flaws conspired to extinguish the lives of "Gus" Grissom, Ed White, and Roger Chaffee. And yet, unexpectedly, the Apollo fire ensured the program's ultimate success. In the wake of disaster, NASA paused to reassess. One by one, numerous potentially disastrous design errors were uncovered and corrected. By December 1968, Apollo 8 had achieved lunar orbit. All that remained was landing on lunar soil.

The Apollo moon mission embodied what scientists call the "three-body problem": What are the trajectories of three mutually gravitating bodies, given their initial positions and velocities? For the initial phase of the moon shot, the three bodies were the earth, the moon, and the Apollo spacecraft. During the moon-landing phase, the three bodies were the moon, the LM, and the command module, which remained "parked" in lunar orbit while the LM descended to the moon's surface.

Newton himself did not solve the three-body problem. He did, however, solve the two-body problem in its full glory. His solution to the two-body problem – that is, to the problem of determining the orbit of the moon around the earth or the orbit of a planet around the sun – exactly satisfied all three of Kepler's descriptive laws of planetary motion. The achievement represents perhaps the single greatest leap in the history of science because Newton had to conjure his tools – the theory of gravitation and the mathematical machinery of calculus – from thin air. It might therefore seem that solving the three-body problem presented a relatively small step beyond Newton's initial leap.

Surprisingly, the problem defied solution for more than two centuries, for good reason: it is insoluble by ordinary methods. This recognition came just as the space race was heating up, but from an unlikely quarter. In 1962, meteorologist Edward Lorenz, while analyzing a mathematical model of weather, discovered *chaos*,[17] the unanticipated tendency of relatively simple systems of equations to yield complicated results. Like Lorenz' weather model, the gravitating three-body system is both simple and inherently chaotic. This means that the trajectory of the third body (for example, the spacecraft) is virtually unpredictable when subjected to a gravitational tug-of-war between the earth and moon. Slight changes to either the initial position or velocity of the spacecraft result in wildly divergent trajectories. When such problems cannot be solved exactly, the only alternative is to approximate their solutions by "number-crunching" on a computer. The success of Apollo, therefore, rested on pushing the technological envelope in two opposite directions: the gigantization of rocketry and the miniaturization of the digital computer.

Newton's *Principia* gave us nearly all the principles that guided Apollo. But in its three volumes, its author gave us much more. Therein, Newton "cultivated mathematics as far as it relates to philosophy."[18] In Book I, he laid out the laws of motion and the fundamentals of theoretical mechanics. Book II, with its emphasis on hydrodynamics (fluid mechanics), constituted what would be called today a text on mathematical physics. "Like the third movement in a supreme symphony," Book III applied the principles established in Books I and II, lest they appear "dry and barren," to the world that we know.[19] In it Newton established the motions of the planets around our sun and determined the planetary masses relative to that of the earth. He estimated the density and the mass of the earth to within twenty percent. He accounted for the flattening at the poles due to the earth's spin, and, most remarkably, he derived the rate of precession of the earth's axis due to the planet's slightly nonspherical shape. He established a theory of tidal action (which had eluded Galileo), calculated the irregularity of the moon's orbit due to the attraction of the sun, and analyzed orbits of the comets so that their returns could be predicted. Each of these feats would have marked Newton for scientific immortality. That one man accomplished them all is beyond comprehension.

Newton's public face differed dramatically from his private one. In public, he claimed assuredly, "I do not deal in conjectures," by which he meant, "I lay down a law and derive phenomena from it."[20] He was critical of "virtuosos" like Descartes who speculated about physical phenomena without a shred of physical evidence. "Tis much better to do a little with certainty, and leave the rest for those that come after you, than to explain all things."[21]

By contrast, the private Newton, far from possessed with dull certitude, was a wildly speculative creature. We see the tip of the iceberg in an appendix to the second edition of *Opticks*, published relatively late in his life. The section entitled "Queries" proffers a series of speculations on numerous aspects of science. Many are astonishing prophecies about the scientific future. Among them:

- The first query – "Do not Bodies act upon Light at a distance, and by their action bend its Rays?" – foreshadows the gravitational bending of light predicted by Einstein's general theory of relativity and confirmed experimentally in 1919.

- Equally astounding, the thirtieth query appears to anticipate the equivalence of mass and energy of Einstein's special relativity: "Are not gross [massive] Bodies and Light convertible into one another?"[22]

Here we glimpse of Newton's alter ego, albeit one he took great pains to conceal from the public. Newton nurtured a lifelong infatuation with alchemy, the pre-science and occult art of transmutation that gave rise to chemistry. To the alchemists, far from being inanimate brute matter,

"Nature was alive with process."[23] The fire in Newton's secret laboratory at Cambridge "burned night and day," and "in the breadth of his knowledge and his experimentation, [Newton] was the peerless alchemist of his day."[24] He kept a private *Index Chemicus*, "a manuscript of more than a hundred pages, comprising more than five thousand individual references to writings on alchemy spanning centuries."[25] Unwittingly, he slowly poisoned himself with quicksilver, the alchemist's most essential element: mercury.

Newton was equally entranced – and equally secretive – about matters spiritual. During bouts of scientific disinterest and sleeplessness, he wrote copiously in Latin on the fine points of theology: the Trinity, the date of creation, the miracles of Christ, and the meaning of Revelations. The private wanderings of his fertile and complex mind were found hidden in a trunk following his death. Among his musings was *Observations upon the Prophecies of Daniel and the Apocalypse of St. John*, a tome of more than a million words.

A few of Newton's musings bridged science and religion, as in this remarkable conjecture on an issue not yet resolved: extraterrestrial life. "And as Christ after some stay in or neare the regions of this earth ascended into heaven, so after the resurrection of the dead it may be in their power to leave the earth at pleasure and accompany him into any part of the heavens, that no region in the whole universe may want its inhabitants."[26]

From whence came the genius of Newton? The prescience of his queries suggests that Newton tapped some mysterious reservoir of intuition as an accomplished writer taps a muse. Once, when asked in a letter from a friend about the source of his inspiration, Newton replied cryptically, "It is plain to me the fountain I draw it from, though I will not undertake to prove it to others."[27] It is also clear that, by nature and by virtue of a solitary existence, Newton held intense powers of concentration far beyond the ordinary. "I keep the subject constantly before me and wait till the first dawnings open little by little into the full light."[28]

Laplace's Daemon

To those of his era, it seemed that Newton had brightly illuminated every dark scientific corner, that nature was now bereft of secrets. The French mathematician Pierre Simon de Laplace, himself an intellectual heavyweight known as the "Newton of France," extrapolated beyond Newton's achievements to envision a universe that is completely deterministic.

> We may regard the present state of the universe as the effect of its past and the cause of its future. An intellect which at a certain moment would know all forces that set nature in motion, and all positions of all items of which nature is composed, if this intellect were also vast enough to submit these

data to analysis, it would embrace in a single formula the movements of the greatest bodies of the universe and those of the tiniest atom; for such an intellect nothing would be uncertain and the future just like the past would be present before its eyes.[29]

In short, Laplace's "daemon," as this hypothetical intellect became known, could conceivably predict the path of each and every particle in the universe for all of eternity! The future state of any particle depended only upon Newton's calculus, his laws of motion, and the initial position and velocity of the particle. To an intellect armed with this information, the future was preordained. Any difficulties were practical, not conceptual. By the mid-twentieth century, humans possessed Laplace's daemon: the digital computer. Apollo hastened its development.

In his enthusiasm, Laplace reduced the universe to a majestic machine. For devout mechanists like Laplace, nature was a clock. Winding the clock, God the Clockmaker set the universe in motion and, having finished his creation, left Newton as the town crier to call out the passing hours.

Dominating scientific thinking for two centuries, "Newtonian determinism" still holds sway in many scientific quarters. And yet, even as Newtonian mechanics guided Apollo toward a successful landing on the moon, its foundations were crumbling. Relativity, quantum mechanics, and chaos theory, each a product of twentieth-century science, chipped away at the Newtonian worldview not by choice but of necessity.

To his credit, it is doubtful that Newton would have embraced determinism so boldly as did Laplace. Problems with radical determinism lurked even in Newton's day. Foremost, determinism is fundamentally incompatible with the free will of sentient beings. Although Laplace's *daemon* might predict with uncanny precision the trajectory of Apollo, it was powerless to anticipate the emotions and actions of its occupants: of Armstrong's fateful last-minute decisions that turned imminent disaster into triumph, or of Anders' wonder at seeing his first earthrise. A strictly deterministic world is, in the words of Alfred North Whitehead, "a dull affair, soundless, scentless, colorless, merely the hurrying of matter, endless, meaningless."[30] By contrast, Apollo's images of fragile spaceship earth were both stunningly beautiful and imbued with meaning for all with eyes to see.

For his monumental accomplishments, Newton was knighted in 1705. In 1727, he died at the venerable age of eight-four. Newton lies buried in London's Westminster Abbey among kings, queens, and poets. At the Royal Society's tercentenary commemoration of Newton's birth in 1946, postponed by four years due to the Second World War, Lord Keynes[31] paid tribute to Newton as a watershed figure in human history whose intellectual contributions ushered in the scientific age. Nevertheless, Keynes' eulogy contained a sage observation about Newton's ultimate place in history: "Newton was not the first of the age of reason. He was the last of the magicians."[32]

However flawed the notion of a clockwork universe, it must have offered some solace to astronaut Michael Collins, Apollo 11's command module pilot. During the lonely hours in lunar orbit following the LM's separation, his job was simply to mind the store while fellow astronauts Neil Armstrong and Buzz Aldrin cavorted on the lunar surface sixty miles below. A quarter-million miles from home, in free fall about the moon, Collins was weightless, as Newton had anticipated. His orbits of the moon were reassuringly predictable, in strict accordance with Kepler's three laws of planetary motion that had spurred Newton's investigations of orbital mechanics. And if all went according to plan, Newton's dependable laws would guide the delicate lunar-orbit rendezvous to reunite Collins with Armstrong and Aldrin. And then the threesome would jettison the faithful *Eagle* and travel home, to that blue marble called Earth, the one truly beautiful object visible in the vast black cosmos.

Two

The Music of the Spheres

There is geometry in the humming of the strings; there is music in the spacing of the spheres.

– **Pythagoras**

Years after his epic journey, Apollo 14 astronaut Edgar Mitchell cast in bold relief what the Apollo moon landings had accomplished: "When [Neil] Armstrong announced to the world that the Eagle had landed, it was tantamount to that unceremonious yet monumental flop of the first sea creature that beached itself on dry land, and lived."[1]

To Mitchell, three daring transitions punctuate the evolution of life on Earth. In the first, sea creatures ambled awkwardly onto terra firma. In the second, land creatures took hesitantly to the air. And in the third, earthlings ventured to an alien world. In each instance, life was extraordinarily ill adapted for its new environment and at great risk. At such evolutionary singular points, the future always hangs in the balance.

In the grand scheme of evolution, we humans are newcomers. Cro-Magnon man, the type specimen of the species, has trod the earth for perhaps 130,000 years. Sharks, by contrast, have ruled the seas for 400 million years. As a species, *Homo sapiens* remains in infancy.

Until quite recently, we humans differentiated our species by its presumably exclusive use of tools and language. But such assumptions about our uniqueness have not survived scrutiny. We now know that many higher animals – not just primates – exploit both tools and language. Crows are canny problem solvers and create tools on the spot, for example by bending a wire to probe for insects. In Vienna, Austria, Betsy, a precocious border collie, understands 340 spoken words and recognizes nearly as many symbols. At Iowa's Great Ape Preserve, Kanzi, a bonobo, grasps 360 keyboard symbols, understands thousands of spoken words, forms sentences, and composes music on a synthesizer.[2] The more closely we study the higher animals, the more our categorical superiority evaporates. Even our claims to unique spirituality fall flat in the face of new observations that elephants, foxes, and even magpies appear to grieve for fallen companions.[3]

And so we humans are not unique, at least not in the ways we thought we were. Our distinguishing characteristic is not so much the use of tools and language per se but the almost infinite variety of ways in which we adapt these apparently ubiquitous abilities.

Abstraction. In a word, it is by the abstraction of tools and language that humans shine. Chief among the grand abstractions of humanity are mathematics and science. Humans have employed mathematics for as long as there has been commerce – at least 5,000 years. In the third century BCE, when Euclid wrote the *Elements*, mathematics was already a well-established discipline.

On the other hand, science, as a formal discipline, is a relatively recent acquisition to the human toolkit. The origins of science are obscure, but the first scientific revolution is not. In 1543, *On the Revolutions of the Heavenly Spheres* was published as Copernicus, its author, lay upon his deathbed. "Revolution" is right. *De Revolutionibus*, its Latin title, literally moved the earth, upended centuries of entrenched mythology, undermined the authority of the Catholic Church, and redefined the place of humans in the cosmos. It laid foundations upon which Newton and others erected the grand edifice of science and thus freed the scientific genie. None of this was the intent of its author, a Polish cleric, astronomer, and mathematician who gazed nightly at the heavens, enthralled.

The Heavenly Spheres

Fascination with the heavens is as old as humanity. Like beholding the eyes of a lover, gazing into the night sky plumbs mysterious depths. It is an act both reassuring and humbling – reassuring because of the faithfulness of the celestial objects. By day, the sun rises and sets, as it has done for recorded history and long before that. By night the moon waxes and wanes with dependability, and Polaris faithfully points our way. The constellations change their orientation but never their patterns.

But the heavens also beckon for humility. How did it all happen? And why? Why is there something rather than nothing? Are we the only ones asking such questions? Each culture on Earth is rich with mythology, feeble attempts to answer questions ultimately too big for comprehension. Nevertheless, in the asking and the pondering we express what is most essentially human: the longing to transcend our limitations.

Our forebears lived close to nature. Studying the heavens, naming the constellations, predicting the equinoxes: these were not simply pastimes. Survival hung on such skills. Hunter-gatherers studied the seasons to anticipate migrations. So did cultivators, whose harvests depended upon planting at the optimal time of year. Those who traveled the seas did so at great peril, greater still unless they could predict the tides and navigate by the stars when far from land.[4]

Many wonders of the ancient world hold astronomical significance, among them Stonehenge, Egypt's Great Pyramid of Cheops, and the Aztecs' Teotihuacan. For at least 5,000 years, humans have celebrated the solstices, paying homage to whatever forces regulate the universe. Primitive observatories abounded in antiquity. The earliest, attributed to the Sumerians, was the *gnomon*: a straight stick set vertically into the ground. Patient following of the shadow cast by the gnomon reveals invaluable practical knowledge.

Over the course of a summer day, the tip of the gnomon's shadow traces out a "frown." In contrast, in winter months, the shadow's locus "smiles," opening upward. Twice a year, and only twice, the tip of the gnomon's shadow walks a perfectly straight line. These days mark the spring and fall equinoxes. At the equinoxes, the perpendicular bisector from the gnomon to the shadow's locus points directly toward true north.[5] Thus, by attentively following the gnomon's shadow, ancient sky watchers inferred the length of year, the cycle of the seasons, and the direction of true north.

Similarly, the nighttime sky holds both mystery and knowledge. Before the telescope, the unaided eye could discern a few thousand stars. Over eons, archaic cultures bestowed names upon the patterns of the "sky people": *Aquarius, Aries, Cancer, Capricorn*, Cassiopeia, *Gemini, Leo, Libra*, Orion, Pegasus, *Pisces, Sagittarius, Scorpio, Taurus*, Ursa Minor and Major, and *Virgo*. Twelve of the familiar constellations, highlighted above in italic, form the signs of the zodiac, the celestial road of the Egyptian sun god *Ra*.

During the course of a night, the constellations rotate *en masse* about a fixed point: Polaris, the pole star. Because the earth rotates 360 degrees in each twenty-four-hour period,[6] the stellar arcs, when captured by modern time-lapse photography, grow by 15 degrees an hour. Some arcs dip below the horizon. Others, those of the *circumpolar* stars, do not. Homer, observing "the Bear never bathes,"[7] meant that the constellation Ursa Major never dips below the horizon at moderate northern latitudes.

Unlike the stars, whose positions remain fixed relative to one another, the sun is a "wanderer." Its position at sunset gradually moves through the zodiac. The path of the sun against the stellar backdrop is known as the *ecliptic*.[8] In one year, the sun traverses a full circuit of the zodiac along its ecliptic, like a toll collector on a merry-go-round.[9] Since antiquity, astronomers have decomposed the motion of the sun into two constituents: diurnal and annual. "Each day the sun moves rapidly westward with the stars; simultaneously the sun moves slowly eastward along the ecliptic through the stars."[10]

The contrasting attributes of the heavens – change and constancy – are embodied in the philosophy of Plato (428-347 BCE). Influenced by Heraclitus, "the weeping philosopher," who lamented, "There is nothing permanent except change," Plato was painfully aware that everything "flows," that is, changes. Like most of his fellow humans, Plato longed for evidence of that which is immutable, eternal, and incorruptible, and he eventually formulated a philosophy that is implicit in much of mythology and religion. To Plato, behind the visible world lay the world of *forms*, those incorruptible ideas that give existence to the things of nature. Mathematics is the quintessence of Plato's ideal world. The things of mathematics – circles, lines, spheres, and triangles – are perfect forms that exist independent of experience and yet make possible organization of the natural world. Although the manifestations of forms in the natural world are subject to change, the forms themselves are changeless.

In his creation allegory *Timaeus*, Plato likened the universe to a creature: "[The creator's] intention was, in the first place, that the animal should be as far as possible a perfect whole and of perfect parts: secondly, that it should be one, leaving no remnants out of which another such world might be created: and also that it should be free from old age and unaffected by disease."[11]

Plato's cosmological biases insinuated their way into the philosophy of his most famous student, Aristotle (384-322 BCE). Aristotle's universe was an amalgam: it combined the ancient two-sphere cosmological model, elements of Platonic idealism, and Aristotle's own whimsical ideas about motion.

The two-sphere model, an early phenomenological model of the universe, very nearly accounted for the cycles of the heavens.[12] For thousands of years, the earth was known to be spherical. This was inferred from the experience of seafarers, who observed that the mountaintops of landmasses appeared on the horizon before their lower elevations became visible.[13] It was inferred also from experiences of sailors who, traveling long distances along the meridians, observed that Polaris resides higher in the night sky the farther north one travels. As early as the third century BCE, Eratosthenes, a librarian in Alexandria, Egypt, estimated the circumference of the earth at 250,000 stadia, a value within five percent of the true measure.

Misnamed, the two-sphere model incorporated *three* concentric spheres, with the interior sphere being the earth herself. Concentric with and surrounding the earth, the *celestial sphere* contained all visible stars. Rotating once daily on an imaginary axis, the celestial sphere ferried the stars on their diurnal journey. Nested between the inner and outer spheres lay a third sphere sporting the sun. Rotation of the middle sphere carried the sun once daily around the earth. Lost in antiquity, the origins of the two-sphere model are obscure, but its motivations are not: it accounted for the rising and setting of the sun, the circumpolar arcs of the stars, and the apparent immobility of the earth. To this day, it is the model employed by planetaria and navigation charts.

The two-sphere model, however, did not account for all observed phenomena. For one, there was the problem of the "wanderers": *planetes* in the Greek. In antiquity, seven "planets" were visible to the naked eye: the sun and moon, Mercury, Venus, Mars, Jupiter, and Saturn. Like the sun, the planets traverse the ecliptic, swimming upstream against the zodiac. But their orbital periods vary wildly, from 88 days for Mercury to 29 years for Saturn. Some wanderers behave at times erratically. Like one who retraces his steps to pick up an object dropped upon the path, Mars appears to reverse course. The two-sphere model failed to account for the wanderers or their *retrograde motion*.

A philosopher, not a scientist, Plato nevertheless articulated the central problem of science: to "save the appearances":[14] the job of a theory is to fit observational data. Further development of cosmological theory fell to a distinguished procession of Greeks: primarily to Plato's students, Aristotle and Eudoxus of Cnidus, and to the geometer Apollonius and the astronomer Hipparchus.

A century earlier, the pre-Socratic Greek philosopher Parmenides accommodated the planets by adding additional spheres to the two-sphere model, one sphere for each planet. Parmenides' collection of *homocentric spheres* suffered a number of deficiencies, among them difficulties with retrograde motion. Eudoxus and Aristotle partially resolved the problem of retrograde motion by introducing yet more concentric spheres, at the cost of a considerable increase in the model's mathematical complexity. But they left unscathed other cosmological difficulties. Chiefly, the brightness of Mars as seen from the earth varies by an astonishing factor of 25. No planetary model centered about the earth could adequately account for observational variations in planetary brightness or apparent size.

Eventually cosmologists abandoned homocentric spheres and superseded them by new mathematical gimmicks: *epicycles*, *equants*, and *eccentrics*. Developed and promoted by the geometer Apollonius and the astronomer Hipparchus, epicycles were circles upon circles, which, by clever construction, could account for some variation in planetary brightness, orbital speed, and orbital direction. Each new epicycle, however, added a layer of complexity.

Meanwhile, paying homage to Plato's notions of the ideal, Aristotle philosophically polished the homocentric model. He presumed the celestial bodies to be smooth spheres and the planetary orbits to be perfect circles, spheres and circles being idealized forms. Similar notions of incorruptibility led Aristotle to conclude that new celestial objects could not be created in a universe already perfected. The Aristotelian universe was therefore flawless, eternal, and static, except for its faithful periodic motions.

Geocentric cosmology attained its zenith in the second century CE thanks to the Alexandrian astronomer Claudius Ptolemaeus (Ptolemy, circa 80-165 CE). From an observatory at Canopus, a coastal town fifteen miles east of Alexandria, Egypt, Ptolemy charted the heavens faithfully, carefully, and, as a general rule, accurately. Ptolemy dramatically advanced the state of the art in astronomy. A gifted mathematician as

well as diligent astronomer, he perfected mathematical techniques that
had eluded other astronomers. His most important contribution, however, was to account for the atmospheric refraction of light, whereby the
atmosphere bends light rays and distorts the apparent positions of stars
that lie low on the horizon.[15]

Ptolemy's masterpiece, the *Almagest* – which translates to "The Greatest" – was a singular accomplishment. "[It] epitomized the greatest
achievement of ancient astronomy," and "was the first systematic mathematical treatise to offer a complete, detailed, and quantitative account of
all the celestial motions."[16] As testament to Ptolemy's mathematical rigor
and his persuasiveness as an expositor, Ptolemaic cosmology reigned as
definitive for fourteen centuries, an enormously long time for any theory
to remain viable.

Ptolemy's model of the cosmos coincided closely with the rise of
Christianity, which first embraced and then absorbed the cosmological
structure of Ptolemy along with its roots in Platonic and Aristotelian
thought. Among the most influential religious scholars to appropriate
Ptolemaic and Aristotelian notions was Thomas Aquinas (1225-1274 CE),
the Scholastic philosopher who contributed most to the systematic formal-
ization of theology for the Roman Catholic Church. In the following passage on matters of cosmology, Aquinas echoes Aristotle, almost parroting
him:

> It is therefore clear that the material of the heavens is, by its intrinsic nature,
> not susceptible to generation and corruption. . . . Furthermore, among the
> sorts of motion which they might experience, theirs is circular, and circular
> motion is the one which produces the very minimum of alteration because
> the sphere as a whole does not change place.[17]

Figure 5 depicts Ptolemy's cosmos as conceived in the sixteenth century. The *Divine Comedy*, "a description of [Dante's] journey through the
universe as conceived by the fourteenth-century Christian," is literally a
voyage through the universe of Ptolemy.

> The journey begins on the surface of the spherical earth; descends gradu-
> ally into the earth via the nine circles of Hell, which symmetrically mirror
> the nine celestial spheres above; and arrives at the vilest and most corrupt
> of all regions, the center of the universe, the appropriate locus of the Devil
> and his legions. Dante then returns to the surface of the earth at a point
> diametrically opposite the one where he had entered, and there he finds
> the mount of purgatory with its base on the earth and its top extending
> into the aerial regions above. At the last he journeys through each of the
> celestial spheres in turn, conversing with the spirits that inhabit them, un-
> til finally he contemplates God's throne in the last, the Empyrean, sphere,
> which nestles Ptolemy's stellar sphere, completing the cosmos.[18]

Like Aristotelian physics, Ptolemaic cosmology suffered from a number of factual discrepancies. In fourteen centuries, the "damned facts," as science terms anomalous data, had accumulated to sizable proportions. In addition to intransigent difficulties with retrograde motion and variations in planetary brightness, the Ptolemaic model failed to account for the precession of the equinoxes. (Completing just one cycle in 26,000 years, the equinoxes drift slowly westward through the ecliptic.) Only by allowing the outer celestial sphere to wobble, in contentious violation of Aristotelian principles, could Ptolemy model this astronomical subtlety.

Predicting the path of a celestial object is one thing; predicting its motion is quite another. To correctly anticipate motion, a theory must place an object at the correct position at the correct time. All the mathematical tricks of the period failed in this regard. Worn threadbare over centuries, Ptolemaic cosmology needed either drastic mending or replacing. Enter Copernicus, the father of the new worldview.

The First Blow to Human Narcissism

Torun, Poland, is postcard picturesque. A UNESCO world heritage site, it lies situated on both banks of the Vistula River at the intersection of ancient trade routes. At night, the town's gothic buildings glow in floodlights and reflect on the Vistula's calm surface. As a member of the Hanseatic League, Torun has always exuded the openness of a trading mecca. It remains a place where new ideas can take root. There, Nicolaus Copernicus was born in 1473, the fourth and last child of a prosperous merchant.

When Copernicus' father died in 1484, care of the family passed to Copernicus' maternal uncle, the scholarly priest Lucas Watzelrode. A few years later, Watzelrode acceded to a position of prominence as bishop of Ermland.[19] Watzelrode's good luck also gilded the fortunes of Copernicus, who, by family consensus, would follow in his uncle's footsteps and be groomed for the priesthood.

To this end, Copernicus' education was paramount, and it began with the study of liberal arts at the University of Cracow. As if on cue, Columbus discovered America, expanding the world's geographic horizons just as Copernicus was poised to expand its intellectual horizons. At Cracow, Copernicus fell under the tutelage of Albert Brudzewski, guiding light of the university's rising humanist faction and author of a commentary on Ptolemaic astronomy. Brudzewski's influence upon Copernicus was profound. Academically, Brudzewski introduced Copernicus to both mathematics and astronomy. Philosophically, Brudzewski lit a flame for humanism. Following Cracow, Copernicus ventured abroad to continue studies at the most prestigious Italian universities: Bologna, Ferrara, and Padua. The disciplines ranged widely as well and included Greek, law, medicine, mathematics, and astronomy.

While at the University of Bologna, Copernicus resided at the home of Domenico Maria de Novara, a Platonist and mathematics professor whose criticisms of Ptolemaic theory completed what Brudzewski had started. There, Copernicus' lifelong passion for astronomy as the "consummation of mathematics"[20] ripened. By the Jubilee year of 1500, Copernicus demonstrated enviable mastery of astronomy and guest lectured on the subject in Rome.

Copernicus' studies continued for a full decade at a leisurely pace. There was no need to rush; a sinecure as canon of the Frauenberg Cathedral, secured for his nephew by Watzelrode, awaited Copernicus whenever he felt ready. Delving into medicine at Padua and furthering the study of law at Ferrara, Copernicus received his doctorate in canon law in 1506, completing his world-class education. A newly minted Renaissance man, "not only a humanist learned in Greek, mathematics, and astronomy, but also a jurist and a physician,"[21] Copernicus returned to his native Poland, where he served as personal physician to Watzelrode until the bishop's death in 1512. Nearly forty years of age, Copernicus had never held a bona fide job.

Situated at a prime spot on the Vistula Lagoon of the Baltic, Frauenberg Cathedral served a wealthy parish. Copernicus' sinecure consisted of a lucrative clerical position without priestly duties. Strictly an administrator, Copernicus neither sought nor took holy orders. Nevertheless, by virtue of his extensive knowledge, he was called upon to perform a variety of important duties: commissary for the diocese, physician to the rich and poor, and arbiter of disputes. In 1522, he advised the government on currency reform. Earlier, in 1514, the pope had sought Copernicus' participation in a commission on calendar reform. Copernicus politely declined and for good reason. By firsthand experience, he understood that observational uncertainties persistently vexed the movements of the sun and moon. Until astronomy could accurately ascertain the celestial movements upon which the calendar is based, it was pointless to reform the calendar.

Undemanding duties afforded Copernicus ample opportunity for making astronomical readings, developing mathematical theory, and marrying the two in cosmological writings. To this end, Copernicus constructed an observatory in the cathedral's eastern tower, where he lived. In 1509, he tested the scientific waters by publishing a short work on astronomy. It met with a favorable review, thanks in part to introductory remarks written by a college friend, who praised the astronomer for exploring "the rapid course of the moon and the changing movements of the fraternal star and the whole firmament of the planets."[22]

Copernicus idolized Ptolemy and esteemed Ptolemy's masterpiece, the *Almagest*, owning not one but two copies, thanks to Gutenberg's recent invention of the printing press. Yet, like Plutarch before him, Copernicus recognized that "the movement of the stars has overcome the ingenuity

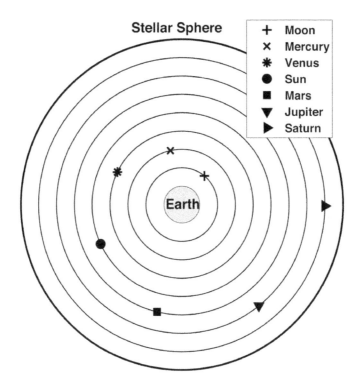

Figure 5: Schematic of Ptolemy's geocentric universe, virtually unchallenged until the sixteenth century. The stellar sphere, which contains all "fixed" stars, rotates daily in a direction opposite that of the planets. (Author.)

of the mathematicians"[23] and set about to correct and update Ptolemy's work. To his surprise, Copernicus found that Ptolemy's geocentric model rested upon fundamental flaws that could not be repaired. Only by a radical change of perspective – setting the earth free of her Ptolemaic moorings – could he salvage cosmology. The sun, not the earth, Copernicus argued, should anchor the cosmos. The earth dances with the planets. As she encircles the sun, she pirouettes on her axis, and the combination of those movements – rotation and revolution – "saves the appearances" of anomalous data.

Shortly after moving to Frauenberg, Copernicus had begun his magnum opus. His situation, however, was delicate. Having no desire to incur the ire of the ecclesiastical authorities, Copernicus hesitated to publish. In 1530, after two decades of delay, Copernicus tentatively released the handwritten *Commentariolus*, a brief outline of the new cosmology. In the form of a letter, the summary circulated among learned friends and acquaintances, to a generally positive reception. Even the ecclesiastics smiled favorably. Cardinal Schönberg encouraged development of the full theory; Pope Clement VII concurred. *Commentariolus* also caught a favorable wind from the Protestants, who, wanting to know more, sent an inquisitive young emissary to Frauenberg in 1539.

The emissary was Joachim Rheticus, whose credentials were sterling. A twenty-five-year-old professor of mathematics at the University of Wittenberg, Rheticus was also the protégé of Philipp Melanchthon, Martin Luther's educational reformer. Having studied Copernicus' planetary system at length, Rheticus offered to publish a full exposition. Within three years, Copernicus, now elderly, lay stricken by stroke and paralysis. He survived just long enough to witness the printing of *De Revolutionibus* in 1543. Presented with an advance copy as he lay upon his deathbed, Copernicus died the same day. Little could he have foreseen his impact upon humankind.

To Copernicus, science, akin to art, was a vehicle "to draw the mind of man away from the vices and direct it to better things."[24] In awe and humility, he thus opens *De Revolutionibus*. Copernicus then turns the first pages of the book of nature, a book that Galileo and Kepler would later champion. "Many philosophers have called the world a visible god on account of its extraordinary excellence."[25] The motivation of the astronomer and cleric is clear: by studying the world we learn about its creator. "For who, ... through diligent contemplation of [the established order] ... would not wonder at the Artificer of all things, in Whom is all happiness and every good?"[26]

Next, Copernicus lauds Ptolemy, "who stands far in front of all others"[27] in the divine science of astronomy. The writer's effusive praise is sincere, as evidenced by the structure of his great work, which closely mimics the *Almagest*. Like the *Almagest*, *De Revolutionibus* is comprised of several books, each devoted to a different facet of the subject. Each book, following Ptolemy, weaves together explanatory commentary, mathematical theorems of a geometric nature, and extensive tables of astronomical data, astronomical projections, and mathematical data.

Revolutionary, but unable to completely abandon Aristotle, Copernicus starts with common ground: the world (universe) is spherical. The earth is spherical too, he asserts in Chapter 2 of Book I, which launches into the historical evidence for such a presumption. In additional concessions to Aristotle, Copernicus assumes that the planetary orbits are "regular, circular, and everlasting."

Not until Chapter 5 does he drop a bombshell in the form of a question: "Does the earth have a circular movement?"[28] Here too, he proceeds modestly and cautiously. The idea is not new to Copernicus; it is in fact ancient. He rightly credits the Pythagoreans Herakleides and Ekphantus for proposing that the earth rotates about an axis. Other ancients were yet bolder. Seventeen centuries before *De Revolutionibus*, Aristarchos of Samos had advocated heliocentrism, eventually earning him distinction as "the Copernicus of Antiquity."[29]

Paying homage to the ancients in Chapter 7, including Aristotle and Ptolemy, Copernicus reviews the arguments for a stationary earth, and in Chapter 8, he gently refutes all. Quoting Virgil, he invokes the principle of relative motion, upon which Einstein will later build the theory of relativity: "We sail out of the harbor, and the land and cities move away."[30]

In Chapter 10, he offers the now iconic diagrams of two competing models of the cosmos: the Ptolemaic model (Fig. 5) and the Copernican model. The earth anchors the former, the sun the latter. Copernicus' cosmos is a *solar* system, and in Chapter 11, Copernicus pulls out the stops. The apparent motions of the sun and stars, and the passing of the seasons – all are explained by three independent attributes of the earth's motion: rotation, revolution, and inclination.

Having set the stage, Copernicus then establishes heliocentric theory on firm mathematical foundations, offering a relentless procession of theorems and geometric diagrams. When the dust settles, the cosmos is wholly altered. At the beginning of Book II, perhaps instinctively softening the psychological blow of his theory, Copernicus apologizes for continuing to speak anachronistically "of the rising and setting of the sun and stars."[31]

As a theory, Copernicanism is parsimonious – it explains so much with minimal mathematical machinery: seasonal variations in climate result from the 23.5-degree inclination of the earth's axis relative to the ecliptic; the equinoxes precess due to a slight wobble in the earth's axis; Mars varies dramatically in brightness because as it and the earth both circumnavigate the sun their separation distance varies sevenfold. Why is the moon's period so much longer than the sun's, given the moon's proximity and the sun's remoteness? The sun's revolution, according to Copernicus, is only apparent; it is due not to the movement of the sun but to the daily rotation of the earth. In contrast, the moon revolves about the earth. There is plenty in Copernicanism to stoke the fires of the science to come. For example, as the moon circumnavigates the earth, her gravitational tug upon the earth, Newton will later explain, activates the tides.

In the Copernican system, the retrograde motions of Mars, Jupiter, and Saturn are virtual rather than actual, optical illusions played by the relative positions of planets in their orbits about the sun. An added benefit of Copernican cosmology, it provides a geometric mechanism by which to determine the relative planetary distances from the sun, both for the inferior planets, those closer to the sun than the earth, and the superior planets, those farther away. Icing the cake, Copernicanism affords a natural explanation for the phases of Venus, inexplicable in Ptolemaic cosmology.

Copernicanism explained much, but to biblical literalists, it blatantly violated scriptures. For example, in the book of *Joshua* (10:12, 13) Yahweh delivered the Israelites from the Amorites in response to Joshua's singular request: "Sun, stand still over Gibeon . . . till the people had vengeance on their enemies."

Indeed, to Aristotelian and Augustinian fundamentalists, Copernicanism opened a Pandora's box of heresy. Heliocentrism dethroned the human race from its self-appointed seat at the center of the cosmos. By implication, if God's highest creation was dethroned, so too was the creator
himself. Within one hundred years of Copernicus, Aristotelian notions
of celestial perfection dissolved as well. Galileo's telescope revealed the
celestial orbs to be blemished by all manner of imperfections, and his discovery of the moons of Jupiter confirmed once and for all that the earth is
not the *axis mundi*.[32] Moreover, "Kepler's star" – the supernova of 1604 –
and the "bearded stars" (comets) undermined Aristotle's naive presumption of the immutability of the heavens.

In an attempt to defuse potential controversy, Copernicus' publisher,
Andreas Osiander, anonymously prepended *De Revolutionibus* with an
unauthorized preface, in the form of a long letter addressed to "The Most
Holy Lord, Pope Paul III." In it, Osiander boldly states Copernicus' noble purpose "to seek truth in all things as far as God has permitted human
reason so to do."[33] Truth, nevertheless, has its consequences, and in words
coined much later, Copernicanism launched a cosmic *paradigm shift*.

Copernicanism not only deposed humans from cosmic centrality, it
rudely awakened them to cosmic immensity. The Ptolemaic universe, finite and cozy, fit within an expanse not appreciably larger than the solar
system. The Copernican universe, in sublime contradistinction, is vast,
possibly infinite. With improved eyesight thanks to the telescope, astronomers began to see previously unnoticed depth in Ptolemy's celestial
sphere and turned their attention to the measurement of stellar distances
by their parallax, from the Greek term for triangulation, *parallaxis*. When a
distant object is observed from different locations, the change of perspective induces an apparent shift of position relative to a fixed background.
The lines of sight from the two vantage points form two sides of a triangle
whose base leg is the distance between the observation points. Trigonom-
etry then yields the distance to the object. Whereas, for the Ptolemaic system, the base leg of the triangle is the diameter of the earth, in the Copernican system, it is the diameter of the earth's orbit, a staggering 180 million
miles. That the stars betray little parallax as the earth traverses her orbit
implies their unfathomable remoteness.

Unwittingly, Copernicus unleashed the Age of Reason, widening a rift
with origins in pre-Christian Greek thought. On one side were Plato and
the Pythagoreans; on the other Aristotle, Democritus, and the sophists.
Whereas the former held that the world could be comprehended by direct insights of an intuitive quality, the latter believed knowledge to be
reliable only if based squarely in the five senses. Sensory knowledge is existential; it is derived from practical human experience, quantifiable, and
lends itself to consensus. In contrast, intuitive knowledge cannot readily
be quantified, is frequently idiosyncratic and is therefore often lacking in
consensus.[34]

The triumph of reason over intuition came at great cost. Disorientation and loss of moorings: these were to become central themes of the Romantic backlash against science. Among the first to count the cost of the ascendancy of science, Goethe mourned:

> The world had scarcely become known as round and complete in itself when it was asked to waive the tremendous privilege of being the center of the universe. Never, perhaps, was a greater demand made on mankind – for by this admission so many things vanished in mist and smoke! What became of our Eden, our world of innocence, piety and poetry?[35]

Epilogue

De Revolutionibus went to press with the imprimaturs of both Catholic and Lutheran theologians. How ironic – within a few years, the Reformation and Counter-Reformation would transform Catholics and Protestants into mortal enemies. As orthodoxy ossified and biblical literalists won the day, Copernicus fell from favor in both camps. "That fool will upset the whole science of astronomy,"[36] blustered Luther. And in 1616, Roman Catholicism deemed heretical all assertions that the earth moves, placing *De Revolutionibus* on its index of banned books, where it remained for more than 200 years.[37]

In 1838, after eighteen months of meticulous observations from the Koenigsberg Observatory, German astronomer and mathematician Friedrich Wilhelm Bessell successfully measured the parallax of a "nearby" star to determine its distance.[38] The star, 61 Cygni, lies 10.9 light-years from Earth. Our nearest stellar neighbor, Alpha Centauri, shines 4.3 light-years distant, and at the outer reaches of the cosmos spin galaxies whose light has traveled for 13.7 billion years to reach us. Staring into the night sky no longer comforts the wonderer quite as it did the ancients. Awe-inspiring still, stargazing is now tinged with an element of fear, as if standing at the edge of a precipitous abyss with no solid perch.

In 2005, archeological digs in the great red-brick sanctuary of the Frauenberg Cathedral unearthed human remains. The excavators sent the skull to Warsaw's crime laboratory for forensic analysis and computer rendering of the "victim's" appearance in life. The digital reconstruction bore a striking resemblance to portraits of Copernicus, convincing archeologists that they had their man. Subsequent DNA analyses corroborated that the final resting place of Nicolaus Copernicus, lost for centuries, had been found, appropriately in the very cathedral where he lived, worked, and studied the cosmos for three decades.[39]

Four centuries after *De Revolutionibus* cast humans from the center of the cosmos as unceremoniously as God cast Adam and Eve from the Garden of Eden, humanity grapples still with the existential vertigo associated with the Copernican loss of place. "What were we doing," lamented Nietzsche in *The Gay Science*, "when we unchained this earth from its sun?"

Three

The Starry Messenger

Who can doubt that it will lead to the worst disorders when minds created free by God are compelled to submit slavishly to an outside will?

– Galileo

In 1600, there existed but a handful of Copernicans. There were good reasons for this lack of converts. For one, that was the year in which the Inquisition burned one of its own at the stake. Among the many heresies that consigned the Dominican friar Giordano Bruno to the flames, he espoused pantheism, the plurality of worlds, and the Copernican cosmology of heliocentrism.

Then, too, there was scant observational evidence that Copernicus was right. His model of celestial motions "saved the appearances" in elegant fashion. But so did others, if not so elegantly. The preeminent European astronomer of the day, Tycho Brahe, aware of the pitfalls of Ptolemy but unable on theological grounds to embrace Copernicanism, proposed a hybrid cosmology. In the Tychonic system, all planets but the earth orbited the sun, while the sun encircled the earth with its planetary flock in tow. It was unwieldy to say the least, but as effective as the unadorned Copernican model in "saving appearances."

In essence, Copernicus had copped out. By publishing *De Revolutionibus* on his deathbed, he left to others the contentious and dangerous task of promoting the theory. But among the early converts to Copernicanism were two whose combined gravitas reset the center of the solar system: Galileo Galilei, a Catholic, and Johannes Kepler, a Lutheran. Neither fared well under the heel of religious authority. But for Bruno, Galileo might not have survived.

The "Galileo affair," as the inevitable showdown became known, drove a wedge between science and faith that persists to this day. The central issue cut to the heart of what it means to be human. Where does the locus of truth reside? Does it lie externally, in religious authority, or can the individual's perceptions and experiences be true guides to ultimate reality? Galileo's "crime" was unabashed belief in the latter, at least in questions of a physical nature. "In questions of science," he steadfastly maintained,

31

"the authority of a thousand is not worth the humble reasoning of a single mind."[1]

In the Galileo affair, the Church won its tactical victory, but at the cost of strategic defeat. The end results included the secularization of thought and the triumphalism of science. Institutionalized religion has never fully recovered.

Galileo never intended injury to the Church. Despite his dangerous mix of naiveté and arrogance, Galileo's motives remained essentially pure: to establish the complementarity of the book of nature and the book of scripture. Truth is one; science and mathematics, as surely as the holy scriptures, reveal the mind of God. To Galileo and to many contemporary scientists, "Mathematics is the alphabet with which God has written the universe."[2]

For its part, the Church behaved abominably, but its self-serving justification carried an element of logic: at all costs, it must spare its flock from additional revolutionary shocks. By 1600, the souls under its care had already suffered two major assaults on their worldview: the discovery of the New World and the Reformation of Martin Luther. Copernican cosmology could be the straw that broke the camel's back. And its chief defender was Galileo.

The Reincarnation of Michelangelo?

Galileo Galilei was born in 1564, the year of Shakespeare's birth and Michelangelo's death. His science rivaled Michelangelo's art. Could it be, as Galileo's followers claimed, that "Michelangelo's spirit had leaped like an inspiration from his aged, failing body to the infant Galileo in the brief span of hours separating the former's death from the latter's birth?"[3]

The surname Galilei – a transliteration of the land of Jesus' birth – suggests a certain pious nobility. Patrician indeed, the family just eked by financially. A musician, Galileo's father Vincenzio played several instruments including the organ and lute. When Galileo was eight his family moved from Pisa to Florence, where Vincenzio joined a coterie of virtuoso musicians and scholars intent upon reviving Greek tragedy through music. Their success gave rise to a lasting art form: opera.[4]

Vincenzio owed his notoriety as a musician to a break with rigid Pythagorean rules of scale, preferring the "sweetness" afforded by more relaxed tuning and contrary to his authoritarian teacher, with whom he parted ways. Like father, like son: Galileo's two greatest works would also challenge the authorities – first Ptolemy, then Aristotle.

In patrician families, education was highly prized. From the age of ten, when Galileo first attended grammar school at a Benedictine monastery, he received excellent instruction. Father and son soon clashed, however, over vocation. Needing Galileo's help to support the family, Vincenzio insisted upon a lucrative medical career for the promising elder son.

In 1581, Galileo obediently left home to take up medicine at the University of Pisa, but new interests quickly side-railed medicine. From the moment he opened Euclid's *Elements*, mathematics hooked him. He found fascination also in the study of motion. The Age of Reason traces its roots to the synthesis of Galileo's two pet interests – mathematics and motion – from which Newton's calculus would rise.

At Pisa, while observing the rhythms of a cathedral's swinging lamp, whose oscillations he timed with his pulse, Galileo conjectured correctly that the period of a pendulum is independent of the amplitude of deflection from the vertical.[5] He was but seventeen years of age. Four years after matriculating at Pisa, without completing a formal degree, Galileo returned home and audaciously set up shop not as a doctor but as a professional mathematician.

Back in Florence, Galileo gathered pupils, constructed proofs, wrote papers, and gave occasional public lectures, a professor without a post. To earn his keep at home, he assisted his father in conducting meticulous household experiments to study the effects of variations in string diameter, tension, and length on the pitch and tonal qualities of strings. Galileo, the father of experimental physics, learned much of his craft at father Vincenzio's side.[6] The originality of Galileo's thought soon attracted notice. By 1589, his notoriety had spread from Florence to Pisa, where he secured a coveted position teaching mathematics at his *alma mater*.

The Alphabet of God

Galileo's tenure at Pisa began shakily. Like Florence, Pisa lay on the Arno River. Unlike Florence, Pisa frequently flooded. Galileo missed his first six lectures, delayed by floods and penalized for his absences. He chafed at wearing formal academic regalia, deeming it an anachronistic nuisance. And his lectures introduced demonstrations that unceremoniously dismantled Aristotelian physics, the core curriculum at every European university.[7] These antics enticed many followers but also garnered enemies.

When his father died at the age of seventy, financial responsibility for the family passed to the eldest son. Although Galileo assumed the burden graciously, a mathematics professor's annual salary was a paltry sixty *scudi*, one *scudo* barely sufficient for two weeks of basic sustenance. By contrast, some Tuscan military commanders received 2,500 *scudi* in compensation annually.[8] Fortunately, Galileo's star was ascending.

The year after burying his father, Galileo left Tuscany to chair the faculty of mathematics at the University of Padua, a post he held for eighteen years. Reflecting in old age, Galileo wistfully recalled those years in Padua as the happiest of his life. There he made friends easily, including well-connected ones. There he perfected his hallmark experiments on motion. And there he established the international reputation that made his name a household word and secured a burgeoning salary that eased the difficulty of caring financially for his siblings and mother.

Galileo's lectures electrified audiences. University students were accustomed to disputations, formal arguments on all manner of subjects. In contrast, Galileo exploited classroom demonstration to powerful effect: "One sole experiment, or concludent demonstration produced on the contrary part, sufficeth to batter to the ground . . . a thousand . . . probable arguments."[9] The combination of compelling argument *and* physical demonstration was breathtaking to behold. Audiences swelled to 2,000, necessitating the construction of a new lecture hall.[10]

In Padua, Galileo met Marina Gamba of Venice, fourteen years his junior, who bore him three illegitimate children: Virginia in 1600, Livia in 1601, and Vincenzio, named for his grandfather, in 1606. Galileo's relationship with his mistress, although informal, was more than hedonistic. By tradition, professors, as agents of universities sanctioned by the Church, seldom married. Galileo regarded Marina as his mate and loved his children. But marriage he avoided. In social conventions, Galileo would fit well in the modern era.

Galileo's science was equally modern. Today's experimental physicist finds a kindred spirit in Galileo, a master in elucidating *how* things work and a virtuoso of "irreducible and stubborn" facts.[11] The tone and style of his writing are familiar. Following the maxim of Francis Bacon "to educe and form axioms from experience [and] to deduce and derive new experiments from axioms,"[12] Galileo establishes the *modus operandi* of contemporary experimental science by blending piercing insight, mathematical formulas and proofs, and physical experiment.

Accounts of Galileo's experiments at Pisa's Leaning Tower are most likely apocryphal. Nevertheless, he did experiment extensively with falling objects to infer general laws. Contrary to the Aristotelian view, Galileo observed that (heavy) objects under the influence of gravity accelerate downward at identical rates regardless of weight. Originally he incorrectly conjectured that a falling object's speed was proportional to the distance it had fallen. Cleverly using inclined planes to "dilute" the effect of gravity, Galileo ultimately concluded that the velocity of a falling object released from rest is proportional to its *time* since release rather than to its distance from the release point.[13] Difficulties in measuring time accurately complicated Galileo's efforts; hence, his genius in diluting gravity to induce "slow motion," which had the effect of rendering his crude clocks more accurate.

Having made fundamental contributions to hydrostatics and fluid mechanics, Galileo turned this knowledge to a practical purpose: the development of a timing device more reliable than the human pulse.

For the measurement of time, we employed a large vessel of water placed in an elevated position; to the bottom of this vessel was soldered a pipe of small diameter giving a thin jet of water, which we collected in a small glass during the time of each descent, whether for the whole length of the channel or for a part of its length; the water thus collected was weighed, after each descent, on a very accurate balance; the differences and ratios of these weights gave us the differences and ratios of the times, and this with such accuracy that although the operation was repeated many, many times, there was no appreciable discrepancy in the results.[14]

The experimenter's fascination with time and motion led, in his later life, to the first pendulum clock.

Experiments with pendulums and rolling balls revealed similarities. A pendulum in motion does not rest at the bottom of its arc; rather it continues its journey past the vertical until attaining the same height at which it was released. Similarly a ball that rolls down an incline continues up a neighboring incline to rest at the level of its original release. Galileo concluded, unlike Aristotle, that objects in motion naturally tend to remain in motion. For this property, called *momentum* in today's parlance, Galileo appropriated the term *impeto*, Italian for the Latin *impetus*, a word first adopted for scientific purposes in the fourteenth century by Jean Buridan of the University of Paris.[15] Newton's first law of motion, in fact, only slightly refined Galileo's law of inertia, which asserts that a body "continues its motion in a straight line until something intervenes to slacken or deflect it."[16]

What is it that "slackens" or "deflects" motion? Galileo was first to grasp the key importance of change in motion: *acceleration*. Unlike Aristotle, who erroneously held that constant force was necessary to sustain motion, Galileo knew that it was acceleration – "the creation or destruction" of motion – that required external force. It was a conceptual sea change.

Through a series of propositions and theorems, Galileo, the facile mathematician and consummate experimentalist, proved what his careful eye had observed: that the distance an object falls from rest varies as the square of the time of its fall. His proof combined uniform lateral motion with accelerating vertical motion to conclude: "A projectile which is carried by a uniform horizontal motion compounded with a naturally accelerated vertical motion describes a path which is semi-parabolic."[17] Since then, projectiles of all types – from baseballs tossed skyward to Apollo 11's lunar excursion module on its descent to the Sea of Tranquility – have obediently followed Galileo's parabolic course.

Occasionally Galileo's thought experiments rivaled in ingenuity those of his successor Einstein. One such conceptual experiment revealed the fallacy in the Aristotelian paradox of the "misbehaving arrow." According to Aristotelian physics, an arrow, having detached itself from the motive force of an archer's bowstring, should fall straightway to earth. Galileo conjured a common shipboard analogy to convince otherwise. As every sailor knew, an object dropped from a ship's crow's nest fell to the foot of the mast. It mattered not whether the ship was in port or under sail. In the latter case, this was possible only if the object inherited the horizontal velocity of the ship. Set in motion, an object remains in motion.

A far deeper thought experiment of Galileo's led Einstein ultimately to general relativity. It originated to counter Aristotle's misconception that objects fall to earth at differing rates according to their weights. By both observation and logic, Galileo knew better: neglecting air resistance, two objects released simultaneously from the same elevation strike level ground at the same instant. Galileo's argument is modernized in *Einstein's Mirror:*

> Imagine that a building is ablaze, and that a man carrying an injured child jumps from the roof onto a stretched out blanket held by firemen below. If each body falls at a rate dependent upon their weight, the child will tend to fall slower than the man. The man will have to hold on tight to prevent the child rising above him as he falls. This suggests that the man and child together will fall at a rate part way between the rate of either. On the other hand, because the man and child together are heavier than either separately, according to Aristotle this combined bundle will fall together faster than either separately![18]

The seeming paradox vanishes only when child and man fall at the same rate, as Galileo correctly inferred.

At Padua, Galileo added "inventor" and "entrepreneur" to the skills of mathematician and experimentalist. He invented a device called the geometric and military compass. With improvements it served both as a geometer's compass and as a rudimentary slide rule capable of calculations including square roots. The first few he made himself, but as their popularity grew, Galileo hired an instrument maker. With canny business acumen, Galileo sold each compass for only five *scudi*, hardly more than it cost to produce. But he charged and pocketed nearly twenty *scudi* for instruction in its use (an object lesson not lost on Bill Gates, who has made billions from the recognition that the money is in software, not hardware). When the Venetian arsenal subsequently adopted the military compass to aid in shipbuilding, Galileo benefitted handsomely.

To further augment his salary, Galileo took in private students from elite families, among them the sixteen-year-old Don Cosimo de Medici, heir to the Duchy of Tuscany. Perhaps his name – Italian for "cosmos" – was a portent. In 1609, Galileo turned his attentions skyward.

Although Flemish eyeglass makers invented the spyglass, early models were but toys. Intrigued by their potential scientific applications, Galileo puzzled out the mathematics of refraction in a single night of intense reflection. Shortly thereafter he fabricated an instrument to rival the Flanders spyglass. Two months later, in a letter to his brother Michelangelo dated August 29, 1609, Galileo wrote excitedly of having perfected a spyglass with an optical power of 10, far surpassing that of the original instrument. The spyglass matured quickly into the telescope, thanks to Galileo.

Soon the Venetian Senate summoned him to demonstrate the instrument to the august body. To "the infinite amazement of all,"[19] from the tops of the highest campaniles in Venice, ships could be seen at sea two full hours before they were close enough to port to be visible to the naked eye. In recognition of his accomplishment, the Senate granted Galileo lifelong tenure and raised his salary five-fold.

Improvement by incremental improvement, Galileo raised the power of the telescope to a factor of 30 and trained it not on the sea but upon the heavens. In March 1610, he printed *Sidereus Nuncius* (*The Starry Messenger*), an illustrated book of his fresh astronomical observations. Intending *Sidereus* for the common man, Galileo wrote in colloquial Italian rather than Latin. He described seeing "stars in number more than ten times those previously observed," and of our nearest celestial neighbor, Galileo effused:

> It is a most beautiful and delightful sight to behold the body of the moon.
> . . . [It] certainly does not possess a smooth and polished surface, but one rough and uneven, and just like the face of the earth itself, is everywhere full of vast protuberances, deep chasms, and sinuosities.[20]

He astonished readers by announcing the discovery of "four planets, neither known nor observed by any one of the astronomers before my time." And in honor of his royal student (and to grease the skids for appointment as court mathematician), Galileo dubbed these moons of Jupiter as the "Cosmian stars." Cosimo, however, wished to share the glory with his siblings, and the moons were renamed the "Medicean stars." Clearly disobeying Ptolemy, these stars orbited a center different from the earth.

Three weeks following a favorable horoscope by Galileo, Duke Ferdinand died from illness. Upon his succession as Grand Duke of Tuscany, Cosimo appointed Galileo court mathematician, a lifetime post that carried a salary of 1,000 *scudi* and ensconced Galileo as chief mathematician at the University of Pisa.

An overnight sensation, *The Starry Messenger* earned Galileo international acclaim. Johannes Kepler, the German mathematician and astronomer, gushed, "I may seem rash in accepting your claims so readily with no support of my own experience. But why should I not believe a most learned mathematician whose very style attests the soundness of his judgment?"[21] But not all eyes on Galileo were admiring. At the height of fame, Galileo had unwittingly set himself on a collision course with the Inquisition.

The Lines of Battle

As court mathematician, Galileo returned to Florence. Upon his departure from Padua, Marina Gamba married a respectable man of her own social status. Graciously, Galileo encouraged the marriage. In Florence, he rented a villa with a roof terrace from which to behold the entire night sky. But to his dismay, Galileo found the "thin air" of Florence to be "a cruel enemy of my head and the rest of my body,"[22] from which he suffered all manner of physical maladies as well as depression and loss of sleep.

Earlier, in 1604, Galileo had witnessed the appearance of a new star in the heavens, now known to have been a supernova. It was a severe blow to Aristotle's premise of the incorruptibility of the heavens. Shorn of all Aristotelian inclinations, Galileo had quickly embraced the whole of Copernican cosmology. Unlike Copernicus, however, Galileo possessed an Olympian ego and formidable weapons with which to make the case for heliocentrism, among them the telescope and firsthand knowledge of the principles of motion. Thus armed, Galileo set out to dismantle the Aristotelian and Ptolemaic worldviews that had reigned for millennia. The Church, however, had founded its theology on Aristotle as well as Jesus.

The seemingly innocuous event that triggered the schism between faith and reason – an intellectual earthquake whose aftershocks continue to rock the modern world – was a dinner party.

As an official court function, the dinner boasted a host of illustrious guests, among them the duke's wife; his mother, Grand Duchess Christina; a number of academicians including Galileo's former student, Pisan colleague, and friend, Father Castelli; and Dr. Boscaglia, a professor of philosophy. The dowager duchess commandeered the conversation, steering it to the newly revealed Medicean stars, named in honor of her progeny. Were they real celestial objects or merely illusions? Castelli and Boscaglia both assured her of their genuineness.

Father Castelli then gracefully bowed out of the conversation and left the party. No sooner had Castelli departed than the duchess summoned him back to continue the discussion, which then ventured to the more controversial topic of heliocentrism. In a letter to Galileo, Castelli described with relish the subsequent events. Boscaglia had cautioned the grand duchess that the motion of the earth is suspect because it violated scripture. Castelli then jumped into the fray, winning over all present to Copernicanism, excepting the duchess herself and Boscaglia. His letter glowed in triumph. A follow-up letter reported that, ultimately, even the grand duchess had embraced the new worldview. It seemed that Copernicus and his ardent disciple Galileo might soon convert the whole world.

Castelli's letter had effects opposite those intended. Hot-blooded and quick to take umbrage, Galileo jumped to arms over Boscaglia's objections to Copernicus. A modern detractor describes Galileo's overreaction with hyperbole: "His counter-blast to the dinner-table chirpings of the obscure Dr. Boscaglia (who is never heard of again) was a kind of theological atom bomb, whose radioactive fall-out is still being felt."[23] By a long and impassioned *Letter to Castelli*, Galileo placed a philosophical stake firmly in the ground. Fully intending the position paper to be widely circulated, Galileo expanded it a year later into his now-famous fifty-page *Letter to the Grand Duchess Christina* of 1615. It begins:

> Some years ago, as Your Supreme Highness well knows, I discovered many things in the heavens which had not been seen before our own age. Both the novelty of these things as well as some consequences which naturally followed from them and which contradicted some physical notions commonly held among academic professors, stirred up a small number of professors against me – as if I had placed these things in the sky with my own hands in order to upset nature and overturn the sciences.[24]

Today, when universities are seen as hotbeds of open intellectual ferment, it may seem strange that fellow academicians numbered among Galileo's severest opponents. But the university then was the educational arm of the Church, and many scholars belonged to religious orders. Of these, the Dominicans and the Jesuits wielded the greatest influence. The Dominicans, deeply suspicious of any challenge to Aristotelian metaphysics, "constituted the rear guard for the ultraconservative and reactionary elements of Catholicism,"[25] while the Jesuits tolerated scientific revelation, provided it could be reconciled with orthodoxy. Regarding Copernicus, the Jesuits adopted a wait-and-see ambivalence.

Objections to Copernicus invariably had roots in theology, "Genesis, Psalms, Ecclesiastes, and Joshua . . . all expounding literally that the sun is in the heavens and travels swiftly around the earth."[26] In his *Letter to the Grand Duchess Christina*, Galileo fought fire with fire. Turning straightway to theology, he appealed to the wisdom of the holy fathers: Saint Jerome, Saint Aquinas, Saint Augustine, and others.

> They [Galileo's detractors] threw various charges and published many writings filled with vain arguments, and they made a serious mistake of sprinkling these with citations taken from places in the Bible which they had failed to understand properly, and which were far from suitable for their purposes.[27]

Accepting the Bible to be devoid of error, *provided its true meaning is understood*, Galileo turned the tables on his detractors. Do *they* who quote scripture to refute scientific evidence truly understand the scriptures' meaning? Surely not every written word was intended as literal. Otherwise "it would be necessary to assign to God feet, hands, and eyes, as well as corporeal and human affections such as anger, repentance, hatred, and sometimes even the forgetting of things past and ignorance of those to come."[28] Thomas Aquinas had acknowledged that the writers of the scriptures recorded "what God, speaking to men, signified, in the way men could understand and were accustomed to."[29] God's language was colloquial. One could miss the point if one were too literal.

Galileo also appealed to history. Copernicus was hardly the first to introduce heliocentrism. The idea was ancient, having originated in the sixth century BCE with the Pythagoreans. It was embraced by such guiding lights as Plato and Aristarchos of Samos and publicized by Archimedes in the *Sand-Reckoner*.

Then, Galileo appealed to the validity of the physical senses, the scientist's ultimate instruments.

> Nothing physical which sense experience sets before our eyes, or which necessary demonstrations prove to us, ought to be called into question (much less condemned) in the testimony of biblical passages which may have some different meaning beneath their words.[30]

To Galileo, seeing was believing. "I do not . . . believe the same God who gave us our senses, our speech, our intellect, would have put aside the use of these, to teach us instead such things as with their help we could "find out for ourselves, particularly in the case of these sciences of which there is not the smallest mention in the scriptures."[31] To the Peripatetics (those who unwaveringly followed Aristotle), faith subjugated the testimony of the senses, to the point that some, like Jesuit father Clavius, denounced what their eyes beheld.[32]

At heart Galileo's impassioned plea was to accept the book of nature as a new window on creation. "Within its pages are couched mysteries so profound . . . that the . . . studies of hundreds upon hundreds of the most acute minds have still not pierced them, even after continual investigations of thousands of years."[33] To censure Copernicus was to forbid access to the book of nature, "a [great] detriment to the minds of man."

Galileo's Catholic theology was sound, and his challenge to Church interpretation might have passed without serious incident but for two things: his timing was bad, and he had acquired too many enemies.

Sweeping Europe, the Protestant Reformation had put the Roman Catholic Church on the defensive. The best defense is often a good offense. Between 1545 and 1563, high Church officials met on two dozen occasions to enumerate and condemn Protestant heresies and to codify the Church's official position on the holy scriptures, traditions, and sacraments. The meetings, known collectively as Council of Trent, laid ideological fortifications from which the Church launched its fearsome Counter-Reformation. High on the list of the heretical proclivities of Protestantism was the propensity of individuals to read, interpret, and arguably misinterpret the scriptures by the light of conscience. The Council of Trent specifically "forbade going contrary to the teaching of the fathers when their teaching on a subject was unanimous and the subject itself was a matter of faith or morals."[34] Church doctrine was thereby considered sacrosanct if it passed two tests: (1) the holy fathers were unanimous on the subject, and (2) the matter pertained directly to faith or morality.

But did this ruling apply to Copernicanism? First, should not Copernicanism be consigned to the domain of science, not faith or morals? Second, since the holy fathers had not studied the issue, how could there be unanimity? It was Galileo's view that Copernicanism lay outside the jurisdiction of the council. With sleight of hand, in the *Letter to the Grand Duchess Christina*, Galileo attempted to shift the burden of proof in scientific matters from the domain of science to the domain of religion. This he did by the contention that "before a physical proposition is condemned, it must be shown to be not rigorously demonstrated."[35] In other words, the Church has no right to condemn scientific ideas until it has studied them and found them lacking. He justified this shift by asserting that fallacies are easier to uncover than certainties. In mathematics, this is generally true. One counterexample suffices to undermine a conjecture no matter how many examples support the conjecture. In science more generally, it isn't necessarily so.

It fell to Cardinal Bellarmine to sort out the Copernican issue. As both chief theologian and chief executive of the Roman Catholic Church, Bellarmine cut a powerful figure. As a Jesuit, he was intellectually and scientifically inclined. However, as the principal defender of orthodoxy, Bellarmine could be ruthless. By sentencing Bruno to the flames, Bellarmine had earned the fearsome appellation "hammer of the heretics." He found the intention of the Council of Trent abundantly clear: "Petulant minds must be restrained from interpreting Scripture against the authority of tradition in matters that pertain to faith and morals."[36]

To the most fundamentally pious, Galileo's despised telescope had uncovered a number of affronts to Aristotle and "illusions of the devil." Besides the moons of Jupiter, the telescope revealed imperfections in the moon's surface: mountains, valleys, and irregularities. In sacred art, the Blessed Virgin was oft portrayed with her feet upon the moon. When an audacious artist painted the newly revealed irregularities into her perch, the sacrilege was attributed to Galileo. The telescope also revealed the sun's blemishes – sunspots – which appeared to circumnavigate the sun, suggesting that it rotated about an axis. Wherever one pointed the blasted telescope, it contradicted Aristotle and provoked blasphemies.

Still, it remained to be seen whether Copernicanism officially touched upon "faith and morals." Fortunately for Galileo, another had sent up a trial balloon. The Carmelite Foscarini, who had written a book defending both Galileo and Copernicus, had requested Bellarmine's imprimatur. Bellarmine was equivocal. To Foscarini, he wrote unofficially:

> I say that it appears to me that your Reverence and Signor Galileo did prudently to content yourselves with speaking hypothetically and not positively, as I have always believed Copernicus did. For to say that assuming

the earth moves and the sun stands still saves the appearances better than eccentrics and epicycles is to speak well. This has no danger in it, and it suffices for the mathematicians.[37]

There then followed an ominous turn: "But to wish to affirm that the sun is really fixed . . . is a very dangerous thing, not only because it irritates all the theologians and scholastic philosophers, but also because it injures our holy faith and makes the sacred Scripture false."[38] To make his point, Bellarmine put belief in the movement of the sun on a par with that of the virgin birth:

> Nor may it be answered that [heliocentrism] is not a matter of faith, for if it is not a matter of faith from the point of view of subject matter [science], it is on the part of the ones who have spoken [the prophets]. It would be just as heretical to deny that Abraham had two sons and Jacob twelve, as it would be to deny the virgin birth of Christ . . .[39]

Bellarmine had ruled Copernicanism to be a matter of faith after all. Galileo and others had the freedom to consider it a hypothetical model of reality, but they skated on thin ice if they ventured beyond hypotheses. Bellarmine's response to Foscarini, intended also for Galileo's eyes, went a step further, placing the burden of proof back in science's camp and forcing Galileo into a corner. There were only two ways out: either supply convincing proof of the validity of the Copernican system or accept it merely as a working hypothesis.[40] Bellarmine's wait-and-see approach rankled the impatient Galileo.

By 1615, the religious world had split into two scientific camps: Galileists and anti-Galileists. The latter reacted as strongly to the messenger as to the message, for Galileo invoked strong passions. "He is passionately involved in this fight of his," wrote a confidant of the grand duke, "and he does not see or sense what it involves, with the result that he will be tripped up and will get himself into trouble."[41] Galileo took claims against Copernicanism as personal affronts and responded with fierce oratory and vehemence. Those incautious enough to publicly challenge him frequently found their arguments crushed by Galileo's tactics and themselves humiliated. His favorite device – *reductio ad absurdum* – assumed the opponent's point of view, followed it to the point of folly, and left the opponent looking imbecilic as the argument collapsed. "It was an excellent method to score a moment's triumph, and make a lifelong enemy."[42] One so mocked was the Florentine philosopher Ludovico delle Colombe. His Italian surname translated as "pigeons."[43] And so it was that the anti-Galileists became known as the "pigeon league." Licking their wounds, the pigeons lay in wait.

Prominent in the "pigeon league" were two Dominicans. Tommaso Caccini, a firebrand priest, denounced Galileo from the pulpit, tarring mathematicians in general and Galileo in particular as "practitioners of diabolical art [and] enemies of true religion."[44] Another, Niccolo Lorini, sent a copy of Galileo's *Letter to Castelli* to the inquisitor general in Rome, hoping for an official injunction against Galileo's apostasies. Rumors flew from Rome that Copernicanism might be banned outright.

Sensing a gathering storm, in late 1615 Galileo took the papal bull by the horns, so to speak. He requested and received permission from the duke to plead his case before the Vatican, where he had both friends and enemies in high places. Galileo's hopes were buoyed by the discovery, so he believed, of a secret weapon.

To satisfy Bellarmine, Galileo needed proof positive of Copernicus. There was much circumstantial evidence for Copernicanism, but no smoking gun. Copernicus had been well aware that, were his theory correct, Venus would demonstrate phases like our moon. Indeed, Galileo's telescope had revealed, among other wonders, the phases of Venus. Maddeningly, however, the Tychonic system, which Galileo considered a detestable compromise, also accounted for these phases, so they were not conclusive. Had Galileo been able to demonstrate stellar parallax, the Copernican issue would have resolved quickly. However, not until 1838 was a successful parallax measurement achieved, the stars being so vastly more remote than anyone could have imagined.

Galileo grew increasingly irate, indignant at "the ignorance, malice, and impiety of [his] opponents who had won the day."[45] To him, the validity of Copernicus was painfully obvious, and yet arguments capable of convincing the skeptics loomed just beyond reach. Had Galileo read Kepler's *Astronomia Nova*, then in his possession, he would have found strong circumstantial evidence for heliocentrism in Kepler's first two planetary laws. Whether constrained by pride or too impatient for Kepler's intricate mathematical calculations, "Galileo did not take refuge in Kepler."[46] Instead, he appealed to an academic strength: fluid mechanics.

Galileo's secret weapon was a theory of the tides. If the earth was stationary, as Ptolemy presupposed, what accounted for the motion of the tides? Galileo correctly attributed the tides to the rotation of the earth about its axis, but as Newton's predecessor, he lacked an understanding of universal gravitation. Therefore, Galileo's theory implicitly linked the tides to the relative motion of the earth and the sun rather than to the gravitational effect of the moon upon the earth.

By failing to account for the correct tidal period of 24 hours, 50 minutes, Galileo's tidal theory was weak, and it failed to convince Bellarmine. Still, he knew Galileo casually and respected his scientific accomplishments. When summoned before the pope, Bellarmine reported that he could find no fault in Galileo save his continued insistence in treating Copernicanism as reality rather than hypothesis. The time had come, however, to draw a line in the sand. At the pope's request and while Galileo was still in Rome, Bellarmine convened a panel of theologians to rule on Copernicanism. The eleven unanimously ruled as "formally heretical"[47] the contention that the sun anchors the cosmos.

Informed of the verdict immediately, Galileo was accorded an audience with Bellarmine at the cardinal's palace. After politely greeting Galileo at the door and extending other courtesies, Bellarmine cut to the chase. He informed Galileo of the panel's edict on Copernican theory and "admonished Galileo to abandon defending this opinion as fact." The very next week, the Congregation of the Index officially proclaimed Copernicanism as "false and contrary to Holy Scripture."[48]

The edict did not fault Galileo by name, nor did Bellarmine officially censure him. Indeed, he offered Galileo some succor in regard to his persecutors: he reassured Galileo of his high esteem among the congregation of cardinals and noted that "they would not lightly lend their ears to calumnious reports."[49] At the end of May 1616, Galileo departed Rome carrying a letter from Bellarmine exonerating him. The letter also firmly established that the ruling of the Sacred Congregation of the Index had been communicated clearly to Galileo: the earth does not move. Galileo returned to Florence with a false sense of security, having misinterpreted Bellarmine's words and letter as allowing him free pursuit of Copernicanism, provided he did so hypothetically.

"Sparkling with Malice"

In October 1616, at the age of sixteen, Galileo's oldest daughter, Virginia, took her vows to become a nun. A year later daughter Livia followed suit. In 1617, Galileo moved from Florence to a picturesque hilltop villa in nearby Bellosguardo, where the air was clear and the view charming. Florence, domed and tiled, sprawled below, on the north bank of the Arno River. To the east and below lay the village of Arcetri, whose Convent of San Matteo sheltered his daughters. And above spread the unobstructed dome of heaven, where Galileo could read from the book of nature whenever health and circumstance permitted.

Health, however, was not the norm. Years earlier, while visiting a cave in Padua, Galileo had contracted a mysterious illness, possibly malaria, which killed two colleagues. Ever since, debilitating illness had periodically confined him to bed. Other physical ailments also tormented him: gout, hernia, eye infections, and cataracts. At least once he visited the Loreto Shrine, ostensibly in search of cures.

His daughters sequestered, Galileo's household was an all-male bastion. Two new students, Mario Guiducci and Niccolo Arrighetti, took up residence with Galileo and his son. During his bouts of illness, Galileo's students assiduously transcribed his notes on motion, preparing them for eventual publication. In September 1618, Galileo's attentions were diverted by a new display from the book of nature, the first comet observed since the birth of the telescope. Bedridden, Galileo missed the show, but others, aided by his telescope, followed the small "hairy star" across the nighttime sky.[50]

The election of Galileo's student Guiducci to the prestigious Florentine Academy included the honor of his delivering a pair of lectures on comets. The paper's combative tone – it was clearly ghostwritten by Galileo – rekindled old animosities and escalated tensions. An angry rebuttal by a Jesuit father, Grassi, elicited Galileo's *Il Saggiatore* (*The Assayer*). This riposte carefully avoided direct mention of Copernican cosmology, but it mocked those who favored majority opinion over direct evidence in matters of science. Making his point in an everyday metaphor, Galileo wrote contemptuously: "But reasoning is like racing and not like hauling, and a single Barbary steed can outrun a hundred dray horses."[51]

In 1619, Marina Gamba died, leaving Galileo sole responsibility for their three illegitimate children. A year later his mother died at the age of 82. The same year, 1620, his early pupil, Grand Duke Cosimo II, passed away at the age of only thirty, leaving the duchy to his eldest son, ten-year-old Ferdinando II. Beyond these personal losses, two more deaths were noteworthy because they appeared to tip the delicate balance of power between Galileo and the anti-Galileists. Cardinal Bellarmine, instigator of the official anti-Copernican edict, died in 1621. Then in 1622 Pope Gregory died, and Cardinal Maffeo Barberini ascended to the papacy as Pope Urban VII.

Barberini and Galileo were old acquaintances. Although Barberini preferred to see the scientist use "greater caution in not going beyond the arguments used by Ptolemy and Copernicus, and finally, in not exceeding the limitations of physics and mathematics,"[52] he admired Galileo's ability. Early in his papacy, Urban went so far as to write in a glowing letter to Galileo's new patron, Duke Ferdinando II, "We embrace with paternal love this great man whose fame shines in the heavens and goes on earth far and wide."[53]

An intellectual and a lover of the arts, music, and fine architecture, the new pope commissioned composers and sculptures. Urban saw himself as progressive. There was every reason to believe his papacy would bring fresh air. Galileo sensed opportunity and seized it. In April 1624, at the age of sixty, he arrived in Rome. Warmly received by the pope, who granted Galileo no fewer than six private papal audiences, Galileo pleaded his case strongly: Italy was losing face in the world, and potential converts to Catholicism, especially in Germany, were scared off by the Church's scientific intolerance. The pope was not an easy sell. Pope Urban had never really supported Bellarmine's edict against Copernicanism, but he remained adamant that the holy book trumps the book of nature. Finally a fragile compromise took shape: Galileo was granted license to write about both Ptolemaic and Copernican worldviews, provided he did not take sides. Pontiff and scientist parted on good terms. Pope Urban even threw in a sweetener: the promise of a position as canon for Galileo's ne'er-do-well son, Vincenzio.

Upon returning to Florence, Galileo threw himself into writing what was to become the first of his two great works: *Dialogue on the Two Chief World Systems*. Into it flowed "all the force that his science, his religion, his life experience, and his flair for the dramatic allowed."[54] Delayed by illness and family obligations, Galileo labored for five long years.

Completed on Christmas Eve 1629, *Dialogue on the Two Chief World Systems*, "sparkling with malice,"[55] unabashedly promoted heliocentrism. Published in Italian in 1632, the book was instantly recognized as a literary and scientific masterpiece. Letters flowed in from a broad and adoring public. A fellow mathematician beamed, "Wherever I begin, I can't put it down."[56] Another reader effused, "You have given [the Copernican system] life, and . . . laid bare the breast of nature."[57]

Despite the book's bias toward Copernicus, Galileo genuinely believed he had fulfilled the letter of the injunction laid down by Bellarmine and Urban, if not the spirit. He had obediently submitted the *Dialogue* to Vatican censors and awaited their verdict for an interminable two years. With its preface lightly edited, the *Dialogue* passed muster, eventually garnering four official imprimaturs. By a shrewd literary device, Galileo believed that he had further protected himself. His insurance policy: the word *Dialogue* in the title.

The *Dialogue* was cleverly constructed as a conversation between Salviati, Galileo's alter ego, and Sagredo, an intelligent but skeptical character. Salviati's powerful voice ultimately overcame Sagredo's doubts, rendering the book strongly supportive of Copernican theory. A third character, Simplicio, played the fool. Into Simplicio's mouth, Galileo placed all the lame arguments of the Peripatetic philosophers. Both sides of the argument were thereby presented, but those of the Aristotelian opposition were simply straw men.

Set in the palace of the illustrious Sagredo, the *Dialogue* progresses over a period of four days. On day one, the characters dive immediately into the controversy, discussing the place of the earth in the solar system. In lengthy day two, the friends pick up the argument and turn to philosophy in general, weighing authority against observation in matters of truth. Day three turns mathematical, with explanations of the apparent retrograde motion of Mars, of the apparent variation in size of the planets, of the changing of the seasons, and of why sunspots migrate on an annual cycle rather than a diurnal one. On day four, the ebb and flow of the tides, Galileo's "secret weapon," is discussed. All in all, it was a rout for Copernicus.

Those who had lain in wait for Galileo to trip up now played their hands. Anonymous sources convinced the pope that he had been played the fool, insinuating "that the scientist had intended to make sport of Urban himself by impersonating him in the ingenuous and ignorant figure of Simplicio."[58] Urban erupted in anger that could not be mollified. Perceiving a double-cross, the Vatican ordered that all unsold copies of the *Dialogue* be bought back. The book, however, had sold out, and the prohibition simply created a thriving black market. Galileo was summoned to Rome to stand before the Inquisition. "May God forgive Signor Galilei," seethed the pope, "for having meddled with these subjects."[59]

Two overarching objectives had compelled Galileo to write the *Dialogue*: "first, to arouse general interest in the problem of Copernicanism among cultured persons, even though they were not versed in astronomy . . . ; and second, to educate the highest Vatican authorities to the dangers that the Catholic Church would encounter if it insisted arbitrarily in maintaining its attitude of 1616."[60] Galileo's success was near total in the first aim. In the second, he failed miserably, largely because the pope, preoccupied by other worries, was tone deaf. It is unlikely that Galileo intended to embarrass the pope. He did intend, however, to demonstrate that the Church and her functionaries were in grave danger of embarrassing themselves, and in this he was absolutely correct.

Galileo seriously misjudged the new pope. Instead of bringing fresh air, his papacy brought dark clouds. Impatient and arrogant, Urban deluded himself, "I know better than all the cardinals put together." Domineering and extravagant, he installed relatives in key posts, using nepotism to diminish opposition. As the Thirty Years' War turned sour, the pope, unable to sleep, silenced the birds in the Vatican gardens by having them destroyed.[61]

It was bad enough to have Catholicism at war with Protestants, but internecine conflict was still worse. In 1632, Rome plunged into political crisis. Directly challenging the pope, the Spanish ambassador, Cardinal Gaspar Borgia, attempted a coup under the pretext that Urban pandered to heretics by failing to create a league of Catholic states that would officially link religion and politics.[62] As the Galileo affair came to a head, the pope was in no mood to be further accused of coddling heretics. Galileo must pay his pound of flesh.

At nearly seventy years of age and incapacitated by ill health, Galileo received official summons to Rome. Through the pope's brother, Cardinal Francesco Barberini, his most influential friend in high places, Galileo pleaded for leniency. He reminded Urban of his age and health and that Vatican censors had vetted the *Dialogue*. To no avail. Galileo could either travel to Rome of his own free will or be dragged there in chains, but to Rome he must go. Galileo completed his will and dutifully obliged.

The accused arrived in Rome in late February 1633, having been underway for two weeks and in quarantine for another two, standard procedure in times of the plague. In a concession to his frail health, the pope permitted him to reside in the comfortable quarters of the Tuscan ambassador. Deeply fond of Galileo, Ambassador Niccolini did all within his power to ensure his guest's physical comfort and soothe his agitated mind. Niccolini too had friends in high places, and he conveyed to Galileo all he knew. The news was not good: forces were conspiring against him.

Summoned to Rome in haste, Galileo was humiliated by an agonizing limbo of nearly two months' duration. Not until April 12 was Galileo finally brought before the inquisitor for questioning.

Once the trial – too generous a word, really – got underway, it was swift and ruthless. Ten judges, all Dominican cardinals, presided over Galileo's case. Two were related to the pope; one was his brother (who was nonetheless sympathetic toward Galileo). Galileo had neither counsel nor access to the evidence against him. With the deck stacked, the verdict was a foregone conclusion.

To Galileo's surprise, the initial questioning focused on evidence dating to his first trip to Rome sixteen years earlier. In self-defense, he recounted truthfully, "Lord Cardinal Bellarmino informed me that the said opinion of Copernicus could be held hypothetically . . . [but] taken absolutely, the opinion could be neither held nor defended."[63] In corroboration, he produced Bellarmine's letter, dated May 16, 1616, which he had saved all these years. The letter and his subsequent audiences with the pope lent him leeway, he felt, to examine Copernicanism as a hypothetical means of "saving the appearances." To this point of the questioning, he remained truthful in maintaining his innocence. Surprise turned to shock, however, when the inquisitor produced an alternate version of Bellarmine's letter containing the harsher injunction, "If he should not acquiesce, he is to be imprisoned."[64] When retrieved by historians from the Vatican's secret archives, the alternate version contained no signatures, neither Bellarmine's, nor witnesses, nor Galileo's.[65] And it added the phrase "in any way whatsoever" to the prohibition against teaching Copernicanism.

Under the gravity of his predicament, Galileo began to dissemble by claiming that he wrote the *Dialogue* to *refute* the Copernican worldview. This low moment did not help his case.

All in all, there were four dispositions. At the second, a team of theologians examined the *Dialogue*, concluding that the book shamelessly promoted Copernicanism. Ironically, the only Jesuit among the three panelists took the most umbrage. Railing against Galileo, he spat, "He declares war on everybody and regards as mental dwarfs all who are not Pythagorean or Copernican."[66]

At the third disposition, on May 10, Galileo tendered his formal written defense, in which he steadfastly maintained "absolute purity of . . . mind" and challenged the alternate version of Bellarmine's letter. Finally, he pleaded for mercy, citing the mental torment he had endured for ten long months.

For the fourth and last time, he faced the inquisitors on June 21, his situation dire in the extreme. At issue was the question of "intent," which if necessary could be extracted using the instruments of torture. Galileo resolutely maintained his innocence but disingenuously claimed to support

Ptolemy's traditional cosmology over Copernicus' radical one. "Do with me what you please,"[67] he offered up in final resignation.

The verdict was pronounced the next day. He was "vehemently suspected of heresy" and subject to "all the censures and penalties [of] the sacred Canons . . . against such delinquents." Galileo well understood that such penalties might include the fate of Bruno.

The Bruno affair had so tainted the Church that, even to stamp out heresy, it could ill afford to execute one so prominent and revered as Galileo. Therefore, the inquisitors enjoined Galileo to confess his errors "with a sincere heart and unfeigned faith, [and] in our presence . . . abjure, curse, and detest the said errors and heresies." That is, by forced confession, Galileo could live and the Vatican could save face. It was hardly a magnanimous offer; even if it spared his life, the sentence consigned Galileo to the Vatican dungeons. As a final indignity, Galileo was required "seven penitential psalms once a week for three years."[68] Of the ten presiding cardinals, seven signed off on the terms.

Even under threat of execution, Galileo found two clauses of the confession so abhorrent he refused to sign. He had not lapsed in behavior as a good Catholic, and he had not used deceit in obtaining the imprimatur for the *Dialogue*. The mighty Inquisition backed down, and, in return, Galileo capitulated. Kneeling, he read his handwritten "confession":

> I, Galileo Galilei, . . . aged seventy years, arraigned personally before this tribunal, and kneeling before You Most Eminent and Reverend Lord Cardinals, Inquisitors-General against heretical depravity throughout the whole Christian Republic. . . . But whereas – after an injunction had been judicially intimated to me by this Holy Office, to the effect that I must altogether abandon the false opinion that the Sun is the center of the world and immovable, and the Earth is not the center of the world, and moves, and that I must not hold, defend, or teach in any way whatsoever, verbally or in writing, the said doctrine, and after it had been notified to me that the said doctrine as contrary to Holy Scripture – I wrote and printed a book in which I discuss this doctrine already condemned . . .[69]

Church authorities minced no words in their condemnation of Galileo, finding his heresies "more scandalous, more detestable, and more pernicious to Christianity than any contained in the books of Calvin, of Luther, and of all other heretics put together."[70] The news of Galileo's humiliation spread like wildfire throughout Italy and beyond. Perhaps the "unkindest cut" of all, the thing that Galileo most feared, was the complete prohibition of his life's work. Sure enough, the 1664 edition of the *Index* listed among its prohibited books Galileo's *Dialogue*. And there it remained for nearly two centuries.

"A Woman of Singular Goodness"

"It is a fearful thing to [face] the Inquisition. The poor man came back more dead than alive."[71] Such was the observation of Ambassador Niccolini, into whose custody Galileo was released following sentencing. For his part, Galileo felt "his name had been stricken from the roll call of the living."[72] Unable to sleep, he rambled about, babbling to himself, ruminating incessantly on the injustice of the sentence.

At the intercession of Cardinal Barberini, Galileo was spared the dungeons of the Holy See and released to the Tuscan embassy in Rome. From there he was transferred to Siena in July 1633. Still a prisoner, he landed in the custody of a sympathetic and kindly soul, Sienese archbishop Piccolomini, who took it upon himself to restore the great man's broken spirit. First, Piccolomini treated Galileo as an honored guest rather than a prisoner. Second, beyond basic charity, he gave Galileo a purpose, a project to utilize his expertise and occupy his mind. The cathedral needed a new bell, and Galileo, with his knowledge of fluid mechanics, was to supervise the casting. With the project's success, Galileo's spirits rallied. Rising like a phoenix from the ashes of circumstance, Galileo owed much to Piccolomini for the return of some vigor. In early 1634, pope Urban, believing Galileo to have been mollycoddled in Siena, re-sentenced him to house arrest in Arcetri.

Age and ill health, the arduous journey, the months of waiting, the unbearable uncertainty, and the unrelenting stress of the Inquisition, individually or in concert, could easily have snuffed out Galileo's candle. What kept him going against the odds?

First and foremost, Galileo steadfastly maintained his innocence, genuinely believing he had committed no crime. "I have two sources of perpetual comfort," he wrote to a supporter, "first, that in my writings there cannot be found the faintest shadow of irreverence towards the holy Church; and second, the testimony of my own conscience, which only I and God in heaven thoroughly know."[73]

To these two "perpetual comforts" could be added a third, so dependable that perhaps Galileo had momentarily taken it for granted: the abiding love of a devoted daughter. In a letter to Galileo, a family friend wrote from Pisa, "I know of no one who in the same way as she remained your unique and gentle comforter in your tribulations."[74]

From the ages of thirteen until their respective deaths, Galileo's two illegitimate daughters remained cloistered at the Convent of San Matteo. The older daughter Galileo had named Virginia after his beloved sister. Upon taking her vows as a nun, Virginia adopted the name Suor (Sister) Maria Celeste in homage to her father's fascination with the heavens. "Livia – Galileo's strange, silent second daughter, . . . also took the veil and vows at San Matteo to become Suor Archangela."[75] Of the two, Suor Maria Celeste accepted the cloistered life with resigned dignity and plunged herself into the daily affairs of the convent. She alone of his three children "mirrored Galileo's brilliance, industry, and sensibility, and by virtue of these qualities became his confidante."[76]

Despite the rigors imposed by the extreme poverty and deprivation of her order, Suor Maria Celeste doted on her father as did he on her. He sent her frequent monetary contributions, without which she might not have survived, while she prepared him treats of marzipan and plums, and medicines for his various ailments. Most striking, however, was her support throughout his trial and imprisonment, when she continually reminded him of his accomplishments and his stature in the world. Suor Maria Celeste's devotion to her order and her father exacted a severe price. As Galileo himself noted, "she had not paid much attention to herself."

In late March of 1634, Suor Maria Celeste Galilei contracted dysentery. Despite her father's daily visits, love, and prayers, she "slipped the surly bonds of earth" in early April. Within a year of his conviction by the Inquisition for heinous crimes, Galileo lost his Rock of Gibraltar.

"Pernicious Fruit"

"Bereft of my powers . . . I am left with no other comfort than the memory of the sweetnesses of former friendships."[77] Since 1628, when Galileo's household of freeloading relatives had returned to Germany, he had lived more or less alone. The solitude that once comforted him was now unbearable to the aged Galileo, broken, disgraced, and a prisoner in his own home. He turned to the only remaining solace he could find: the world of ideas. And against all odds, in the final eight years of his life, by sheer willpower and the strength of a still potent intellect, Galileo transformed the world and became the first man of the modern age.

Following his encounter with the Inquisition, Galileo had arrived in Siena the shell of a man. Under Archbishop Piccolomini's solicitous care, Galileo's thoughts returned to science as his psyche healed. Forbidden to write about celestial matters, Galileo turned attention to his first scientific love: motion. "To be ignorant of motion is to be ignorant of nature,"[78] Aristotle had espoused. As a keen student of the book of nature, Galileo had early recognized the fundamental importance of Aristotle's dictum. Twenty-five years previously, as a young professor at Pisa, Galileo had first cogitated about motion and had drafted a treatise on the subject, which was never published. Motion preoccupied him during his meteoric rise and happy days in Padua, when he laid the foundations for experimental physics by mastering pendulums and inclined planes. Now back in Arcetri mourning Suor Maria Celeste, Galileo found comfort by losing himself in his second great work: *Discourses Relating to Two New Sciences*. Soon he knew he was onto something big.

In a self-assessment confirmed by history, *Two New Sciences* was "superior to everything else of mine hitherto published."[79] In today's parlance,

Galileo's *Two New Sciences* were strengths of materials and dynamics, still foundational subjects for modern engineers. Set in a shipyard, *Two New Sciences* was intended to be practical. It was also rock solid, chock-full of mathematical propositions, theorems, and corollaries. The modern scientist or mathematician recognizes *Two New Sciences* as a textbook in mathematical physics.

For the great work, Galileo reconvened the familiar cast of characters featured in *Dialogue*: Salviati, Sagredo, and Simplicio. As before, the dialogue took place over four days, the first two devoted to the properties of materials and the latter two to the dynamics of motion. At the onset of day three, Galileo reveals his primary purpose: "To set forth a very new science dealing with a very ancient subject."[80]

Aristotle's whimsical views of motion had persisted for nearly two millennia. His confusion, which contaminated all subsequent thinking, lay in his failure to separate *how* from *why*. By entangling physics with metaphysics, Aristotle muddied the waters of understanding. Seeking clarity, Galileo focused exclusively on the how.

Aristotle had legitimately distinguished two categories of motion, which he termed "natural" and "violent." "Free" and "forced" motion, as we classify these categories in today's lingo, occupied days three and four, respectively, of *Two New Sciences*. Galileo was first to recognize that objects subjected to unbalanced forces accelerate. Falling objects, for example, experience natural downward acceleration. Ignorant of a theory of gravitation, however, Galileo contented himself with descriptions rather than explanations. Early on day three, he places prescient words into the mouth of Salviati: "The present does not seem to be the proper time to investigate the cause of the acceleration."[81] It took the combined genius of Newton and Einstein to explain the why.

By combining free and forced motions, Galileo could describe complex albeit commonplace phenomena. For example, he correctly inferred that velocity decomposes into horizontal and vertical components. On earth, only the later, influenced by gravity, is subject to acceleration. Recombining free motion in the horizontal direction with forced motion in the vertical yielded the now familiar parabolic trajectory of thrown objects. This Galileo phrased as Theorem I, Proposition I, on day four:

> A projectile that is carried by a uniform horizontal motion compounded with a naturally accelerated vertical motion describes a path that is a semi-parabola.[82]

Age and an "unquiet mind" hampered Galileo's progress on *Two New Sciences*. "I find how much old age lessens the vividness and speed of my thinking,"[83] Galileo wrote to a benefactor. Nevertheless, he persevered, and by demolishing one by one the Peripatetic arguments of Simplicio, Galileo, through Salviati, weeded a garden in which the seeds of modern science could grow. The first to sprout was physics, earning Galileo universal acclaim as the father of experimental physics.

In 1637 Galileo completed the last sections of *Two New Sciences*. Publication was problematic. On the one hand, the pope and the anti-Galileists continued their machinations to ensure that no further works of Galileo saw the light of day. "Under the lying mask of religion, this war against me . . . continually restrains and undercuts me in all directions,"[84] he fulminated. On the other hand, Galileo had attained international fame. In France, Galileo's admirers included the Jesuit educated philosopher René Descartes (himself a Copernican), the world-class mathematicians Marin Mersenne and Pierre de Fermat, and the astronomer Pierre Gassendi. In Germany, astronomer Johannes Kepler fell squarely in Galileo's camp.

Restrictions on Galileo gradually eased, perhaps a case of "out of sight, out of mind." Still forbidden to receive visitors, Galileo nevertheless entertained a procession of luminaries in his twilight years, among them the English philosopher Thomas Hobbes and Hobbes' countryman, the poet and philosopher John Milton. In a polemic defending freedom of the press, Milton recalled his visit to Italy and put his finger on the reason for the demise of Italian glory: "There it was that I found and visited the famous Galileo, grown old, a prisoner of the Inquisition."[85]

Surreptitiously, Galileo's admirers launched an international effort to locate a publisher. Heading the search was Elia Diodati, a Geneva-born Parisian. After considerable intrigue, Louis Elzevir of Holland stepped forward, visiting Galileo clandestinely at Arcetri in 1636. Section by section, the manuscript was smuggled to Elzevir and a cover story crafted to give Galileo a fig leaf of deniability should the book be intercepted.

By 1638, the terms of his confinement had relaxed sufficiently for Galileo to attend mass on feast days, provided he avoided personal contacts. If only the authorities had known. In the spring, *Two New Sciences* rolled off the presses in Leiden. Galileo, however, could no longer read his masterwork; his vision had succumbed to glaucoma and cataracts.

> This universe, which I with my astonishing observations and clear demonstrations had enlarged a hundred, nay, a thousand-fold beyond the limits commonly seen by wise men of all centuries past, is now for me so diminished and reduced, it has shrunk to the meager confines of my body.[86]

Into Galileo's shrinking and desolate world stepped Vincenzio Viviani, a sixteen-year-old boy, whom the enfeebled Galileo welcomed as a live-in companion and amanuensis. In return, Galileo mentored Viviani, who showed uncommon prowess in mathematics. Their hours together passed congenially, their relationship more like grandfather and grandson than teacher and pupil.

And then there were three. In 1642, Evangelist Torricelli – the precocious student of Galileo's old friend and colleague, Father Castelli – joined Galileo and Viviani. Within a month of Torricelli's arrival, however, Galileo took ill, confined to bed by pain and fever. His condition deteriorated, and on the evening of September 8, the great man slipped quietly away, lovingly attended by his son Vincenzio and his two pupils. Perhaps confidence in the legacy that the publication of *Two New Sciences* would ensure eased Galileo's transition from this world to the next.

Throughout the long years of his sentence, Galileo's supporters had been kept under surveillance. Counted among the loathsome sympathizers was the Sienese archbishop whose kindness had knitted together Galileo's brokenness. When the archbishop made too public his candid assessment of the Galileo affair, a spy was quick to report the infraction. "And since such seeds sown by a prelate might bear pernicious fruit, I hereby report them."

> Archbishop [Piccolomini] has told many that Galileo was unjustly sentenced by this holy congregation, that he is the first man in the world, and that he will live forever in his writings, even if they are prohibited, and that he is followed by all the best modern minds.[87]

And so it was. The seeds of "pernicious fruit" have sprouted the world over. As Galileo envisioned, "There will be opened a gateway and a road to a large and excellent science, into which minds more piercing than mine shall penetrate to recesses still deeper."[88]

In the year of Galileo's death, Isaac Newton was born, ushering in the modern age of which we remain a part. "If I have seen further than others," Newton acknowledged in a humble moment, "it is by standing upon the shoulders of Giants."[89] No doubt, Copernicus, Galileo, and Kepler were among the giants Newton had in mind.

Epilogue

Pronouncing Galileo "the greatest light of our time," his patron, Grand Duke Ferdinando II, assumed responsibility for funeral arrangements and a memorial worthy of the great man. It was not to be. Pope Urban, whose vengeance even Galileo's death could not abate, quashed the duke's plans for public orations and a mausoleum. Instead, Galileo was interred in a nondescript, unmarked tomb.

Two years later, in 1644, Pope Urban VIII died. Within forty-five minutes of his last breath, mobs demolished his statue at Collegio Romano, venting their spleen over the havoc wreaked by the Thirty Years' War.

In 1647, Galileo's devoted pupil Viviani succeeded him as court mathematician. Privately Viviani vowed to properly commemorate his teacher's burial site when the time was ripe. Viviani died in 1703, his obligation to Galileo frustrated by politics and circumstance. Having no children, Viviani willed to a nephew his earthly possessions and his unfulfilled responsibility. From the nephew, the responsibility to Galileo passed down to yet another generation.

The wisps of smoke that announced a new pope in 1730 brought also a new climate. Pope Clement XII was Florentine. The time had come to repatriate Galileo, a duty Viviani's heirs had not forgotten. In 1737, while Galileo's body was being exhumed for relocation to its new marble sarcophagus, excavators made a stunning find: two adjacent biers contained three coffins. In Galileo's tomb, two identical coffins were stacked. The remains from adjacent coffins were quickly identified as those of Galileo and Viviani, who had requested burial alongside his master. The third coffin, directly beneath that of Galileo, contained the remains a young woman.

In the riveting conclusion to *Galileo's Daughter*, Dava Sobel recounts: "The disciple [Viviani], driven to despair by his failure to pay the tribute [due] his mentor, had given Galileo something dearer than bronze or marble to distinguish his grave."[90] The young woman in the coffin was Suor Maria Celeste. In death father and daughter had been bound as closely as they had been in life. And together they remain.

In 1835, Galileo's *Dialogue on the Two Chief World Systems* was officially dropped from the *Index of Prohibited Books*, where it had lain since 1664.

On November 17, 1979, Pope John Paul II addressed the Pontifical Academy of Sciences to encourage open-minded reexamination of the Galileo case, and in 1982 the Galileo Commission was formally charged to investigate the "grave incomprehension" of the "Galileo affair."[91] In October 1992, 350 years following Galileo's death, Pope John Paul II officially apologized for the Church's harsh treatment of Galileo, conceding at last that the earth moves.

In 1995, NASA's space probe *Galileo* arrived at Jupiter following a six-year voyage, its delicate interplanetary maneuvers based upon Newtonian mechanics, which in turn rested upon principles of motion first articulated by Galileo. For two years *Galileo* relayed extraordinary photographs of the largest of the solar system's gaseous giants. Its primary mission accomplished, *Galileo* enacted a number of daring fly-bys of the "Medicean stars," the four of Jupiter's eleven moons first observed by Galileo nearly four centuries ago. In 2002, the camera of the aging spacecraft failed. Like its namesake, after opening our eyes to other worlds, *Galileo* went blind.

Four

The Star-Crossed Prophet

> I was almost driven to madness in considering and calculating this matter. I could not find out why the planet would rather go on an elliptical orbit (rather than a circle). Oh ridiculous me!
>
> – Johannes Kepler

Nestled between Bavaria to the east and the Black Forest to the west lies the old German state of Württemberg. Graced by low mountain ranges dotted with castles, Württemberg is charmingly picturesque. On the periphery of the region's *Hauptstadt*, Stuttgart, nearly twenty miles west of the city center, lies a medieval village, Weil-der-Stadt.

In the sixteenth century, Weil-der-Stadt was a free imperial city, a remote outpost of the Holy Roman Empire. And on this small island of Catholicism in a sea of Lutheranism, the astronomer Johannes Kepler was born in 1571. The timing, for one of Kepler's religious faith, ability, integrity, and temperament, could hardly have been worse.

The "Peace"of Augsburg

The religious partitioning of Europe into a patchwork quilt of religious enclaves stemmed from the Peace of Augsburg in 1555. Almost thirty years earlier, the Protestant Reformation had swept Europe like a firestorm, ignited when the Augustinian monk Martin Luther pinned his *Ninety- Five Theses* to the door of the Wittenberg Cathedral.

Luther's litany of complaints railed against the excesses of unchecked Catholicism, principally the selling of indulgences. However, at the heart of the Reformation lay a theological matter deeper than mere ecclesiastical malpractices: the matter of authority. Who has the authority to interpret the scriptures? Is the individual free to interpret holy writings according to the dictates of conscience, or is an intermediary between God and man necessary to prevent errors of theology? The latter point of view (and the Catholic monopoly on Christianity) was weakened by the serendipitous invention of the printing press, which placed bibles into the hands of the laity. With bibles widely distributed, individual interpretations of scripture flourished. With its temporal and spiritual power quickly eroding, the Roman Catholic Church unleashed the Counter-Reformation. Spiritual and physical battles raged for the souls of the lost and the allegiance of the saved.

The Peace of Augsburg legally rent Christendom asunder by codifying an uneasy truce between Catholicism and its principal rival, Lutheranism. Signed by the Holy Roman Emperor Charles V, the Peace of Augsburg incorporated the compromise *cuius regio, eius religio*: whoever rules sets the religion.

Before the era of the modern nation state, the Holy Roman Empire was a far-flung sprinkling of principalities ruled by an assortment of noblemen: princes, dukes, counts, and barons. Each had autonomous rule of his own territory, and each, following the Peace of Augsburg, established the official religion of his domain. By the standards of the day, the Peace of Augsburg contained a magnanimous provision. Although it forbade the practice of Catholicism in Lutheran territory and vice versa, it incorporated a grace period during which families could migrate to religiously compatible villages or regions, protected by law from economic penalty or persecution. As it happened, the duchy of Württemberg was "most contentedly Lutheran" in 1571 because the ruling duke was a Protestant.[1] Weil-der-Stadt, however, remained Catholic.

The Protestant Galileo

Unlike his predecessor Copernicus or his contemporary Galileo, Johannes Kepler sprang from the hinterlands of nobility. This accident of birth predisposed him to a lifetime of grinding poverty and devastating personal misfortunes. His alcoholic father Heinrich had a "vicious and inflexible nature."[2] When attempts at innkeeping failed, as they did frequently, Heinrich became a mercenary soldier, abandoning the family. Kepler's mother, Katharina, had a wicked tongue, an excess of curiosity, and an inclination to gossip that earned her many enemies.[3] When Kepler was only three, Katharina left her children in the care of their grandparents and ran off in search of her derelict husband. While she was away, Johannes contracted smallpox. Miraculously he survived. But the pox left its toll: permanently crippled hands, damaged eyesight, and poor immunity to disease.

When Kepler was six, a brilliant comet lit the nighttime sky. One evening, Kepler's mother grabbed him by the hand and led him up a hill, where the two watched the spectacle long into the night. In later life, Kepler retained precious few fond memories of his mother. Yet this one tender act of motherly instinct almost certainly changed the course of his history and ours, for it sealed Kepler's destiny as an astronomer.

A year later, the Keplers fled Weil-der-Stadt. Johannes' paternal grandparents, pious Lutherans in Catholic Weil-der-Stadt, held enough prestige that their religious leanings were initially overlooked. As the Counter-

Reformation heated up, however, their position became untenable, and the family relocated to Leonberg, ten miles east in Lutheran Württemberg. The move was fortuitous for Kepler, for it spurred his education. There he was enrolled in a Latin school, the formal educational conduit for true scholars. Though his studies were frequently interrupted by either illness or the need to help support his family, he finally finished the course of study, placing so highly on the *Landesexamen* that he earned a scholarship from the duke of Württemberg. From then on, Kepler's continued education was a gracious gift of the duke's, and Kepler would feel a lifelong obligation to him.

Although family provided him little comfort or solace, Kepler owed to his grandmother the foundations of the faith that would sustain him throughout life. Lutheranism appealed to him immensely for the freedom of conscience it afforded, though as a sensitive child he was troubled by the viciousness Christians could inflict upon one another through doctrinal rigidity.

After graduating from Latin school in 1584, Kepler entered the seminary at Adelberg. There Kepler thrived, immersed in the love of study and the world of ideas, some unorthodox. When warned by a fellow student that he was skating close to heresy, Kepler responded, "My beliefs are my beliefs. I will make no secret of them."[4] This trait would later serve him well in the world of science and nearly destroy him in the world of religion.

The fault line between faith and science had been established by St. Augustine (354-386 CE), the most influential thinker of the early Church, in *Enchiridion*, a handbook for Christians. In it Augustine admonishes the faithful: "When . . . asked what we are to believe in regard to religion, it is not necessary to probe into the nature of things, as was done by those whom the Greeks call the physici. . . . It is enough for the Christian to believe that the only cause of all created things, whether visible or invisible, is the goodness of the creator."[5]

To Augustine, while it was permissible to be curious about nature and to have a passing interest in natural philosophy, it was also unnecessary. All one really needs in this life is faith. Faith trumps science.

Thus, the Church's view toward science was subject to the whims of fortune. In the early Middle Ages, as the Church struggled against paganism, it held a dim view toward probing the nature of things. Early astronomy, in particular, was too closely allied with astrology and considered pagan. By the twelfth century, with paganism largely defeated, the Church relaxed and universities began to crop up throughout Europe, officially sanctioned by the Church and populated by professors who were agents of the Church. Among these jewels were Oxford and Cambridge in England and the universities of Heidelberg and Tübingen in Germany. When the Duchy of Württemberg transitioned to Lutheranism after the Peace of Augsburg, so did the town of Tübingen and its university.

Tübingen

In 1589, Kepler matriculated at Tübingen to study Lutheran theology at its seminary. There his passion for natural philosophy awakened. Years later, in his great work *Astronomia Nova*, Kepler reflected upon this awakening: "When for the first time in my life, I tasted the sweetness of philosophy, I was taken by a [strong] passion for it in general, not yet for astronomy in particular."[6]

The natural philosophy that seduced Kepler was not the disinterested variety established in Enchiridion. In the educational reforms instituted by Philipp Melanchthon, a compatriot of Martin Luther's, moral law and natural law were seen as coequals. Nature herself was a window into the mind of God. In contrast to Augustine, Melanchthon held that, to truly know God, one must read the scriptures and the book of nature. And as powerful tools of natural philosophy, both mathematics and astronomy were legitimate paths for seeking God.

Kepler quickly distinguished himself at Tübingen. By the end of his master's education he ranked second of fourteen in his cohort and was assigned the best instructors for his follow-on studies in theology. Among these was the great mathematician and astronomer Michael Mästlin, who influenced Kepler's life and direction dramatically, inadvertently diverting him from the study of theology. Galileo was similarly diverted by a mathematician, as was the Lion himself, Isaac Newton. Mathematics is a highly subversive discipline, it would seem, and mathematicians singularly predisposed to foster independent thinking.

In Kepler's day, astronomy and astrology were virtually indistinguishable. Having taken up the study of astronomy under Mästlin, Kepler naturally tried his hand at astrology and developed quite a reputation. Like his mentor Mästlin, Kepler was an ambivalent astrologer. On the one hand, he thought misguided and bordering upon superstitious attempts to predict specific events in the lives of particular individuals. On the other hand, as a student of the book of nature, it seemed apparent to him that correlations existed between celestial and terrestrial happenings. Do not, for example, the tides, the laying of eggs by sea creatures, and the menstrual cycles of the female human all follow the phases of the moon?

Tongue in cheek, Mästlin instructed the budding astrologer to always predict disaster, for disaster eventually came to everyone with certainty. Nevertheless, Kepler's horoscopes were cautious. He maintained a habit of incorporating what could be observed directly about the individual into what could be inferred from the heavens, to the chagrin of many a rogue.

His mentor Mästlin first introduced Kepler to the dangerous notion of Copernicanism by risking to loan the boy (who seldom returned borrowed books) a rare copy of *De Revolutionibus*.[7] In Kepler's day, at least four different cosmological models floated around the hallowed halls of Tübingen, each vying for legitimacy. Copernicanism fared little better in

Lutheran circles than it had under Catholicism, and none of the professors fully embraced Copernicus' radical model. Kepler, on the other hand,
quickly gravitated toward the Copernican view, recognizing "all the mathematical advantages that Copernicus has over Ptolemy."[8]

Integral to the education of the seminarians at Tübingen was disputation, a form of public debate in which students appropriated opposing
views and engaged in lively discussions before an audience. Kepler, who
could not espouse a point of view he did not truly hold, argued force-
fully for Copernicanism, raising the eyebrows of skeptical professors. In
1593, Kepler authored a short dissertation on Copernicanism, which tackled head-on one of the principal objections to the theory, namely that if the
earth moved, its inhabitants should have some sensation of that motion.
His intention was to present his work *Somnium* (*Dream*) as a disputation.
The head philosopher would have nothing of it, and *Somnium* remained
unpublished until after Kepler's death.

Kepler's attachment to Calvinist doctrine on the matter of "ubiquity"
further rankled his professors. Uncertain times were not compatible with
freethinking, and the Lutheran theology of the period was as inflexible as
any rigid fundamentalism of today. Ubiquity, which concerned the sacra-
ment of communion, was Luther's answer to Aquinas' transubstantiation,
wherein the host and the wine are literally transformed into the body and
blood of Christ at the moment of the Eucharist. Luther denied this, claiming that Christ is everywhere already present (ubiquitous) in the host and
wine. Kepler, who could find no mention of ubiquity in the scriptures, discounted Luther's doctrine, favoring the Calvinist interpretation that the
bread and wine remain just that. Lutheranism eventually abandoned the
doctrine of ubiquity, but not before Kepler had suffered dearly for his independence of thought.[9]

In addition to the standard fare of Plato and Aristotle, philosophy
classes at Tübingen occasionally introduced Kepler to less orthodox
thinkers. There Kepler discovered the geometric mysticism of Nicholas
of Cusa and the number mysticism of Pythagoras. He resonated with both
and became "number-intoxicated."[10]

As he neared graduation, it became increasingly clear to his professors
that Kepler was ill suited for the Lutheran ministry by temperament and
theological unorthodoxy.[11] Fearing embarrassment, the faculty sought a
face-saving alternative. As luck would have it, a way opened. Word came
of a Lutheran school in Graz, Austria, in need of a mathematics instructor.
Kepler was the man.

Graz

At the age of twenty-two, full of misgivings, Kepler borrowed fifty *gulden*,
bought a horse, and embarked with his cousin on the difficult journey to
Graz by horseback, barge, and wagon.[12]

Unlike Württemberg, which was comfortably Lutheran, Graz was a house divided. The seminary where Kepler was to teach had been founded in 1574 as a Lutheran alternative to the local Jesuit college. Both coexisted in Graz. Kepler's assignment included teaching mathematics and astronomy and serving as district mathematician. In this latter capacity, he was required to maintain calendars to facilitate astrological prognostications. His outspokenness quickly earned him the enmity of the school's rector, but his forecasting ability redeemed him. In 1595, when three of Kepler's astrological predictions proved true, the young man was transformed into something of a celebrity despite his apprehensions that astrology "nourished the superstition of fatheads."[13]

In Graz Kepler was soon introduced to Barbara Müller, a twenty-three-year-old with whom he was smitten despite the fact that Barbara was already twice widowed and had a child by her first marriage. Her father, a practical man of considerable success, was not eager to marry his daughter to a dreamer, and it would be four years before they were finally married.

At the seminary Kepler began to carve a niche as a scientist. In the midst of a lecture, Kepler had an epiphany that was to place him on the scientific map. His initial attraction to Copernican cosmology was partly mystical. On the one hand, as a man of deep faith, Kepler believed that the plan of creation must be elegantly simple. Of the Ptolemaic and Copernican cosmologies, the latter more simply described the universe. On the other hand, Kepler was always on the lookout for occult numerical patterns wherever they could be found. In Kepler's day, just six "planets" were known, counting the sun and the moon. "Why six?" asked Kepler. Could there be six planets because six is a *perfect number*, that is, an integer for which the following decomposition holds: $1 + 2 + 3 = 1 \times 2 \times 3$.[14] Was there also, for example, an occult relationship among the orbital radii of the planets?

Of a Platonic and Aristotelian bent, Kepler examined mathematically perfect geometric objects for answers to his queries. Following Aristotle, Ptolemy and Copernicus had both presumed that the planetary orbits were perfect circles. Kepler also initially accepted circular orbits without question. Regarding the relationships of the orbital radii, armed with a naive belief in the geometrical perfection of the solar system, Kepler examined all sorts of geometrical arrangements among the planets. For example, inscribe within a circle many equilateral triangles. Their sides form tangents to an inner circle (Fig. 6). What is the ratio of the diameters of the two circles? The same question may be asked for inscribed regular polyhedra, polygons of four, five, six, or more congruent sides. Try as he might, inscribed polygons did not result in ratios that correlated with the approximate orbital radii of the planets. Then Kepler remembered the Platonic solids, of which, curiously, there are exactly five: the tetrahedron, the hexahedron (or cube), the octahedron, the dodecahedron, and the icosahedron, with 4, 6, 8, 12, and 20 identical faces, respectively. He suspected that

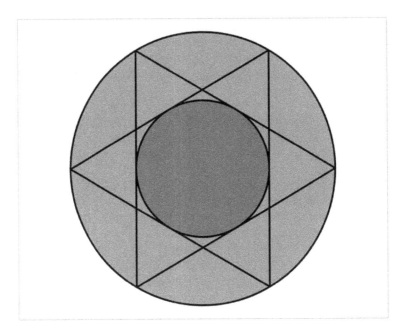

Figure 6: Triangles inscribed in circle; circle inscribed in triangles. (Author.)

the six planets could be related to the five Platonic solids. To his joy, when these objects were inscribed within or circumscribed by perfect spheres in a particular arrangement, the successive radii of the concentric spheres bore some correlation to the apparent radii of the orbits of the six known planets (Fig. 7).

Kepler's delight at his "discovery" knew no bounds. "The intense pleasure I have received from this discovery can never be told in words. . . . I shunned no toil of reckoning, days and nights spent in calculation, until I could see whether my hypothesis would agree with the orbits of Copernicus, or whether my joy was to vanish in air."[15]

On the basis of this discovery (more imagined than actual, it turned out), Kepler completed in 1595 his first major astronomical work, *Mysterium Cosmographicum* (*The Cosmographic Mystery*) and set out to find a publisher. He sought the endorsement of his professors back in Tübingen. Mästlin was moderately enthusiastic about *Mysterium*; others were less so. Kepler's geometric model of the cosmos had the sun at its center, and his unabashed promotion of Copernicanism offended the staid sensibilities of his conventional reviewers as unnecessary and unwise. Still, there was grudging recognition for Kepler's accomplishments.

In *Mysterium* Kepler also speculated on an *anima movens*, a motive power that compels the planets to orbit. He envisioned the mysterious force as emanating radially from the sun and sweeping the planets along like a rotating paddle wheel. To account for the longer periods of revolution of planets more distant from the sun, he conjectured that this force diminished inversely with distance from the sun. Today, some of Kepler's notions seem almost silly; nevertheless, his speculations contain the seeds of Newton's theory of gravitation.

Mysterium was accepted for publication in late 1596. When the first copies of his book arrived a few months later, Kepler sent copies to all to whom he was indebted, whom he needed to placate, or who shared his passion. The recipient list included his benefactor the duke of Württemberg, the renowned astronomer Tycho Brahe, and a still obscure mathematics professor in Padua, Italy, by the name of Galileo Galilei. Kepler's star was rising, but his joy was short-lived.

Two months after his birth, the Keplers' first child, Heinrich, died. A year later, their second child, Susanna, also died just a month after her birth. The household plunged into despair. Meanwhile, storm clouds gathered over Graz. In 1596, with the arrival of the young Catholic archduke Ferdinand, the Counter-Reformation entered the religiously schizophrenic city.[16] That Kepler's former professors at Tübingen were among the most vociferous at baiting the pope both disgusted him and undermined Kepler's chances of peaceful coexistence in Graz.

In September of 1598, Ferdinand decreed that all servants of Lutheran churches and schools must evacuate his territory within fourteen days, under the penalty of death. Kepler fled, temporarily leaving his family. A month later, unlike most other exiles, Kepler was exempted from the decree and allowed to return to his post as district mathematician. Ferdinand, it seems, had recanted after recognizing the advantages of having an eminent scholar in his domain, confident that Kepler would eventually convert to Catholicism.

Back in Graz, Kepler was mired in a catch-22. Deeply rooted in Graz, Barbara was loath to relocate. However, without the sermons and sacraments of the Lutheran soil that sustained him, Kepler felt isolated and spiritually malnourished. Converting to Catholicism would have spared Kepler and his family a great deal of misery, but conscience did not allow it. In the autumn of 1600, the archduke gave him an ultimatum: accept Catholicism or be expelled from Graz. In the presence of a thousand witnesses, Kepler stood firm in Lutheranism, sealing his fate and that of his family. With six weeks to vacate Graz, his predicament was desperate. Personal and financial ruin, or worse, loomed on the horizon.[17]

Fortunately, the publication of *Mysterium* had produced, among other things, a three-way flurry of letters between Kepler, his former teacher Mästlin, and the preeminent European astronomer, Tycho Brahe. In June 1599, Tycho had moved from Denmark to Prague to assume the position of imperial mathematician under Emperor Rudolf II. Tycho, remember, had proposed a hybrid cosmological model that incorporated elements of Ptolemy and Copernicus. Although Kepler rejected this theory as in-

Figure 7: Five Platonic solids, nested as Kepler envisioned the solar system: Kepler Museum, Weil-der-Stadt, Germany. (Ulrich Rist.)

elegant, he was painfully aware that his own model of the solar system based on the platonic solids did not quite fit observational data. Second to none, Tycho's observational data just might help Kepler correct his own cosmological system. Naively Kepler longed for a faculty position at his *alma mater* Tübingen, unaware that he had long ago burned those bridges. When Tübingen failed to provide its controversial alumnus a safety net, Kepler played his ace in the hole. He wrote Tycho for permission to visit Prague, requesting to examine Tycho's data to perfect his theory. With characteristic magnanimity, Tycho replied, "Come not as a stranger, but as a welcome friend."[18]

As Rudolph's imperial mathematician, Tycho had recently begun what would ostensibly be his life's work: the meticulous compilation of the Rudolphine Tables of astronomical data, named to honor his patron. Tycho's planetary observations were by far the most accurate in existence, phenomenally accurate in fact. Astronomical tables compiled prior to the introduction of the telescope typically held margins of error as gross as five degrees of arc. Tycho's wizardry with mechanical devices reduced the margin of error of naked-eye measurements to just ten *seconds* of arc, that is, 1/360 of one degree, a stunning 1800-fold decrease in measurement error.[19]

Following Kepler's initial visit, Tycho offered Kepler a position as mathematical assistant. Kepler hesitated for legitimate reasons, citing that "for observations his sight was dull, and for mechanical operations his hand was awkward."[20] Secretly, he also bridled at the thought of being Tycho's assistant, feeling himself every bit Tycho's equal. Yet on September 10, 1600, when the clock ran out on Kepler's welcome in Graz, Kepler accepted Tycho's offer. Accompanied by his wife and stepdaughter, Kepler set out for Prague with all the family's belongings in two small wagons.

Prague

At the time, Kepler could not have known that his fortunes were about to improve or that Prague would become his home for twelve productive years. Tübingen had rejected him in his time of need. He and his family were refugees. On leaving Graz, his wife had relinquished most of her inheritance. How Tycho would receive him and under what terms was uncertain. Kepler took some comfort by seeing himself as a martyr for Lutheranism, but the stress of uprooting was enormous. Falling ill along the way, he arrived in Prague sick and virtually destitute.

Happily Tycho received Kepler warmly. However, the two men could not have been more different. Tycho was wealthy, robust, and a mechanical genius. Kepler was nearly indigent, sickly, mechanically awkward, and of poor eyesight. Nevertheless, in mathematical ability, Kepler was Tycho's superior. At first it was a marriage of convenience. Kepler needed employment, and he needed Tycho's data, while Tycho needed Kepler's mathematical prowess to turn his data into theories. Both men were touchy and proud. Tycho jealously "protected his observations like a wolf guarding her cubs,"[21] relinquishing them only gradually to Kepler as their mutual trust grew. Kepler's fragile ego and Tycho's big ego often clashed. At bottom, however, both men were essentially good-hearted, and eventually their egos were chipped away until the marriage of convenience transformed into one of genuine affection. Tycho eventually embraced Kepler as a full partner.

Their collaboration might have been long and productive, but fate took an ironic twist. Within the year, Tycho died suddenly and somewhat mysteriously, having barely begun the task that could have occupied a lifetime. On his deathbed, while repeating his dying mantra, "Let me not seem to have died in vain,"[22] Tycho entrusted the completion of his astronomical tables to his new partner. Kepler dutifully accepted the task, remaining faithful to Tycho's memory and work for the remainder of his own life. Immediately following Tycho's death, Kepler was appointed his successor as imperial mathematician. In just over a year, Kepler had gone from indigent refugee to a position of highest esteem. Letters of congratulation poured in from across Europe.

Still, Kepler lacked Tycho's gregarious personality and his clout. His official salary of 500 *gulden* annually was only a sixth of Tycho's. But the official numbers were largely irrelevant, as the imperial coffers were often depleted. The superstitious Emperor Rudolph maintained imperial mathematicians not for their astronomical and mathematical abilities but for their astrological prognostications, upon which he increasingly relied as he sank ever deeper into madness. The prince's weak and erratic leadership left Bohemia insolvent and in chaos. Kepler's pay remained in arrears; often he received only promises or irredeemable notes. It seems that the stars to which he devoted his life had predestined Kepler to live constantly on the edge of poverty.

Meanwhile, the religious strife that had always dogged Kepler followed him to Prague, whose religious affairs were more convoluted than those of Graz. Graz had had but two factions, while Prague boasted four. In addition to Catholics and Lutherans, Prague was home to the Ultraquists, a schism representing liberal Catholicism, and the Bohemian Brethren, a radical Protestant sect. Just two years after Kepler's arrival in Prague, Emperor Rudolf decreed that only Catholics and Ultraquists were welcome there. Fortunately, preoccupied with the toys of his imperial museum and hounded by mental instability, Rudolph failed to follow through on his pronouncements. For the moment, Prague remained religiously integrated, and while the sun of tolerance shone, however feebly, Kepler worked feverishly.

Kepler's first two mature works were published during the Prague years: *Astronomiae Pars Optica* (*The Optics of Astronomy*, 1604) and *Astronomia Nova* (*The New Astronomy*, 1609). These were as scientifically sound as *Mysterium* had been naive and fanciful. In the first, Kepler established the science of optics by elucidating the rules of refraction. In the second, he laid the foundations of modern astronomy by specifying the first two laws of planetary motion.

The laws of planetary motion had revealed themselves to Kepler only grudgingly. When Kepler first arrived in Prague, one of Tycho's assistants had been working for some time without success on the irregular (retrograde) motion of Mars. With uncharacteristic bravado, Kepler bragged that he could solve the problem in weeks. Ultimately, it took years. Numerous Aristotelian and Ptolemaic misconceptions had to be dislodged one by one. Particularly problematic was the assumption that the planetary orbits are circular. Using this assumption, Kepler could never quite match the velocity of Mars to any of a myriad of empirical laws. For some of these laws, the discrepancy between the predicted and observed position was as small as eight minutes of arc (approximately 1/8 of a degree),[23] but confident of the reliability of Tycho's measurements and possessed of an integrity that disallowed force-fitting data to theory, Kepler knew that nature had not yet relented.

Kepler continued to search for the key. Ultimately he abandoned uniform circular motion to consider time-varying circular motion. To this end, he examined eccentric circular orbits, with the sun displaced from the center. Perhaps motivated by earlier musings regarding motive force, Kepler hit upon the construction of rays from the sun to the orbit, which divide the disc into (nearly) triangular pieces (Fig. 8). When the "triangles" have equal areas, a pattern emerged, that when explicated became Kepler's famous second law: equal areas are swept out in equal times. Whenever nature yielded a glimpse of the mind of God, however fleeting, Kepler was beside himself with joy.

The joy was often short-lived. More detailed examination of eccentric circular orbits revealed that the theory still retained small errors.

By unflinching integrity and sheer determination, Kepler compensated for what he lacked in natural gifts and physical health. Abandoning circular orbits, he examined oval ones, of which there are infinite varieties. After countless failed attempts and six years of unremitting labor, he turned to a particular oval orbit now familiar to every student of high school geometry: the ellipse.

In hindsight it seems strange that Kepler had not first tried the ellipse, which had been known to Greek geometers for centuries. The Greeks were well versed in the conic sections, the mathematical shapes that result from slicing a right circular cone at various angles to its axis. These canonical shapes include circles, ellipses, parabolas, and hyperbolas. Recall that Galileo had uncovered nature's predisposition for the parabola in the paths of projectiles. But the ellipse was unfamiliar to Kepler.

Having stumbled upon the ellipse as a possible orbital path, Kepler naturally took the next step: placing the sun at one of the ellipse's two foci. Nature's lock yielded immediately to the elliptical key. When its path was presumed to be elliptical, Mars matched Tycho's precise observations exactly. Moreover, unlike the other oval-shaped orbits Kepler had examined, elliptical orbits exactly preserved his second law: equal areas in equal times. Another Keplerian law had been born, later renamed the first: the planetary orbits are ellipses, with the sun at one focus.

Kepler's delight was unbounded. Now that Mars had been conquered, Kepler half-jokingly appealed to Rudolph to fund the war on "Mars' other relations, father Jupiter, brother Mercury, and the rest."[24] But it was not to be. While Kepler was preoccupied with extracting order from the heavens, forces on earth were allying against him.

First, his home life was anything but smooth. He, an unabashed intellectual who loved to discuss religion and ideas, and Barbara, a woman of

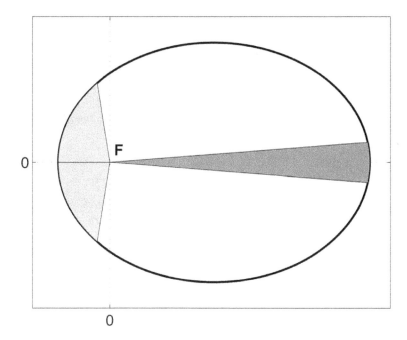

Figure 8: Illustration of Kepler's first and second laws: 1) that the orbits of the planets are elliptical, with the sun at one focus (F) of the ellipse; and 2) that rays from the sun to a planet traveling an elliptical orbit sweep out equal areas in equal times. More specifically regarding the second law, because the light and dark gray sectors of the ellipse shown have identical areas, the planet must travel faster when nearer the sun (sweeping out the light gray sector) than when farther from the sun (sweeping out the dark gray sector). (Author.)

simple piety, had little to talk about. But ill-suited as they were to each other, they managed to produce three more children, Susanna, Friedrich, and Ludwig. Of the three, there was something about Friedrich that particularly tugged at the heartstrings of both mother and father.

Meanwhile, the fragile peace that had temporarily graced Prague began to unravel. In 1608 Protestants formed the Protestant Union. Responding in kind, Catholics founded the Catholic League. Both conscripted armies, mostly of mercenaries unconstrained by any scruples. In 1611, when a Catholic army attempted to occupy Prague to enforce the religious order that Rudolph had failed to impose, a Protestant militia lay in wait. After routing the Catholics, the Protestants embarked upon a rampage of burning monasteries and assassinating monks. When their bloodlust went unfulfilled, mobs entered Jewish ghettos and indiscriminately killed every Jew they could find. Any hope that Kepler had of remaining in Prague vanished in 1612 with the death of his patron Rudolph, which ended the golden age of Prague.

As the outer world crumbled, so Kepler's life disintegrated from within. Barbara took chronically ill. The three children contracted smallpox, and Friedrich died. Barbara's recovery faltered, a casualty of unspeakable grief. "The boy was so close to his mother, people would not simply see their relationship as merely love, but as a deeper, more lavish bond," wrote Kepler of his wife's anguish.[25] Shortly thereafter, Barbara too slipped away, succumbing in 1611 to the triple ravages of illness, depression, and grief.

Linz

By 1610, the handwriting was on the wall: Kepler could not remain in Prague. Prior to Barbara's death, Kepler had initiated negotiations for a teaching post in Linz, a moderate-sized Austrian town similar to Graz, where Barbara had been happy. When Prague descended into religious warfare, Kepler accepted the safe harbor at Linz, where he moved with his remaining two children. Were there justice upon the earth, Kepler's fourteen years in Linz, the autumn of his days, should have been filled with tranquility and prosperity. Instead, they were mostly a hell on earth.

It began well enough. In 1613, Kepler remarried, choosing his second wife for her practical qualities: "not a trace of pride, no extravagance, patience at work, an average knowledge of how to keep house, middle-aged and enough sense to learn what is missing."[26] Her name was Susanna, like that of his eldest daughter, and she was relatively well educated for a woman of the time. The two were well suited. Still in her childbearing years, Susanna bore Kepler six children in fourteen years, the first three in rapid succession. But soon all three succumbed to tuberculosis or pox. On the heels of these grievous losses, fate cast still more blows. Kepler's stepdaughter Regina, whom Kepler had raised as his own, died suddenly, leaving behind a devastated husband and children. Then, in 1619, Kepler was excommunicated by the Lutheran Church, to which he had remained faithful at great cost to himself and his family. His crime: doctrinal errors and "fantasies," among them his Calvinist leanings on the Lutheran doctrine of ubiquity, now well documented in exchanges of letters with rigid Württemberg authorities. Then his sister Margaretha sent word that their aged mother Katharina had been accused of witchcraft.

Kepler's work on the Rudolphine Tables demanded tedious calculations and periods of uninterrupted concentration. When life's vicissitudes made attention to them impossible, Kepler turned elsewhere for diversion and solace. Tossed about like a shipwrecked sailor on a raging sea, Kepler clung to the one piece of driftwood that had always borne him to safety: the idea that there is order in the universe. Focusing on the order rather than the maelstrom: that was the secret to Kepler's resilience. For such times, Kepler kept an unanswered question to gnaw on: Was there a simple relationship between the temporal period of an orbit and its spatial dimension?

In Linz, nine years after the publication of *Nova*, beaten down but not yet broken, Kepler hit upon the elusive mathematical relationship: The ratio of the cube of a planet's mean distance from the sun and the square of its period (i.e., its "year") is a constant; that is, the ratio is the same constant for all planets.

In 1619, he published *Harmonice Mundi* (*Harmony of the World*). The title is testament to Kepler's undergirding faith. The final section contained his now-famous third law of planetary motion. His sense of accomplishment is best captured in his own words:

> What I prophesied two-and-twenty years ago, as soon as I discovered the five solids among the heavenly orbits . . . that for which I joined Tycho Brahe, for which I settled in Prague, for which I have devoted the best part of my life to astronomical contemplations, at length I have brought to light, and recognized its truth beyond my most sanguine expectations.[27]

On the heels of this success, Kepler set out to record all his discoveries in a single volume. What emerged was the first clear and popular textbook on Copernican astronomy, entitled *Astronomiae Copernicanae* (*Epitome of Copernican Astronomy*). Published in 1621, it immediately joined Copernicus' Revolutions on the *Index* of banned books, depriving Kepler of much-needed royalties.

The same year, Katharina's trial for witchcraft came to a head. A troublesome old crone, Katharina Kepler had many enemies. From a simple domestic dispute, forty-nine charges were trumped up, more than sufficient to convict her.[28] On September 28, 1621, she was, according to the verdict, shown the instruments of torture in order to extract a "voluntary" confession, so that her immortal soul might be saved before her body was offered to the flames. But Katharina, unlike almost anyone else who had faced the *territio verbalis*, resolutely refused to confess. Against all odds, she was released, thanks in part to the numerous intercessions of her famous son, by letter and in person. Little good freedom brought her. Broken by age, harsh mistreatment, and imprisonment, she died shortly after her release at the age of nearly eighty.[29]

Also in 1621, Kepler finally completed the *Tabulae Rudolphae*, fulfilling his twenty-year-old promise to Tycho. As the first truly accurate astronomical tables, the *Rudolphine Tables* were a boon not only to astronomers but to seafarers as well, who could now navigate by the night sky. They deserved publication, but conditions in Linz made that impossible, as they also made Kepler's continued residence there unsafe.

Countryless

In 1626, a peasant revolt overtook upper Austria. Mobs burned monasteries and cloisters and plundered the castles of the well-to-do. The Linz home of Kepler's printer, commissioned to print the *Rudolphine Tables*, was burned to the ground. For the last time, Kepler packed up his household, loaded it and his family onto a barge, and struck out in the middle of winter for another safe haven. But this time, there was nowhere left to turn and no benefactor to turn to.

Kepler found temporary quarters for his family in Regensburg, the trip farther upriver impeded by ice. He then struck out on his own on errands deemed so important they could not wait until spring. In Ulm, he found a printer, and in 1627 the *Rudolphine Tables* were finally published. Kepler paid for the printing out of his own pocket, with what it is hard to say.

The completion of the *Tables*, a goal so long stymied, marked a crisis point for Kepler. With his major task fulfilled, Kepler feared that his vocation as imperial mathematician would evaporate. With foreboding, he journeyed to Prague to see Emperor Ferdinand, who had succeeded his uncle Rudolph and who, as a boy, had forced Kepler to flee Graz. To Kepler's relief, Ferdinand received him graciously. The emperor offered Kepler back pay and a university faculty position. It seemed too good to be true, and it was. The offer was conditioned upon Kepler's conversion to Catholicism. He refused, on the bedrock principle that he could not submit to the pope in matters of conscience. His stand angered both Lutherans (because he considered conversion) and Catholics. Kepler left Prague empty handed: without a country, a church, or a position and with friends of both persuasions praying for his salvation.[30]

If religion could not save Kepler, astrology might. A new and unlikely patron emerged: Ferdinand's charismatic general, Albrecht von Wallenstein. Ironically, Wallenstein had earned his status as national hero by crushing the Protestant rebellion. His star rising, Wallenstein needed an astrologer to feed his ego. Kepler reluctantly accepted the general's patronage but never pulled his punches in his predictions. Among these, he forecast "a horrible disorder" that coincided eerily with Wallenstein's assassination in 1634.[31]

The few years in Wallenstein's employ provided Kepler temporarily with a salary, a home in Sagan in Silesia, some domestic joys, including the marriage of his daughter Susanna, and a degree of protection from the atrocities of the Thirty Years' War. However, when Ferdinand sacked Wallenstein, Silesia erupted in warfare. Kepler, now distrusted by all, left Sagan on horseback in a final quixotic search for a nonexistent refuge in a world insanely committed to religious conflagration.

On November 2, 1630, exhausted by an arduous journey in bitter weather, Kepler rode for the last time into Regensburg, the small town on the Danube through which all his journeys seemed to pass. While resting at the inn of an acquaintance, he contracted a fever, which passed into delirium. For several days Kepler drifted in and out of consciousness while a procession of Lutheran ministers came and went, each refusing to offer him communion.

Johannes Kepler, the "Protestant Galileo," breathed his last labored breath in Regensburg on November 17, 1630. His dying words expressed resolute faith in Christ, "from whom comes all solace, welfare, and protection."[32] Despite what Kepler had suffered at the hand of religion, he never lost sight of the pearl inside that dark oyster. Not even death, however, could grant Kepler immunity from the religious carnage that hounded him about Europe. During a clash between Lutherans and Catholics after his death, cavalry overran the Regensburg cemetery containing Kepler's remains. In the melee, Kepler's gravestone was obliterated and lost to history.[33]

Against formidable odds and crushing personal misfortune, Johannes Kepler teased from nature three descriptive laws of celestial motion, from which Newton derived predictive laws. Of Kepler's unique place in history, American astronomer and scientific icon Carl Sagan observed, "He was the first astrophysicist and the last scientific astrologer."[34] In other words, Kepler was a scientific amphibian, singularly equipped to claw his way intellectually from the dark and superstitious world of his birth into the dawn of the Enlightenment.

Epilogue

On March 6, 2009, in a spectacular nighttime launch, a Delta II rocket boosted NASA's newest space telescope into an irregular, Earth-trailing, heliocentric orbit. Unlike the jack-of-all-trades Hubble Space Telescope, the Kepler Telescope was commissioned for a specialized mission: *Kepler* is a planet hunter. Its main scientific instrument a precision photometer, *Kepler* sits far from Earth patiently trained upon a small swath of sky containing somewhat more than 100,000 stars. When an exoplanet – a planet outside our solar system – passes in front of one of those stars, the light that reaches *Kepler* dims imperceptibly to human eyes. *Kepler*'s task is to detect the faint changes in each star's signature brightness that announces an orbiting planet. Like its namesake, *Kepler* is a faithful observer. On average, an exoplanet is discovered daily. By the end of its first year in orbit, *Kepler* had identified more than 400 exoplanets. Seventy percent of stars, it now seems, have rocky planets, a few earth-like. Perhaps on one of these "Goldilocks" planets, where conditions for life are just right, other beings are peering back at us with technologically enhanced eyes, also wondering whether or not they are alone in the universe and determined, as are we, to find the answer.

Five

The Jewish Saint

If you would know the secrets of nature, you must practice more humanity than others.

– **Thoreau**

In late 1933, the most famous living scientist on earth immigrated to the United States from his native Germany. Despite worldwide fame, as a Jew, Albert Einstein was *persona non grata* to the Nazis, and his days were numbered.

Across the Atlantic, preparations for the prize catch proceeded at fever pitch. The Institute of Advanced Studies had just been created at Princeton University to showcase the originator of the theory of relativity. Dignitaries, festivities, and marching bands awaited his arrival. They were not to be disappointed. As the ship bearing Einstein pulled into dock in New York, the epicenter of physics shifted from Berlin to the United States, and here it has remained for seventy-five years.

Grateful but unimpressed, Einstein wrote irreverently of his new home to an old friend: "[Princeton] is a wonderful little spot, a quaint and ceremonious village of puny demigods on stilts. Yet, by ignoring certain social customs, I have been able to create for myself an atmosphere free from distraction and conducive to study."[1] There, in an unpretentious home on Mercer Street, he spent the rest of his days.

In his adopted country, Einstein was adored by ordinary folk, watched by the FBI, and occasionally revered as a Jewish "saint." For a rare few, simply meeting Einstein turned mystical. Such was the experience of Hiram Haydn, editor of the *American Scholar*. En route to a tea in Einstein's honor, Haydn inexplicably perceived Einstein's face as belonging to a being of a wholly different species. Almost as if he read Haydn's mind, Einstein smiled. Haydn later recounted. "This act constituted the most religious experience of my life." Still visibly shaken after the tea, Haydn was comforted by his companion Christian Gauss, Princeton University's dean. "Such moments," Gauss appreciated, "tear a rent in ordinary perceptions, and cut a hole in the fabric of things, through which we see new visions of reality."[2]

Haydn's glimpse through perception's veil proves fittingly symbolic, for no other scientist, living or dead, has altered human perceptions of reality more radically than Einstein.

New Theory of the Universe – Newtonian Ideas Overthrown
(*The Times of London*, Nov. 7, 1919)

Following the first experimental validation of general relativity, as alluded to in the headline above, Einstein became an instant cultural icon. The scientific equivalent of Princess Di or the Beatles, Einstein enjoyed rock-star status. No living scientist before or since has received the acclaim lavished upon Einstein.

In *The Ascent of Man*, science historian Jacob Bronowski likened Newton and Einstein to mythological gods. "Of the two," he took pains to distinguish, "Newton is the Old Testament god; it is Einstein who is the New Testament figure." The post-Einsteinian universe is as different from the Newtonian one as the god of love is from the god of wrath. Einstein, like Jesus, "was full of humanity, pity, [and] a sense of enormous sympathy."[3]

A Life in Review

Born to an easygoing father and a domineering mother in Ulm, Germany, March 14, 1879, Albert Einstein was something of an underdog. His father, Hermann, an engineer and entrepreneur, frequently relocated the family following failed business ventures. Albert's musically inclined mother, Pauline, forced violin lessons upon him at a tender age. Tantrums of protest ultimately gave way to an appreciation of Bach, Beethoven, Mozart, and Schubert. In later life, Einstein could charm family and friends with his accomplished playing.

The gift of a small magnetic compass, presented to five-year-old Einstein by his father, ignited a passion greater than music. In later years, Einstein recalled that he "trembled and grew cold"[4] in excitement while holding the compass, which revealed that "there had to be something behind objects that lay deeply hidden."[5] Something in this experience never left him.

> I sometimes ask myself how it came to pass that I was the one to develop the theory of relativity. The reason I think is that a normal adult never stops to think about problems of space and time. These are things which he has thought of as a child. But my intellectual development was retarded, as a result of which I began to wonder about time and space only when I had already grown up.[6]

Although the family housekeeper referred to him as *der Depperte* (the dopey one),[7] "retarded" is too strong a word. Einstein was slow to talk. Easily frustrated as a child, Albert was prone to outbursts that bordered on the violent. Quipped his sister Maja, "A sound skull is needed to be the sister of a thinker."[8] Awkward and out of place in school, Einstein showed little promise in his early years. His distaste for rote learning irritated his Prussian teachers, and his disinterest in sports provoked the taunts of other boys.

Einstein's curiosity first found nurture at the family's table, where on Thursdays at noon the family regularly set a place for a poor Jewish medical student by the name of Max Talmud. Talmud, sensing Einstein's intellectual hunger, engaged him in conversations that ranged from science to mathematics to philosophy, treating the boy as an equal. By the age of ten, Einstein was something of a math whiz, quickly solving Uncle Jakob's tricky problems. At the age of twelve, Einstein discovered Euclid, which he cherished like a sacred text.

Einstein's twelfth year proved pivotal in another regard. He found a devout faith, to the point of criticizing his parents for their lack of religious devotion. Piety, however, was short-lived. Reading more and more widely of science and philosophy, Einstein

> soon reached the conviction that much in the stories of the Bible could not be true. The consequence was a positively fanatic [orgy of] freethinking coupled with the impression that youth is intentionally being deceived by the state through lies; it was a crushing impression. Suspicion against every kind of authority grew out of this experience . . . an attitude which has never left me . . .[9]

Disdain of authority would prove indispensable in challenging a god as large as Newton, but first it would cause Einstein considerable misery.

An early act of rebellion was characteristic. Just as Einstein became eligible for the German draft, his father's latest business venture failed. Relatives in Italy came to the rescue, with the condition that the Einsteins move to Milan. But German law forbade men to leave the country prior to completing their military service. Thus, at the age of fifteen, Einstein was left in a boarding school in Munich. Without family or friends and subject to authoritarian teachers, Einstein grew increasingly despondent. A family doctor, while treating Einstein for a routine illness, recognized that the high-strung teen was headed for emotional breakdown. His letter and the support of a sympathetic math professor provided Einstein's ticket to escape. With these two endorsements, he was granted a passport and happily renounced his German citizenship.

Reunited with family in the freer atmosphere of Italy, his spirits soared. Einstein's sister, Maja, marveled at the rapid transformation of withdrawn dreamer into amiable young man. In Italy, Einstein helped with the family business and argued with his father over his future. Einstein thought of teaching philosophy, but his father demanded he study something practical. His father prevailed, and Einstein agreed to study electrical engineering at Zuericher Polytechnikum, Switzerland's M.I.T. Alas, he promptly flunked the entrance exam. Einstein's ability in math and science, however, did not escape the notice of the school officials, nor did his youth. The prestigious Polytechnic agreed to admit him at seventeen, provided he could secure a high school diploma in the intervening year. For that, the family packed him off to Aarau, near Zurich.

There Albert thrived. Swiss teachers were all that the German ones were not: relaxed, informal, open to free discussion and student invention. One instructor, Jost Winteler, invited Einstein to lodge with his family. The Wintelers, warm people with four children, made Einstein feel as one of their own. There he endeared himself to the young Winteler children by creating and flying homemade kites, and there he fell in love with their oldest daughter, Marie. In the fall of 1897, Albert left the Wintelers for Zurich Polytechnic, and Marie went off to a nearby town to teach first grade. They were deeply in love, but it was not to last.

In Zurich, Einstein's rebelliousness flared once again. Accustomed to the Aarau's relaxed atmosphere, Einstein was informal with his professors, an attitude unappreciated in the Polytechnic's columned halls. Worse, Einstein had developed the habit of focusing only on what interested him to the exclusion of all else. Even mathematics was sidelined, as he cut classes to make more time for hands-on laboratory experiments. In later life, Einstein offered this rationale:

> The fact that I neglected mathematics to a certain extent had its cause not merely in my stronger interest in the natural sciences than in mathematics but also in the following strange experience. I saw that mathematics was split into numerous specialties, each of which could easily absorb the short lifetime granted us. Consequently I saw myself in the position of Buridan's ass, which was unable to decide upon any specific bundle of hay.[10]

At the Polytechnic, Einstein met Mileva Maric, a Serbian woman four years his senior. The two were ill-suited: Einstein outgoing, striking in appearance, and ready with a boisterous laugh; and Mileva reserved, humorless, and unattractive. Still they shared a passion for physics. Eventually Mileva dislodged Marie as the love interest of Einstein's life. Unable to bear the thought of informing Marie that the romance was over, he simply stopped writing her. Devastated, Marie eventually married but never fully recovered from Einstein's sudden rejection. Einstein's romance with Mileva fell under a cloud from the start, when his mother, who could not tolerate Mileva, set out to undermine the relationship by any means.

By the time of his graduation from the Polytechnic, Einstein had earned the adoration of his classmates for his "uncompromising honesty, complete lack of pettiness," and "inability to ingratiate himself to those in authority."[11] Those same qualities had also earned him the contempt of his professors. "You're a clever fellow, Einstein, but you have one fault," said Heinrich Weber, the physics professor who admitted him to the Polytechnic: "You won't let anyone tell you a thing."[12] By the summer of 1900, one hurdle stood between the tight-knit class of five and graduation: the final exam. University positions awaited only those with the highest scores. Einstein finished fourth, beating only Mileva, who failed outright.

The decision of the two classmates to marry softened the blow to Mileva. But first they must tell and appease Einstein's mother. During a vacation with the family in Italy, Einstein broke the news. Pauline burst into tears, launching an angry tirade against Mileva. Even Hermann Einstein got into the act. Penniless, unemployed, and without support from his family, Einstein postponed marriage until he could land a job. With his poor showing on the final and the animosity of professors, finding employment would prove far more arduous than he could have imagined.

At twenty-two, Einstein accepted a temporary teaching post in Winterthur, seventeen miles north of Zurich, where Mileva remained. Thanks to a train connection, Albert and Mileva could rendezvous in Zurich on Sundays. But their meetings soon produced another complication; Mileva became pregnant. Then her academic hopes were dashed forever when she failed the Polytechnic exam for the second and final time.

As Mileva made plans to return to her family in Novi Sad for the birth of her child, Einstein's prospects of employment rose and fell. Open positions kept going to others. Former classmate Marcel Grossman came to the rescue by convincing his influential father to intercede on Einstein's behalf to the director of the Swiss Patent Office in the capital city of Bern. Assured by Grossman that the job was his for the plucking, Einstein moved to Bern. The same day he received a letter from Mileva's father, writing for his daughter, still weak from childbirth, that Albert was now a father. Anticipating a girl, Albert and Mileva had settled in advance upon the name Lieserl.

The wheels of bureaucracy turn more slowly than the gears in Bern's famous clock tower. The promised position still in limbo and money nearly depleted, Einstein resorted to tutoring university students in mathematics and physics. An advertisement in the Bern newspaper netted just two takers, but both were great catches. Maurice Solovine and Conrad Habicht shared Einstein's intellectual loves and his disdain for material possessions and those who made fame or fortune their life's goal. To Solovine, Einstein was "a great liberal, a very enlightened spirit."[13] Formal tutoring was soon suspended in favor of freewheeling discussions of philosophy, physics, and ultimate reality. As the inquiries deepened, so did the comradeship. The trio formed the core of the Olympia Academy, so named in derision of academic stuffiness. By the time the long awaited position finally opened, Einstein's financial plight bordered upon life threatening. Too proud to appeal to family for support, Einstein was close to starving, worried one friend. At last, on June 23, 1902, Einstein secured the lowly position of patent examiner, second class.

Though of far lower status than a professorship, the position afforded Einstein some amenities, including a steady if modest income. Moreover, the undemanding job left Einstein free to pursue his own investigations after his relatively short eight-hour workdays (which included Saturdays) and on Sundays. On good days, he could steal moments at his desk for scientific contemplation. Although his colleagues were easygoing and congenial, his supervisor was demanding. But even this had a silver lining. "More severe than my father, [he] taught me to express myself correctly."[14] So ideally suited was his situation to his needs that Einstein came to think of the patent office as his "worldly monastery."[15]

Just as it seemed the clouds had lifted, Einstein's father died of heart failure at the age of fifty-five. It was the "deepest shock" of Einstein's life, which left him "dazed" and "overwhelmed by a feeling of desolation."[16] The time at his dying father's bedside was apparently tender. The two reconciled, and Hermann granted his son permission to marry Mileva.

After more than a year of separation, Mileva joined Einstein in Bern. In January of 1903, the two married in a simple civil ceremony attended only by Solovine and Habicht. The marriage got off to a rocky start. Mileva was depressed, grieving the loss of Lieserl. To this day, it is uncertain whether the child was given up for adoption or died from a bout of scarlet fever. Einstein, having the time of his life, offered scant support to his wife. The members of the Olympia Academy congregated in one another's homes most evenings to discuss physics, and on weekends they hiked in the mountains. Of those days Solovine reminisced, "Our joy was boundless."[17] Gloomy by nature and excluded from vigorous hiking by a physical handicap, Mileva remained outside her husband's tight circle.

To Einstein, the years of the Olympia Academy were the best of his life, before the world expected him "to lay golden eggs." But lay them he did. In 1905, his *Wunderjahr*, Einstein published four papers, three of which revolutionized physics. So radical was Einstein's thinking that the physics community was slow to pick up the scent. First to take notice was Max Planck, Germany's preeminent physicist. In a 1906 letter to Einstein, Planck conveyed that the elite physics faculty of the University of Berlin was warming to his unorthodox ideas. In particular, they wanted to know more about relativity.

In 1908, Einstein, still a lowly patent examiner, was asked to deliver a lecture at 7 a.m. at Bern University. Expecting no one at that hour, Einstein was astonished by the overflow crowd. Although his performance that day was less than dazzling, his ideas were catching fire.

On July 6, 1909, finally offered a professorship at the University of Zurich, Einstein resigned from the comfortable patent office that had served as his intellectual monastery for eight years. The very next day he left for a conference in Geneva to mingle freely with the greatest physicists of his era. Einstein had just turned thirty. He and Mileva were again parents. Hans Albert had been born in 1904, and their second son, Eduard, was due any day.

Two years later, lured by a prestigious professorship in Prague, the Einsteins moved from Zurich. Prague's outwardly cold people and dingy apartments stood in marked contrast to Zurich's light-heartedness and cleanliness. Sensing their unhappiness, Marcel Grossman again came to Einstein's rescue. In 1912, Grossman, by then dean at Zurich Polytechnic, offered Einstein a position at his alma mater. Einstein returned to Zurich a conquering hero. Only the darkening of Einstein's relationship with Mileva marred the exuberance of their return.

Escapes from Prague had taken Einstein to Berlin, where he renewed acquaintance with his recently divorced first cousin Elsa. The two began a correspondence in which Einstein confided that he felt trapped between Mileva's unlikeableness and his pity for her. Einstein wrote Elsa, "I have to love someone, otherwise life is miserable. And this someone is you."[18]

Simultaneously, Einstein's growing stature as a physicist increasingly netted prestigious offers. By 1913, he had been wooed by the University of Berlin, the mecca of physics. A strong selling point: of the twelve persons in the world who understood relativity, eight resided in Berlin. As the final sweetener, he would be relieved of teaching duties to devote his energies purely to thinking. Einstein accepted, and in April 1914 the peripatetic family moved once again. The Einsteins had just arrived in Berlin when Mileva returned with the boys to Switzerland. Berlin terrified her for two sound reasons: proximity to Einstein's mother, who despised her, and to Elsa, whom she suspected of encroachment upon her husband.

Mileva's return signaled the de facto end of the marriage, though not its formal end. Einstein, although relieved by the separation from Mileva, wept over the loss of his sons. That August, Europe erupted into war. In addition to all its other cruelty, the war restricted travel between Germany and Switzerland, limiting Einstein's contact with his sons to letters.

Freed of the encumbrances of family, Einstein plunged feverishly into his work. In one year, 1916, he published ten papers and the popular treatise *Relativity*. One monumental paper successfully culminated a decade of effort on the general theory of relativity, the greatest achievement of synthetic thought in human history. Exhausted by work, worn down by conflict with Mileva, and distressed by the constraints of a war he opposed and despised, Einstein collapsed. Believing he had cancer, he resigned himself to die.

If poorly suited for marriage, Einstein was worse suited to bachelorhood, which nearly killed him twice. Elsa convinced him to move into an adjacent apartment where she could care for him. When his loss of 56 pounds was correctly attributed to an ulcer rather than cancer, Einstein began to recover, thanks to bed rest and Elsa's cooking. Worried that scandal would damage the reputations of her daughters Ilse and Margot, Elsa broached the subject of divorce and remarriage. Einstein appealed to Mileva for divorce, promising her the entirety of the money he was virtually certain to receive from a Nobel Prize. She reluctantly accepted but plunged into mental illness. In February 1919, Einstein divorced Mileva, married Elsa, and moved next door to the spacious apartment of his "harem": Elsa and two new stepdaughters.

Later that year, when an astronomical expedition led by Astronomer Royal Arthur Eddington confirmed the gravitational lensing of starlight predicted by general relativity, Einstein was thrust into the international spotlight, where he remained until the end of his life. Many have been destroyed by such fame, but Einstein, who had sought only truth, was singularly disposed to accept the yoke of fame with an uncommon grace. "A hundred times every day I remind myself that my inner and outer life are based on the labors of other men, living and dead, and that I must exert myself in order to give in the same measure as I have received and am still receiving."[19]

Unlike many brilliant scientists and mathematicians, Einstein remained accessible to the common man. He carried on voluminous correspondence, much of it with ordinary mortals. When begged by a distraught mother to speak to her deranged son, who was living as a hermit on a mountaintop, Einstein complied. Despite recognition of the man's obvious delusions, Einstein remarked afterward, "Yet when I was with him I felt the peace of high places and wondered if he was sane and the rest of us mad."[20]

On the occasion of his fiftieth birthday, Einstein received token gifts from around the world, most from common folk. A seamstress sent him a poem; a pouch of tobacco from an unemployed man brought tears to his eyes, provoking Einstein to write a heartfelt thank-you note.[21]

Einstein could suffer fools, but he could not tolerate bigotry, injustice, oppression, or authoritarianism, whether from the right or left. On matters outside his own household, Einstein's unerring moral compass often reoriented those who considered themselves moral. In 1914, Einstein risked the kaiser's reprisals by signing a petition in opposition to the war, one of only four scientists in Germany to do so. When, years later, overt anti-Semitism erupted throughout Germany, Einstein embraced Zionism but never lost sight of the rights of Arabs in a Jewish state. During the Second World War, as a naturalized U.S. citizen, Einstein aided scores of Jews in immigrating to the United States, whose policy toward Jews was far from open-armed. Appalled by the use of the atomic bomb to end World War II, Einstein championed international organizations that sought to limit the proliferation of nuclear weaponry and ban their use. After the war, he denounced McCarthyism, only to become a target of FBI witch hunts.

In 1920, Einstein's world began to crumble. His mother succumbed to cancer. On the scientific front, some of the implications of quantum mechanics, which Einstein's 1905 paper on the photoelectric effect had helped launch, were profoundly disturbing. Hitler, sensing opportunity in empty German stomachs, maneuvered to seize power in war-ravaged Germany. On June 24, 1922, a right-wing mob assassinated Walter Rathenau, the

Weimar Republic's foreign minister, a friend of Einstein and fellow Jew. As an outspoken Jew and pacifist, Einstein was a marked man.

Periodic travel out of country initially lowered Einstein's national profile, affording some protection against Nazi machinations. In 1921, he traveled to the United States for the first time. In 1922, Einstein received the Nobel Prize in physics after having been nominated unsuccessfully numerous times, the first more than a decade earlier. The award had been delayed for legitimate and illegitimate reasons. Before the advent of precise atomic clocks, special relativity was difficult to validate experimentally. But as evidence gradually mounted in its favor, some crusaded relentlessly to discredit "Jewish science," chief among them Einstein's nemesis, Philipp Lenard, who later became chief of Aryan science under Hitler.[22] So effective was the opposition that the Nobel committee elected to award no prize in 1921 rather than to bestow it upon Einstein. Level heads eventually prevailed. In an act of redemption, the committee awarded the 1921 medal to Einstein simultaneously with the 1922 medal to Niels Bohr, thereby recognizing their joint contributions to quantum mechanics. Ironically Einstein's award was not for relativity but for the explanation of the photoelectric effect, which Lenard had first observed experimentally. True to his word, Einstein forwarded Mileva the entire $32,500 prize, a small fortune in its day.

In late 1927, Einstein attended the Fifth Solvay Conference in Belgium, where began a strange and epic debate. Of the thirty attendees, more than half were eventual Nobel laureates. They included Werner Heisenberg, who formulated the uncertainty principle of quantum mechanics; Niels Bohr, whose collaboration with Heisenberg to comprehend the wave-particle duality of light ultimately coalesced into the Copenhagen interpretation of quantum mechanics; and Louis de Broglie, a young Frenchman who had caught Einstein's attention by showing that, like light, matter itself manifests wave-particle duality.

Presided over by Dutch physicist H. A. Lorentz, a freewheeling debate over quantum mechanics raged at Solvay, with Einstein in the unusual position as defender of the classical order. "Oh, it was delightful," Austrian physicist Paul Ehrenfest reported to his students.[23] At breakfast, Einstein would pose a *Gedanken* experiment designed to poke a hole in quantum mechanics, which would agitate Bohr. He, Heisenberg, and Pauli would then go to work, and by dinnertime, they "could usually prove that [Einstein's] thought experiment did not contradict uncertainty relations," recalled Heisenberg.[24] The next morning, it would start all over, and so it went for days, Einstein serene and Bohr so vexed he muttered to himself. "It was one of the great ironies of the debate that, after a sleepless night, Bohr was able to hoist Einstein by his own petard."[25] By consensus Bohr won the debates, but Einstein stole the show. For Einstein, de Broglie had only reverence: "I was particularly struck by his mild and thoughtful expression, by his general kindness, by his simplicity and his friendliness."[26]

The Einstein-Bohr debates continued for decades. "In all the history of human thought, there is no greater dialogue than that which took place over the years between Niels Bohr and Albert Einstein about the meaning of the quantum."[27] For at stake was nothing less than the very heart of physics.

In 1928, just two months after the Solvay Conference, Lorentz died. At his funeral, Einstein delivered a brief but deeply touching eulogy to his beloved mentor. Einstein later admitted, "he meant more to me personally than anybody else I met in my lifetime."[28] That same year, one hundred Nazi scientists collectively published a diatribe against Einstein's "Jewish physics" under the German title *Hundert Autoren gegen Einstein* (*One Hundred Authors against Einstein*). Unfazed, Einstein responded with wry humor that were he wrong, one would suffice.

In 1928, lugging a heavy suitcase up a hill, Einstein collapsed physically for the second time. Diagnosing an enlarged heart, doctors confined Einstein to bed rest and ordered him to give up smoking, a habit he had greatly enjoyed. Although Albert was a model patient, the additional responsibility of nursing him exhausted Elsa. Einstein needed a personal secretary. From his bed, Einstein greeted Helen Dukas, who had come for the interview, with the endearing words, "Here lies an old child's corpse."[29] From that moment until after his death, Helen Dukas played Einstein's Cerberus,[30] never leaving his side. She so faithfully guarded his privacy and later his reputation that only recently have biographers had access to his shadow side.

Einstein used his illness and respite from the limelight to full advantage by turning his attentions to the task of developing a grand theory to unite all the forces of nature, including gravitation. For the rest of his life, the quixotic search for a unified field theory occupied Einstein's genius. Though he would ultimately fail in this, Einstein felt a duty to tackle the thornier problems that younger physicists, reputations at stake, couldn't afford to address. He weathered setback after setback in the belief that "the search for the truth is more precious than its possession."[31]

A Bird of Passage

By 1930, the former lowly patent clerk had offers of professorships from all over the world: Oxford, Holland, Cal Tech, Madrid, and Princeton, to mention a few. Quipped Einstein, "I now have more professorships than rational ideas in my head."[32] In December of that year, Albert and Elsa Einstein visited America for the second time. On the voyage, Einstein confided to his diary, "I decided today that I shall essentially give up my Berlin position and . . . be a bird of passage for the rest of my life."[33]

While in California, Einstein visited Cal Tech at the request of its president, Robert Millikan, who had received the Nobel Prize in physics the year after Einstein. There, in addition to scientific lectures, Einstein spoke widely on issues of peace and justice, much to the chagrin of Millikan, a militant patriot. Indeed, Millikan had been admonished in a letter from a like-minded military general to avoid "aiding and abetting the teaching of treason to the youth of this country by [hosting] Dr. Albert Einstein," who branded himself "not only a pacifist, [but] a militant pacifist."[34] Sure enough, Einstein encouraged voluntary resistance to the draft, believing such civil disobedience could be an effective way to eliminate the scourge of warfare. In California, Einstein found supporters, among them author Upton Sinclair and silent film star Charlie Chaplin, with whom Einstein happily attended the premier of *City Lights*.[35]

Despite their antithetical political views, Millikan eagerly recruited Einstein for Cal Tech. At Cal Tech's behest, Einstein made a third trip to the United States in 1931, during which he met American educator Abraham Flexner. Unbeknownst to Millikan, Flexner had an agenda at cross-purposes with Millikan's. With philanthropic support, Flexner had been charged to set up a "scholar's paradise" in the United States to attract the scientific elite of the world. The location of the site had not yet been chosen, but Flexner made it clear that Einstein was at the top of the new institute's wish list. *"Ich bin Feuer und Flamme dafür,"* responded Einstein enthusiastically: "I am fire and flame for it."[36]

In April 1932, Einstein returned to Germany to his summer home in Caputh. Shortly after Einstein turned fifty, he and Elsa had used most of their savings to build a modest lake house in Caputh, on the Havel River south of Potsdam. "Absurdly happy as soon as he reached the water,"[37] Einstein quickly mastered the twenty-three-foot sailboat *Tümmler* (Dolphin) that friends had given him for his birthday. At Caputh, Flexner again visited Einstein to say that Princeton University was the chosen site for the new Institute for Advanced Studies (IAS). But Einstein had changed his mind, feeling that he could not bear to leave home, family, and friends despite the jeopardy in which he lived. "He does not know the meaning of fear," complained Elsa.[38] A few days later, however, through an intermediary, Einstein's family received this chilling message from the commander of the German army: "His life is not safe here anymore."[39] If indifferent to his own well-being, Einstein was not indifferent to that of his family. He accepted Flexner's offer while leaving open the door for an eventual return to Germany.

On March 29, 1933, Einstein resigned from the Prussian academy to join the ranks of the diaspora that included fourteen Nobel laureates and twenty-six of Germany's sixty professors of theoretical physics. Ilse and Margot had already left Germany, just ahead of Nazi thugs. When Max Planck appealed personally to Hitler about the decimation of German science, Hitler exploded: "If the dismissal of Jewish scientists means the annihilation of contemporary German science, then we shall do without science for a few years."[40] The wider list of European refugees included Edward Teller, Victor Weisskopf, Hans Bethe, Lise Meitner, Niels Bohr, Enrico Fermi, Otto Stern, Eugene Wigner, and Leo Szilard. Given that nearly all contributed in one way or another to the development of the atomic bomb, Einstein's earlier prediction was ultimately realized: "If and when war comes, Hitler will realize the harm he has done Germany by driving out the Jewish scientists."[41] Upon leaving their cottage at Caputh, as if by premonition, Einstein told Elsa, "Take a very good look at it; you will never see it again."[42] Indeed, following their departure for America, Nazis raided the cottage.

Albert and Elsa Einstein arrived in Princeton in October 1933, where Albert set up shop at the newly created IAS. When asked what he needed, Einstein politely requested a desk, a chair, paper, and pencils. Then, as an afterthought, he added a large wastebasket, "so I can throw away all of my mistakes."[43] Joined shortly thereafter by Margot, the Einsteins settled comfortably in Princeton.

An idyllic location affords no guarantee of an idyllic life. In the next several years, first Ilse and then her mother Elsa died. Einstein put work aside to tend Elsa's bedside faithfully until her death in December 1936. He never remarried. For the remaining two decades of his life, Einstein lived on Mercer Street in the company of his secretary Helen Dukas and stepdaughter Margot.

An aneurysm of the abdominal aorta hospitalized Einstein at the age of seventy-six. Not believing in the artificial prolongation of life, Einstein refused surgery. At 1:15 on the morning of April 27, 1955, Albert Einstein passed away peacefully. A few hours earlier, confident of success, he had sat up in bed to review his latest notes on the unified field theory, which Helen Dukas had brought at his request. To this day, a unified physical theory remains elusive.

In the fifty-plus years since his death, Einstein's ashes, strewn upon the Delaware River, have dispersed to remote corners of the globe, a fitting tribute to the earth's first citizen, who "no longer belongs to any one country, any one group, any one age . . . but to all friends of the future."[44]

Light and Shadow

The lives of all men and women are laced with light and shadow. The greater the life, the greater the contradictions. It is not surprising then, that Einstein, whom many regarded as a radiant creature, should also cast a shadow. As troubling as his imperfections are to those who would canonize him, they serve also as reminders of Einstein's ultimate humanness.

By virtue of his birth in Ulm, Einstein was culturally Swabian. In contrast to their more rigid and authoritarian compatriots, the Swabs are fun-loving, relatively carefree, and ready with a laugh, particularly to a bawdy tale. During the youthful and impetuous days of the Olympia Academy, Einstein lightheartedly referred to himself as the "Valiant Swabian," a nickname lifted from a poem.[45] "Long live impudence;" he asserted, "it is my guardian angel in this world."[46]

Indeed, Einstein owed his worldly success to impudence, believing to his core that "blind respect for authority is the greatest enemy of the truth."[47] A young Einstein had once wept at a military procession, sensing that blindly submitting to military or other authority is to sell one's soul. Later, the comrades of the Olympia Academy gorged themselves upon the skepticism of Don Quixote, Hume, Mach, Spinoza, and Poincaré. And from them, Einstein summoned the courage to challenge Newton's preconceptions of space and time. In the opening pages of *Principia Mathematica*, Newton had laid out the bedrock principle of his mechanics: "Absolute space, in its own nature, without relation to anything external, remains similar and unmovable."[48] Similarly, he assumed that time is absolute; that is, it flows at the same rate independently of the observer. But German physicist Ernst Mach, whom Einstein revered, derided Newton's absolute time as metaphysical nonsense, as did French mathematician Henri Poincaré, who famously declared, "Absolute space, absolute time, and Euclidean geometry are not conditions to be imposed on mechanics."[49] Einstein also took Mach's positivism to heart. The positivist's credo: *Esse est percipi* – only what is observed exists.[50]

Despite its role in freeing Einstein from intellectual conformity, positivism was not his *modus operandi*. The philosophical bent behind inductive science, positivism builds theories by accreting facts. In contrast, Einstein most often worked deductively. He began with basic principles and deduced how the universe must behave if it remains true to those principles, "guided not by the pressure from behind of experimental facts, but by the attraction in front from mathematical simplicity."[51] Inviolate to Einstein were three foundational principles: (1) the existence of an objective reality independent of the observer; (2) strict Newtonian causality, with its concomitant certainty and determinism; and (3) locality.[52] By the latter – the principle of local causes – it is meant that events that are widely separated in space cannot instantaneously interact.

Unlike relativity, which was consistent with Einstein's metaphysical assumptions, quantum mechanics violated all three cherished principles. Einstein fought it tooth and nail, conceding in old age, "All these fifty years of pondering have not brought me any closer to answering the question, What are light quanta?"[53] As quantum theory gradually gained acceptance, Einstein the iconoclast was seen more and more as a "calcified shell."[54] With characteristic self-deprecation, he admitted, "To punish me for my contempt of authority, fate made me an authority myself."

The image conjured up by the description "brilliant theoretical physicist" is perhaps a tall and lanky caricature, something like a Paul Dirac – aloof, cerebral, and antisocial – or a Robert Oppenheimer – ascetic, ill at ease, and chain-smoking. Einstein was the antithesis of such stereotypes. As a young man, he was strikingly handsome. Graceful in speech, slow but articulate in both German and English, he never needed to grasp for the right word even in his second language. Upon first meeting Einstein,

Cambridge-trained physicist and novelist C. P. Snow was awed at Einstein's "massive body, very heavily muscled" and "powerful sensuality."[55] It is not surprising then that the sensual, intelligent, attractive, and disarmingly unpretentious Einstein should find himself surrounded by the attentions of women. Time and again, his moral compass, unerring in matters universal, failed in matters domestic. Following the death in 1987 of Einstein's close friend and executor Otto Nathan, unsettling aspects of Einstein's personal life, theretofore carefully guarded, surfaced. Among recent disclosures from the Einstein archives were revelations about the existence of Lieserl, his illegitimate daughter, and his many liaisons with women. Her husband's unfaithfulness distressed Elsa deeply, driving her alternately into tears or fits of rage. Ultimately electing not to leave her wandering husband, Elsa reconciled herself to Einstein's flaw, confiding to a friend that one of Albert's genius could not be expected to be above reproach in all aspects of life.

Perhaps the most disconcerting aspect of Einstein's shadow was the callousness he sometimes inflicted upon those closest to him. When his first marriage turned sour, he confided to Elsa, soon to be his second wife, "I treat [Mileva] as an employee I cannot fire. I have my own bedroom and avoid being alone with her."[56] His divorce ultimatum to Mileva was so brutal she collapsed both physically and emotionally.

The kindly and compassionate genius, who promoted the worthwhile causes of humanity and welcomed the stranger, could be distant to the point of neglect from his own children. In the early days of fatherhood, Einstein was a devoted if not exemplary father. As an adult, Hans Albert marveled, "Out of a little string and matchboxes and so on, [my father] could make the most beautiful [toys]."[57] Of his son, the father effused, "The boy gives me indescribable joy."[58] However, following his divorce from Mileva, Einstein's relationships with both sons deteriorated, partly because Mileva poisoned the well. The breach between Einstein and Hans Albert was particularly painful for both and took a lifetime to repair. To the credit of each, and to Mileva, whose bitterness softened, the father-son relationship survived. Eduard – intelligent, quick, and possessed of his father's passion for the violin – plunged into schizophrenia, blaming his father's desertion for casting a shadow over his life. Committed to an asylum in Switzerland at age twenty, Eduard remained there for his final two decades. Einstein visited Eduard for the last time in 1933, just prior to fleeing to America. As a comfort to both of them during their final visit, Einstein took along his violin. Following the move to America, Einstein steadfastly refused to reestablish contact with Eduard, reasoning, to himself at least, that doing so would further destabilize Eduard's fragile mental state.

Of Einstein, a close associate said, "his heart never bleeds."[59] Similarly, a biographer concluded that he had little gift for empathy. Yet this cannot be the whole story. When their own marriages failed, both his sister

Maja and his stepdaughter Margot chose to live out the remainder of their lives with Einstein. When a stroke confined Maja to bed, Einstein, who loved his sister dearly, doted upon her, reading aloud to her each evening. When she passed on, Einstein was desolate, staring into space for hours, missing her "more than can be imagined." Still, he summoned the compassion to comfort his stepdaughter regarding Maja's death: "Look into nature," he said, "and then you will understand it better." The remaining two, the genius and the stepdaughter, grew ever closer in Einstein's final years, enjoying long walks together, often in silence. Of his stepdaughter, Einstein paid the utmost compliment, "When Margot speaks, you see flowers growing."[60]

Of Einstein's interactions with others, a close friend observed: "I never saw him lose his temper, never saw him angry, or bitter, or vain, or jealous, worried, impatient, or personally ambitious. He seemed immune to such feelings. But he had a shy attitude toward everybody. Yet he was always laughing, and he often laughed at himself. There was nothing stuffy about him."[61] Perhaps Einstein really *was* of a different species.

For most of his life, Einstein remained a devoted, fearless, and unflagging pacifist. When Hitler fanned the flames of nationalism in his inexorable march toward a second world conflagration, Einstein countered in a public forum: "Nationalism is an infantile disease. It is the measles of mankind."[62] Yet, in an ironic twist of fate, the gentle pacifist helped to conceive and spur development of the atomic bomb. The dark promise of the atom lay nascent in Einstein's famous formula relating mass and energy. Einstein's Hungarian friend Leo Szilard saw it first. While crossing the street in 1933, Szilard, also a theoretical physicist, had an epiphany both sublime and terrifying: theoretically, a nuclear chain reaction was feasible. Such an event would liberate far more energy than it consumed. Just in case he were right, Szilard filed a patent for an atomic bomb. By 1939, Szilard and Italian physicist Enrico Fermi had designed an atomic pile of uranium and graphite to produce a controlled chain reaction. The design looked promising, and Szilard began to worry that an atomic bomb was not only possible, it was inevitable. Worse, he fretted that Germany's Heisenberg might have reached the same conclusion. Unsure what to do, Szilard approached Einstein, only to be dumbfounded when Einstein responded, *"Daran habe ich gar nicht gedacht,"* meaning, "I never thought of that."[63]

Einstein, however, grasped immediately the ramifications of an uncontrolled chain reaction. A personal friend of Queen Elizabeth of Belgium, Einstein drafted a letter to the Belgian ambassador in an attempt to block German acquisition of uranium ore mined in the Belgian Congo, the world's principal source of nuclear material at that time. Meanwhile, American economist Alexander Sachs, who knew Franklin Roosevelt personally, convinced Szilard to take the urgent matter directly to the president. Einstein drafted the letter. Revised and then signed by Einstein,

Szilard, and Eugene Wigner, it was hand-delivered by Sachs to Roosevelt. After politely listening to Sach's summary, Roosevelt jumped to the point. Summoning his aide, Roosevelt noted, "This requires action." Those three words set in motion the Advisory Committee on Uranium. Unbeknownst to all attendees, the German bomb project was already in progress under Heisenberg's direction. In 1942, the committee recommended establishment of the top-secret Manhattan Project. Three years and $2 billion later, what was deemed "moonshine" by some physicists in 1933 materialized as lethal reality.

Clandestine reports from German physicists of Hitler's systematic elimination of the Jews converted Einstein temporarily from pacifist to ardent war worker. Never involved directly in the Manhattan Project, Einstein nevertheless put his talents to use to evaluate the designs of underwater explosive devices. Meanwhile, the German bomb project fizzled, and the Americans, with unilateral nuclear capability, brought the Second World War to a close by obliterating Hiroshima and Nagasaki, each with a single atomic bomb. The day following the destruction of Nagasaki, officials in Washington released a detailed history of the bomb's development, compiled by Henry DeWolf Smyth, a Princeton University physics professor. The Smyth Report, as it became known, laid much of the credit for the bomb at Einstein's doorstep. First, his formula brought to light the prodigious quantities of energy latent in ordinary matter. And second, his letter to FDR had incited government action. Einstein felt a heavy burden for his role, however unintended, in liberating the nuclear genie. With renewed vigor he re-embraced his hallmark pacifism, taking both to the press and the airwaves to call for a supranational organization to manage nuclear weaponry. "The release of atomic energy has not created a new problem," he wrote the November 1945 issue of *Atlantic Monthly*. "It has merely made more urgent the necessity of solving an existing one."[64]

Critical of his adopted country's monopoly on nuclear weaponry, Einstein, the quintessential democrat, found himself under FBI surveillance under suspicion of being "red." Unbowed, Einstein publicly challenged Senator Joe McCarthy's witch hunts for communists. Acquainted firsthand with the dangers of fascism, Einstein warned against trading one evil for another: "America is incomparably less endangered by its own Communists than by the hysterical hunt for the few Communists that are here."[65]

A week before he died, still trying to atone for the bomb, Einstein signed a manifesto from which sprang the international Pugwash disarmament conferences. In response to a question regarding how the next war would be fought, Einstein sadly replied: "I do not know how the Third World War will be fought, but I can tell you what they will use in the Fourth – rocks."[66]

If Einstein was not consistent in his pacifism, neither was he entirely consistent in his view toward nationalism. Reluctantly at first, and then passionately, Einstein championed Zionism. In 1921, following his rise to superstar status, the "wandering Zionist" stumped around America to push for the creation of a Jewish nation-state. Thousands lined the motorcade route. Thousands more attended his lectures. Decades later, following the creation of Israel in 1948, Israel's new prime minister, David Ben-Gurion, offered Einstein Israel's presidency, a largely ceremonial position that expressed nevertheless the deepest admiration of the Jewish people. Wisely recognizing that genius and leadership do not necessarily go hand in hand, Einstein graciously declined. Still Einstein presciently anticipated Israel's greatest moral test: how would it treat its Arab neighbors?

That Einstein's life was fraught with inconsistencies may offer some consolation to lesser mortals. In one regard, however, he was resolutely steadfast from youth to death. "He bristled at all forms of tyranny over free minds, from Nazism to Stalinism to McCarthyism."[67] When, in 1918, socialist student revolutionaries in Berlin overtook academic buildings and jailed the university's rector and deans, Einstein and two colleagues put their lives at some risk to confront the agitators and free the administrators. Although sympathetic to the students, Einstein gently reprimanded them. He explained his actions as follows: "All true democrats must stand guard lest the old class tyranny of the Right be replaced by a new class tyranny of the Left."[68] Years later, as a naturalized American citizen, Einstein took a remarkably personal stand against America's most insidious form of tyranny: racism. When in 1937 Princeton's Nassau Inn refused a room to the black American contralto Marian Anderson, Einstein invited her to stay at his Mercer Street home. She accepted and thereafter continued to reside with Einstein on subsequent visits to Princeton, her last just two months prior to Einstein's death in 1955.[69]

Subtle Is the Lord

During a visit to New York City in late 1930, Einstein received a celebrity tour of the majestic Riverside Church by its renowned senior minister, Harry Emerson Fosdick. Einstein was surprised to find a life-sized statue of himself above the church's west portal amid a dozen stone carvings of history's greatest thinkers. Never one to take himself too seriously, Einstein quipped to Fosdick, a Baptist, "I might have imagined that they could make a Jewish saint of me, but I never thought I'd become a Protestant one!"[70]

To some, Einstein truly was a "Jewish saint," the greatest Jew since Jesus. To others, he was an agnostic, or worse, an atheist who did not believe in a personal God. In religion as in science, Einstein was both eclectic and independent. Certainly he drew strength from his Jewish roots, to which he attributed the best of his defining qualities: "The pursuit of knowledge for its own sake, an almost fanatical love of justice, and the desire for personal independence."[71] But a practicing Jew he was not. "There is nothing in me that can be described as a 'Jewish faith,' " he once proclaimed.[72] Embracing wisdom wherever it manifested, Einstein read both the Old and New Testaments and considered Jesus a "luminous" and "enthralling" figure.

When accosted by a boorish dinner-party guest who blurted, "Professor, I hear you are . . . deeply religious," Einstein characteristically responded with grace and thoughtfulness:

> Yes, you can call it that. Try and penetrate with our limited means the secrets of nature and you will find that, behind all the discernible concatenations, there remains something subtle, intangible and inexplicable. Veneration for this force beyond anything we can comprehend is my religion. To that extent I am, in point of fact, religious.[73]

Of all religious thinkers, Einstein most revered Spinoza, the seventeenth-century Dutch and Jewish philosopher, who epitomized Einstein's undying faith in the harmony, beauty, and comprehensibility of nature. Above all, Einstein marveled that the most incomprehensible thing about the world is that it is comprehensible. But when asked "Do you believe in the God of Spinoza?" Einstein resisted categorization: "I can't answer that with a simple yes or no. I'm not an atheist, and I don't think I can call myself a pantheist. . . . [Childlike wonder], it seems to me, is the attitude of even the most intelligent human being toward God."[74]

However, an aspect of Spinoza to which Einstein clung tenaciously was determinism. In *The Ethics*, Spinoza posited, "Nothing in the universe is contingent [a matter of chance], but all things are conditioned to exist and operate in a particular manner by the necessity of divine nature."[75] In his eulogizing sketch of Einstein, American physicist John Wheeler observed that Einstein clutched determinism "in his mind, his heart, his very bones."[76] Nowhere was this expressed more explicitly than in Einstein's poetic response to an interviewer's question: "Everything is determined . . . by forces over which we have no control. It is determined for the insect as well as the star. Human beings, vegetables, cosmic dust, we all dance to a mysterious tune, intoned in the distance by an invisible player."[77]

Einstein's bedrock determinism lay at the root of his inability to accept quantum indeterminacy. Of Einstein's "duet" with Bohr over quantum theory, it was said, "Never in recent centuries was there a dialogue between two greater men over a longer period on a deeper issue at a higher level of colleagueship, nor a nobler theme for playwright, poet, or artist."[78] At times, Einstein attempted to gain the upper hand by bringing God into the conversation. In German, he referred to the deity with familiar respect as *Der Alte*, the Old One. "God does not play dice," said Einstein, and "Subtle is the Lord, but malicious he is not," the latter of which is engraved above the mathematics faculty lounge at Princeton University. Exasperated by Einstein's frequent references to the deity, Bohr once uttered the most famous retort of their long-standing debate, "Einstein, stop telling God what to do."

Einstein, who saw no conflict between science and religion, viewed the two as complementary. With disarming candor, Einstein confessed what would seem as anathema to many hard-boiled positivists: "I believe that any true theorist is a tamed metaphysicist."[79] Einstein acknowledged unabashedly that many of science's greatest advances had sprung from deep philosophical convictions. In an article for the *New York Times Magazine*, he wrote, "What a deep conviction of the rationality of the universe and what a yearning to understand, were it but a feeble reflection of the mind revealed to this world, Kepler and Newton must have had to enable them to spend years of solitary labor in disentangling the principles of celestial mechanics!"[80] Far from a hindrance to scientific enquiry, religious impulses, properly channeled, motivate the scientist. "I maintain that the cosmic religious feeling is the strongest and noblest motive for scientific research."[81] And in an address at Princeton Theological Seminary in 1939, Einstein argued gently that any conflict between science and religion is artificial; each has its appropriate domain. "For science can only ascertain what is, but not what *should* be, and outside of its domain value judgments of all kinds remain necessary."[82] Ultimately, the two spheres of influence need one another.

Einstein's harmonious view of the outer world matched an interior equanimity that gave equal voice to intuition and reason. When asked if he relied on intuition or inspiration to develop his theories, he credited both. "I believe down here," placing his hand over his heart, "what I cannot explain up here," putting his hand on his head.[83]

Did Einstein believe in the immortality of the individual soul? His answer, while downplaying the importance of the individual, left open a place for a world soul. "I look on mankind as a tree with many sprouts. It doesn't seem to me that every sprout and every branch possesses an individual soul."[84]

To a friend who had sent him a birthday greeting, Einstein confided: "I am a deeply religious nonbeliever."[85] Einstein adhered to no religious creed, nor did he pray to a personal God. His religious views would have resonated with those of many of America's founding fathers, among them Jefferson and Franklin. Nevertheless, Einstein's chief motivation was to know the mind of God, and it angered him when people, including many scientists, denigrated religious inclinations. "There are people who say there is no God. But what makes me really angry is that they quote me for support of such views."[86] Well aware of the negative potential of blind faith and of the atrocities promulgated throughout history in the name of religion, he still had harsh words for those who had lost their sense of wonder in the face of the mystery of creation. "The fanatical atheists are like slaves . . . still feeling the weight of their chains. They are creatures who –

in their grudge against traditional religion as the 'opium of the masses' – cannot hear the music of the spheres."[87]

Einstein's inspiration, like Kepler's, often bordered upon the mystical. Most are born with this sense of wonder, which is lost in adulthood. Einstein, however, never lost the ability to be awed. In an article entitled "The World as I See It," he wrote, "The most beautiful experience we can have is the mysterious. It is the fundamental emotion which stands at the cradle of true art and true science. Whoever does not know it and can no longer wonder, no longer marvel, is as good as dead, and his eyes are dimmed."[88]

Einstein's eyes were anything but. Friends, colleagues, doctors, and acquaintances all spoke of his eyes. To psychoanalyst Erik Erikson, Einstein had the ability "to look into cameras as if he were meeting the eyes of the future beholders of his image."[89] But to those who knew him personally, "no picture could reproduce the shining glow of his eyes."[90] Large, deep, and dark brown, Einstein's eyes, recalled a friend, "would light up in a way that filled me with awe."[91]

Einstein's effects upon others could turn profound. As a close family friend put it, "People sometimes glanced at Einstein as if something mystical had touched them."[92] For example, when in 1946 the Russian writer Ilya Ehrenberg met Einstein, then sixty-seven years of age, he was "struck dumb . . . like a child who for the first time witnesses some extraordinary natural phenomenon."

> I . . . found myself face to face with Einstein. I spent only a few hours with him, but my memory retains those long hours better than many an important event in my life. One can forget joy and trouble, but . . . never . . . amazement. His eyes were astonishingly young, by turns sad, alert, or concentrated, then suddenly full of mischievous laughter like a boy's. He was young with the youth that years cannot subdue.[93]

Despite his status as an icon, Einstein placed no value on fame or fortune. In "The World as I See It,"[94] he wrote, "The ideals which have lighted my way, and time after time given me the courage to face life cheerfully, have been Kindness, Beauty, and Truth. The trite objects of human efforts – possessions, outward success, luxury – have always seemed to me contemptible." The only "success" for which Einstein strived was to be liberated from the self into something larger, to rise above the "merely personal,"[95] and he defined the standard against which we may all take measure: "The true value of a man is determined primarily by the [degree to which] he has attained liberation from the self."[96]

In that regard, Einstein was uncommonly successful. For this reason, he was often misunderstood and occasionally vilified. Fearless, he could confront student protestors, the kaiser, Hitler, and the FBI without cowing. Indifferent to his own safety, he entertained risks unacceptable to others. When sailing, alone or with others, Einstein frequently got into trouble. Possessed of an innate sense of direction and an almost spooky ability to forecast storms, Einstein did not read nautical charts, look at a compass, or carry life preservers. Several times, when his masts blew down, Einstein required rescue. Once, at a conference in Switzerland, he invited Madame Curie to sail with him on Lake Geneva. Far from shore, she commented, "I didn't know you were such a good sailor," to which Einstein impishly replied, "Neither did I." Taken slightly aback, she asked, "But what if the boat should overturn? I can't swim," to which Einstein responded honestly, "Neither can I."[97]

This anecdote highlights a facet of Einstein's nature that set him apart from the main. Einstein's indifference to his own safety and that of his companion was not what it seemed: callous bravado. Rather it arose from an inner tranquility that stemmed from a sense of being resolutely at home in the universe. Indeed, Hedi Born, the wife of physicist Max Born, came to believe that Einstein's mastery of life exceeded even his mastery of physics. While seriously ill, with the outcome in doubt, Einstein had confided to Hedi, "I feel so much a part of every living thing that I am not in the least concerned with where the individual begins and ends."[98]

At home in the universe, Einstein lived in a world of his own, paradoxically separated by some invisible barrier from others, even those closest to him. And yet he "was full of harmony," as Margot observed of her iconic stepfather. "He lived in his own world, and this saves people."[99] A self-professed "lone traveler," Einstein belonged more to the whole realm of nature than to the species into which he had been born. Perhaps this explains Einstein's willingness to sit for artists. Artists understand detachment. At the height of his powers, Einstein posed for artist Winifred Rieber. Her stunning portrait conveys Einstein "as a force of nature." "Everything about him is electric," she said. "Even his silences are electric."[100] How curious, then, that the man who taught us most about electrodynamics and light struck those around him as both electric and radiant.

Epilogue

Einstein's first wife, Mileva, paralyzed by a stroke, died in 1948. His youngest son, Eduard, died in a Swiss psychiatric hospital in 1965, having sunk ever deeper into schizophrenia. Hans Albert, his oldest son, had a distinguished career as a civil engineer at the University of California, Berkeley, and died of a heart attack in 1973. Stepdaughter Margot lived a long life and died in 1986.

Each year since 1927, when "Lucky Lindy" flew solo across the Atlantic, *Time* magazine has featured a Person of the Year, chosen because of their impact, for good or ill, on humanity. The millennial end-of-year issue of 1999 was particularly noteworthy, for it honored the Person of the Century. Winnowing the field of contenders – which included Franklin Delano Roosevelt; Martin Luther King, Jr.; Hitler; Gandhi; and Einstein; among others – was arduous. To break the stalemate, *Time*'s editors pondered the question: "How will the 20th century be remembered?" And so it was that *Time*'s year-end cover bore the lion-maned visage of an aged and tired physicist, from whose large eyes emanate a mixture of penetrating intelligence, compassion, and child-like wonder. In the final analysis, *Time* selected Einstein because he "was the pre-eminent scientist in a century dominated by science."[101]

Perhaps Einstein had unwittingly contributed to his election as Person of the Century by the quip: "Politics is for the present, while our equations are for eternity."[102]

Six

A Wrinkle in Time

All our knowledge is but the knowledge of schoolchildren. The real
nature of things, that we shall never know, never.

– **Albert Einstein**

When he was nearing seventy years of age, Einstein sent a birthday
greeting to his eighty-year-old friend Otto Juliusburg, a psychiatrist. Its
heartfelt message remarked how neither man had grown old in the usual
manner. "What I mean," Einstein elaborated, "is that we never cease to
stand like curious children before the great Mystery into which we are
born."[1]

Einstein was master of the *Gedanken* experiment, *Gedanken* being German
for "thought." *Gedanken* experiments require only the laboratory of
the mind; their only apparatus is imagination. One might reasonably
presume that the thought experiments of humanity's greatest theoretical physicist would lie beyond the ken of ordinary mortals. On the contrary, Einstein's thought experiments often sprang from "immensely simple questions" of childlike innocence. One such question, posited when
he was in his teens, would forever alter human perceptions of physical
reality:

How would the world appear were I riding on a beam of light?

Stopping the Clock

The year was 1905. Like Clark Kent and Superman, Einstein was living
a kind of double life. By day mild-mannered, he examined patent applications at the patent office of the Swiss capital of Bern. In the evenings,
he engaged in irreverent critiques of physics and philosophy with likeminded comrades of the Olympia Academy.

On a fateful spring evening, Einstein cornered his friend and fellow
Olympian Michele Besso. Einstein had been pondering the interplay of
motion, light, and time; he needed a sounding board for ideas all in ferment. At issue was the propagation of light. Light waves do not behave
like sound or water waves. Light passes by an observer always at the same

velocity, regardless of the motion of the observer. The unconventionality of light didn't square with classical mechanics.

On the morning after the intellectually cathartic session with Besso, Einstein found himself in a state of great agitation. "A storm broke loose in my mind," he recalled. As the storm cleared, Einstein glimpsed a way through the conceptual impasse, and the thought was both electrifying and unnerving. The propagation of light could be reconciled with conventional ideas of motion only by reexamining fundamental assumptions about the flow of time.

Six weeks after the meeting with Besso, Einstein submitted an understated paper to the respected German journal *Annalen Der Physik*. In it the unknown physicist laid the foundations for the special theory of relativity, paving the way for a revolution in physics that would play out over much of the twentieth century. The first shoe to drop concerned the nature of time. Einstein had cast "a wrinkle in time."[2]

Time was central to Einstein's Clark Kent routine. Day after day, he rode the tram to and from Bern's patent office, and day after day, it stopped faithfully at the city's famous *Glockenturm*. Mechanical marvels from medieval times, such clock towers bejewel many European cities. To the delight of tourists and children, passing hours are heralded by the *Glockenspiel* – a mechanized pageantry of whimsical characters accompanied by a chorus of clanging bells.

Time, Newton had believed, was absolute. A second in Bern was a second in London and a second on Jupiter. It was presumed that time flowed, like grains of sand in an hourglass, at the same leisurely pace in all corners of the cosmos. By a simple thought experiment, Einstein put the lie to Newton. One need not wield Einstein's genius to grasp his reasoning. It is necessary only to imagine oneself aboard Bern's tram as it leaves the *Glockenturm*. In the words of historian of science Jacob Bronowski, "Suppose the tram were moving away from [the clock tower] on the very beam of light with which we see what the clock says. Then, of course, the clock would be frozen. I, the tram, and this box, riding on the beam of light, would be fixed in time. Time would have to stop."[3]

Time flows differently for different observers. This is the quintessence of the special theory of relativity. The pedestrian on the street and the traveler on the tram see the same clock ticking at different rates. The passage of time is somehow related to the motion of the observer. To ferret out the why and the how, Einstein, like his predecessors Galileo and Newton, returned to square one regarding motion. He began by asking the seemingly banal question: "How does the traveler know when he or she is moving?"

The Principle of Relativity

Virtually everyone in the modern world has had the momentarily disorienting sensation of phantom motion, a phenomenon common when traveling by train. For example, at the station, a traveler's train and another sit adjacent on parallel tracks. At its scheduled time of departure, the other train pulls forward to leave the station. The traveler, who has been absent-mindedly peering out the window toward the now departing train, senses fleetingly that his or her train has begun to inch *backwards*. Occasionally one also experiences phantom motion in automobiles at traffic lights, but the smooth, slow acceleration of electric trains and the lack of auditory clues make the sensation much more startling in a railway setting.

Human beings are physiologically instrumented to detect *acceleration*, defined as changing speed or direction. In response to a rapid tap on the accelerator pedal when the light turns green, one is mysteriously "pushed" backward into the seat. In response to changes of direction, fluid in the inner ear disturbs cilia that signal the brain accordingly. So long as one's visual cues and those of the inner ear are consistent, the passenger, whether driving along a winding road or flying a jet fighter in combat, is unlikely to experience motion sickness. On the other hand, when the cues are out of sync – as when reading in the back seat of an automobile on a mountainous road, watching an IMAX film about flight from the vantage point of a motionless seat, or orbiting the earth in a spacecraft – most humans experience mild to debilitating motion sickness.

Humans are equipped by nature to detect acceleration, but the original question remains unanswered: how does one infer whether or not one is moving at all? Clarity comes by distinguishing *uniform* motion from accelerating motion. Here, and in physics, *uniform* implies "at constant velocity," by which it is meant that both speed and direction remain unchanged. Uniform motion, by definition, precludes acceleration, and thus it precludes the attendant sensory clues previously discussed. For example, at cruising altitude and cruising airspeed on a transcontinental flight, the modern traveler is bemused by the sensation of being utterly motionless. In the absence of turbulence, only the drone of the engines and the scudding of the clouds below provide any sensory clues of motion. Upon lowering the shade and putting on headphones, one can no longer infer whether one is enjoying the in-flight movie at 33,000 feet or watching while at rest on the ground.

Newton's first law recognizes the "naturalness" of uniform motion. Objects in uniform motion remain in such motion – on the same linear path and at constant velocity – unless acted upon by an unbalanced force. Commonsense reasoning forces the conclusion that, without external references, there exists *no* means whatsoever to infer whether one is moving uniformly or at rest. Indeed, fallacious arguments against the Copernican worldview presupposed that the earth could not be hurtling through space and rotating all the while because of the utter lack of sensory clues. Such Aristotelian arguments failed to recognize that motion is relative or that the earth's atmosphere rotates with us.

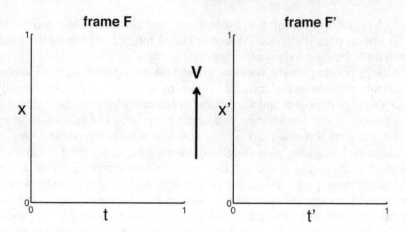

Figure 9: Inertial frames of reference F and F'. (Author.)

By common analogies we arrive at the first tenet of special relativity: all motion is relative. I know that I am moving uniformly only in reference to something else, which either is considered "stationary" or is itself moving uniformly. In more precise language, the principle of relativity (as distinct from the theory of relativity) states that all *inertial frames of reference* are equivalent for expressing the laws of physics. An inertial frame of reference, by definition, is a Cartesian coordinate system, free of rotation, whose origin undergoes translational motion at a fixed velocity. Two such inertial frames of reference, F and F', are compared in Fig. 9. Relative to F, the origin of F' translates parallel to the spatial (x) axis at constant velocity V. By the principle of relativity, it is equally correct to consider frame F' at rest and frame F translating at a velocity of $-V$ relative to F'.

Though a simple notion, the principle of relativity created philosophical waves. In the Ptolemaic world order, the earth held a privileged place at the center of the cosmos. Copernicanism delivered "the severest shock man's interpretation of the cosmos ever received," observed Einstein, "[because] it reduced the world to a mere province so to speak, instead of it being the capital and center."[4] The relativistic worldview is essentially egalitarian: all points in the universe are of equal status. Any point could be designated the "center of the universe." Equally any could serve as the origin of our coordinate system. Moreover, because motion is relative, we cannot state which inertial frame of reference, F or F', is stationary and which is moving. Each is moving in relation to the other; thus, inertial frames of reference are equivalent.

But here we are backed into a corner. The natural philosopher holds dearly the notion of an underlying order of nature that, when understood, can be expressed as physical laws in mathematical form. It is presupposed that those laws are inviolate; that is, that the same laws apply in every

$$\nabla \cdot \mathbf{E}' = 4\pi\rho'$$
$$\nabla \times \mathbf{E}' = -\frac{1}{c}\frac{\partial \mathbf{B}'}{\partial t'}$$
$$\nabla \cdot \mathbf{B}' = 0$$
$$\nabla \times \mathbf{B}' = \frac{4\pi}{c}\mathbf{J}' + \frac{1}{c}\frac{\partial \mathbf{E}'}{\partial t'}$$

Figure 10: One of several equivalent forms of Maxwell's equations, expressed in coordinate system F'. Derived in 1861 and 1862 by Scottish mathematical physicist James Clerk Maxwell (1831-1879), Maxwell's equations form the theoretical basis of electrodynamics. As such, they relate – through a system of partial differential equations – the interactions of the electrical field (**E**), the magnetic field (**B**), and the current density (**J**). (Author.)

quadrant of the universe. And thus, regardless of where we affix the origin of our coordinate system, irrespective of which inertial frame of reference we chose in which to express our laws, those laws should remain the same. Otherwise, if each region of the universe has its own laws, there is no fundamental order to the natural world and no such thing as physics.

The principle of relativity must apply to any physical law. Suppose we apply it to Maxwell's equations (Fig. 10), which express the relationship between electrical (*E*) and magnetic (*B*) fields as they oscillate in space (*x*) and time (*t*). In the figure, Maxwell's equations are written for reference frame F'. By the principle of relativity, which asserts the equality of all inertial frames of reference, the same equations must hold also in frame F, with the sole exception that all primed quantities lose their primes. The only quantities that remain identical in the primed and unprimed coordinate systems are constants such as π and c, the former being a fundamental constant of mathematics and the latter being a fundamental constant of nature: the velocity of light *in vacuo*.

Such frame invariance exposes a fundamental attribute about light: the velocity of light is *absolute*. This is the second tenet of special relativity, and from just two tenets, numerous astounding revelations about the nature of reality spring forth.

Surprisingly, no physicist before Einstein fully appreciated the bedrock principle of relativity on which physics rests. Because of this collective mental block, the state of physics at the beginning of the twentieth century could only be described as schizophrenic. The principle of relativity was assumed at the time to apply for mechanics but not for electromagnetism. Why the inconsistency?

From the time of Newton, natural philosophers (including Newton himself) were uncomfortable with idea of gravitation's action through the nothingness of space. The level of discomfort deepened when physicists realized that electromagnetic waves propagate through empty space. Sound requires a medium – air or water, for example – in which to propagate. How could electromagnetic waves propagate where there was no medium? And so physics invented one, termed the *ether*. This hypothetical intermediary was envisioned, somewhat paradoxically, as a weightless and frictionless fluid that filled all of "empty" space. In addition to presumably transporting electromagnetic radiation, the ether served another purpose. Assumed to be stationary, the ether, if it existed, provided an *absolute* frame of reference for motion. The earth's velocity during its orbit of the sun, for example, could be measured, presumably, by reference to the stationary ether.

Ether or no ether, light travels at a stupendous, but finite, velocity, a fact known since 1676. From eight years of observing eclipses of the moons of Jupiter, the Danish astronomer Olaus Römer noticed a 22-minute discrepancy in the intervals between the eclipses, dependent upon the relative positions of Jupiter and the earth. He attributed the discrepancy to the differences in time required by light to traverse the enormous distance from Jupiter to earth at different times of our year. Römer deduced that "light takes about ten minutes to travel from the sun to the earth";[5] he was off by just one minute.

Two years later, using Römer's data and a crude approximation of the radius of the earth's orbit, Dutch mathematician and astronomer Christiaan Huygens estimated c at the fantastic value of 200,000 kilometers per second. In the intervening years, c has been established to extraordinary accuracy at a considerably higher value: nearly 300,000 kilometers per second. First to measure the velocity of light precisely was the young German-American experimental physicist Albert Michelson.

The Constancy of c

As a twenty-five-year-old physics instructor at the U.S. Naval Academy, Albert Michelson conducted the first of several classic experiments that determined the velocity of light to extraordinary precision. By 1878, he had reduced the experimental error in c to less than 0.02 percent. To reduce measurement error still further, Michelson designed and built an interferometer, which exploits the wave-like properties of light for purposes of measurement. Like water waves or sound waves, light waves combine constructively or destructively to enhance or reduce, respectively, the amplitude of the combined wave. Michelson's interferometer recombined two parts of a single beam of monochromatic light that had been split into two beams that followed different paths. When the two beams rejoined, the resulting interference pattern depended sensitively upon the lengths of the two paths and the speed of light along the two paths. Michelson's early experiments exploited interference patterns to derive the wavelengths of various colors of light. Later, he cleverly adapted the method to accurately measure the speed of light itself. For these achievements, Michelson received the Nobel Prize for physics, the first American to do so.

In 1887, Michelson teamed with American chemist Edward Morley in an experiment designed to detect the "ether wind." If outer space were filled with stationary ether, then the earth's relative motion through the ether should generate an ether wind, just as a bicyclist feels a breeze while riding through still air. Michelson and Morley were steeped in classical physics. Thus, the Michelson-Morley experiment relied on combining velocities using vector addition. For example, moving sidewalks expedite pedestrian traffic along airport concourses because the velocity of the sidewalk relative to the concourse, say, 2 miles per hour (mph), and the velocity of the traveler relative to the sidewalk, say, 3 mph, add to yield 5 mph, the traveler's velocity relative to the concourse. At the combined speed, kiosks and sports bars seem to whiz by.

Similarly, vector addition explains why, between the same two airports, the time required for eastbound flights is typically less than that for westbound flights. In one direction, groundspeed benefits from the jet stream, which generally flows west to east in the Northern hemisphere; in the other direction, groundspeed is retarded. In principle, by timing a round-trip flight between two points whose separation distance is known, one can extract both the wind speed and the airspeed of the aircraft. What happens when an analogous experiment is conducted not with an aircraft but with a beam of light in the ether?

Just as the vector sum of wind speed and airspeed gives groundspeed, so should the sum of the velocity of the earth through the ether and the velocity of light relative to the earth yield the velocity of light c relative to the ether. By making measurements along different directions, it was thought possible to extract both the velocity of the ether wind and the velocity of light, both relative to the presumed ether.

The enormity of c presented Michelson and Morley with practical difficulties. Detection of the ether wind, anticipated to be about 30 kilometers per second, but a tiny fraction of c, necessitated extreme measurement accuracy. By 1887, Michelson had improved the precision of the interferometer well beyond that necessary for success. To do so required a clever variation of Michelson's earlier experiment. The Michelson-Morley experiment was designed not to measure the speed of light *per se* but to detect *differences* in its measured velocity along two paths that were mutually perpendicular. If the direction and magnitude of the ether wind varied during the year, as expected, the interference patterns generated by the interferometer should vary over the course of a year.

To the astonishment of the scientific world, the experiment failed. No change whatsoever occurred in the interference patterns observed. The velocity of light proved maddeningly predictable: the same regardless of time of year, time of day, or orientation of the apparatus. Michelson and

Morley concluded tersely, "the result of the hypothesis of stationary ether is thus shown to be incorrect." It was a bad day for the American experimentalists, but a great one for a maverick theorist in Switzerland.

So deeply ensconced was the presupposition of the ether, and so deeply perplexing were the implications posed by abandoning it, that nineteenth-century physics was left grasping for straws to explain the negative result of Michelson-Morley. Paradoxically, by its failure, the experiment set the stage for the new physics of the twentieth century. Today, Michelson-Morley is the most celebrated physical experiment in history. The experiment inadvertently established the constancy of c, independent of the frame of reference, thereby confirming the second tenet of Einstein's special theory of relativity.

Einstein was but vaguely aware of the Michelson-Morley failure and did not presume the constancy of c from their results. Rather, he inferred it directly from Maxwell's equations, presented earlier (Fig. 10). The fundamental constants of nature must not vary in different frames of reference; otherwise they are neither constant nor fundamental. Because c appears in Maxwell's equations as a fundamental constant, then c must be invariant so long as the frame of reference is inertial. Michelson-Morley served only to confirm experimentally what Einstein knew instinctively.

That c proved constant in all inertial frames of reference strongly suggested that the principle of relativity encompassed electromagnetism as well as mechanics. However, extending the principle of relativity to Maxwell's equations posed a tangle of thorny questions. For all objects of everyday experience – baseballs hurled on the run, apples tossed from bicycles, or passengers walking through a train compartment – the principle of vector addition applies. But to the propagation of light, vector addition apparently does *not* apply because two observers, in uniform motion relative to one another, measure exactly the same velocity for light. How can this be? There was one obvious but rather ludicrous explanation for the null result of the Michelson-Morley experiment, conducted at Case Western. Perhaps, some wagged, Cleveland, Ohio, was the exact center of the universe, in which case the ether wind would completely vanish there.

When all of the facts were sorted out, a perplexing conundrum remained. One of the following statements must be true, but the truth of any doomed some long-cherished scientific presupposition:

1. The earth does not move relative to the ether.

2. The principle of relativity is flawed.

3. Human perceptions of space and time are flawed.

Science had painted itself into a tight corner. It had taken 1,500 years to dislodge the geocentric cosmology of Ptolemy. Few legitimate physicists wished to open door number 1: abandon Copernican theory and re-embrace Ptolemaic cosmology, with the earth absolutely stationary, once again at the center of the universe.

On the other hand, behind door number 2 lurked the unthinkable: give up the principle of relativity, in which case we must abandon physics because there are no universal laws.

The option behind door number 3 seemed most disquieting of all, for it reached beyond physics to human perceptions of reality. Because Einstein had in his early years questioned human perceptions of time, he alone dared to open door 3.

Spacetime

On YouTube[6] is a fascinating video by a billiard expert who combines billiard tricks with domino chains. The video's allure has netted more than one million visits. As the chains topple, billiard balls are released either to drop into prescribed pockets or to initiate the toppling of other chains. It's all over in a matter of moments. The hours of practice and experimentation, the infinite patience, the skill and creativity invested: all like fireworks, gone in a grand but ephemeral display.

However, when the dominoes are cherished scientific notions, watching their demise elicits something closer to terror than to delight. Through neither malevolence nor intent, Einstein toppled one by one, like dominoes, many of the pillars of classical physics. His method was a sequence of clever *Gedanken* experiments, and his intent to know the mind of God.

The first notion to fall was simultaneity. In his trade book *Relativity*, Einstein admirably attempted to render the concepts of special and general relativity accessible to the layperson, and he generally succeeded. Some analogies seem quaint by modern standards because Einstein's primary modes of transportation, by which he illustrated relative motion, were train, tram, and ship. To address the concept of simultaneity, Einstein asks his reader to imagine a train speeding along parallel to an embankment. In the scenario exist two observers. One rides on the train; the other sits on the embankment midway between two telephone poles. At the precise instant at which the traveling observer passes the stationary one, both telephone poles are struck by lightning, which the observer on the embankment sees as simultaneous events.

What does the traveler observe? As the light from each of the flashes moves toward him at the fantastic but finite velocity c, he continues to be carried toward one pole and away from the other at the train's velocity V. In the short interval of time that it takes for the light flashes to reach him, he will have moved slightly closer to the forward pole and slightly farther from the rearward one. As a result, the flash from the forward pole is seen first. To the traveler, the lightning strikes are *not* simultaneous. Einstein thereby concluded that the notion of simultaneity carries meaning only in the context of a particular reference frame.

The experiment, although conceptually simple, exposes something profound about the fundamental role played by light in nature's grand scheme. Light, like the Greek god Mercury, is a messenger. Light ferries information between points separated in space. In the vacuum of deep space, where sound waves have no medium, light is the *only* reliable messenger.

Time flows differently for the traveler and the one who stays put. But how so? In "Einstein's mirror," as another thought experiment has become known, Einstein quantified the relativistic ticking of clocks. In a burst of insight, he proposed the mental construction of a "light clock," comprised only of two parallel mirrors. A single photon of light bounces back and forth between the mirrors. Each round-trip circuit of the photon signals one tick of the clock.

Suppose we place two identical light clocks in inertial reference frames F and F' of Fig. 9, each arranged with its mirrors parallel to the direction of the moving frame. What does an observer in each frame see? An observer in frame F sees the photon in his clock bouncing vertically, up and down, as does the observer in F'. But when the observer in frame F looks at the clock in frame F' that is receding from him, he observes the photon in F' traveling a W-like path, the resultant of its horizontal and vertical motions.

The second tenet of special relativity requires the speed of light, c, to be identical in both frames. Consequently, the observer in F observes the clock in F' ticking more slowly than his own because the path traveled by the photon in F' appears to be longer. With no more mathematics than the Pythagorean theorem,[7] one can derive the relationship for the times t and t' measured by a "stationary" observer in frame F, namely, $t' = t/\beta$. The factor $\beta = \sqrt{1 - (V/c)^2}$, which varies between 0 and 1 depending upon V, appears frequently in the special theory of relativity and is termed the *time-dilation factor*. Whenever the relative velocity V is not zero, β is less than 1.

The implications of this simple relationship are astonishing: an observer in *either* frame of reference sees the two light clocks ticking at different rates. How had pre-Einsteinian scientists so completely overlooked time dilation?

Prior to the twentieth century, no traveler had exceeded the velocity V of a steam locomotive at wide-open throttle, about eighty kilometers per *hour*. Such speed may have seemed breakneck to travelers of a previous era, but it pales in comparison to light's 300,000 kilometers per *second*. In this example, β differs from 1 only in the fifteenth significant digit, far too small a deviation for detection. To observe time dilation, one needs either to travel at velocities that are a significant fraction of c or to measure the passage of time with extraordinary precision. In 1905, neither was possible. Confirmation of special relativity had to wait. Nevertheless, even the possibility of the time's relativity spurred bizarre speculations. One such possibility, the so-called "twins paradox," sparked debate that raged into the 1950s.

Consider identical twins born into a hypothetical future in which interstellar space travel is reality. One twin travels to a distant star aboard a spacecraft at nearly the speed of light while the other remains on Earth. Upon return, the traveling astronaut is shocked to find that her stay-at-home twin has aged dramatically, whereas she retains the flower of youth.

True paradoxes defy resolution. The "twins paradox" is no longer paradoxical because a variant of the experiment has actually been performed. In 1971, two physicists, Joseph Hafele and Richard Keating, flew around the world carrying atomic clocks of extraordinary accuracy. When the times recorded by the traveling clocks were compared to those of atomic clocks on the ground, the results agreed with predictions of special relativity to within 10 percent. More recent and accurate experiments have verified the time-dilation predictions of special relativity to better than $1/100$ of a percent.[8] Today, the phenomenal accuracy of the global positioning system (GPS) is possible only by fully accounting for relativistic time dilation.

Once Einstein tipped the domino of absolute time, other sacred dominoes were soon to fall. Next in succession was the measurement of distance. Suppose we file a photon of light along the x axis of inertial frame F and plot its position as a function of time t (Fig. 11). For convenience, let's measure time in years and distance in light-years, the extraordinary distance light travels in exactly one year. In these convenient units $c = 1$; that is, one light-year of distance per year of time. Thus, the graph of the photon's position with respect to time – its *worldline* – is a straight line emanating from the origin $(0, 0)$ with a slope of 1. What is the slope of the worldline of a light photon in frame F'? Also 1, of course, because the value of c is absolute. The rub comes when a stationary observer in F views the photon in F' traveling at velocity c. To the observer in F, time flows more slowly in F' than it does in F, which is indicated by increased separation between the tick marks along the t' axis in F'. But to preserve the slope of the worldline, the tick intervals along the x' direction of F' must also increase commensurately. Einstein concluded that, to an observer at rest, lengths appear contracted in the direction of motion by the time-dilation factor β.

Again, Einstein's counterintuitive prediction has been verified by experimental observations. Bunches of particles injected into particle accelerators and accelerated to nearly the velocity of light appear to contract in the direction of motion in strict accordance with relativistic predictions.

Special relativity packs at least two psychological punches. First, human perceptions of time and length depend upon one's frame of reference. No two humans perceive exactly the same universe. And second, space and time, which Newton perceived to be independent, are intricately entangled.

The relativistic entangling of space and time was first quantified not by Einstein but by the esteemed Dutch physicist Hendrik A. Lorentz. Lorentz

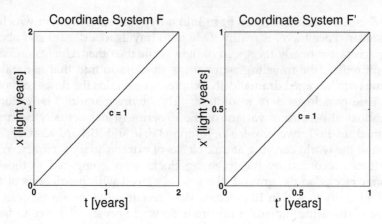

Figure 11: Constancy of the velocity of light, *c*, in two inertial frames of reference. (Author.)

derived a coordinate transformation that accounted for the null result of the Michelson-Morley experiment. But as one ensconced in the tradition of nineteenth-century science, Lorentz incorrectly interpreted his own result. As a proponent of the ether hypothesis, he conjectured that lengths contract in the direction of motion as an object "pushes" against the ether. Einstein, antiauthoritarian to his core, possessed the audacity to challenge conventional wisdom, whether from Newton or Lorentz, and corrected Lorentz' interpretation. No ether is necessary. Length contraction in the direction of motion is a necessary consequence of time dilation and the invariance of *c*.

Despite uncommon intellectual daring, Einstein was not innately arrogant – confident, yes; arrogant, no. With characteristic understatement, he wrote in *Relativity*, "It appears more natural to think of physical reality as a four-dimensional existence, instead of the evolution of a three-dimensional existence."[9] Ironically, relativity landed an early proponent in Hermann Minkowski, the mathematics professor at Zurich Polytechnic who had once called Einstein – who cut his math classes – a "lazy dog." Captivated by the beauty of the theory, Minkowski forgave Einstein his youthful excesses and helped the former wayward student cast relativity in mathematically elegant form. Time, space, ether: other dominoes were soon to fall.

$E = mc^2$

In 2005, the world celebrated the centennial of Einstein's *Wunderjahr*, the year in which the twenty-six-year-old upstart published his dissertation and four scientific papers. The articles appeared in rapid succession in *Annalen der Physik*, an old and venerated journal. Their collective impact was nothing less than the overthrow of Newtonian mechanics. The first, on the photoelectric effect, earned Einstein the Nobel Prize for physics in 1921 and helped launch the revolution in quantum mechanics. In the second, Einstein exploited the molecular theory of heat to explain Brownian motion, the observed random motions of small dust or pollen particles suspended in water. The third laid the foundations for special relativity and became the most highly acclaimed scientific paper of all time. As if he had not upset the scientific apple cart sufficiently, Einstein wrote a fourth paper as a short "footnote" to the relativity paper. Introducing a formula that is at once beautiful and terrifying, he established the equivalence of mass (m) and energy (E): $E = mc^2$.

How Einstein derived the iconic formula was sheer genius. In fits and starts, by the end of the nineteenth century, physics had settled upon three laws of conservation that were foundational to both classical mechanics and to the relatively new discipline of fluid mechanics: conservation of mass, momentum, and energy. Conservation laws are mathematical statements that, in the aggregate, certain quantities are neither created nor destroyed by physical processes. For example, the sum of the momenta of billiard balls remains unaltered by collisions, allowing the billiard shark to accurately predict post-collision trajectories. Conservation of energy explains the utility of fire-making drills long used advantageously by Indians. Friction converts the mechanical work expended on the drill into an equivalent amount of heat. Conservation of mass, on the other hand, presupposes that matter may change form, as for example through a chemical reaction, but that matter cannot be created nor destroyed.

Surprisingly, from thought experiments based upon conservation principles, the relativity of time and length implies the relativity of mass as well. That is, the value that one measures for the mass of an object depends upon the velocity of the object in one's frame of reference. By the term "rest mass," we descriptively connote the mass of an object as it sits at rest in its own reference frame.

Suppose that an object is accelerated from rest to velocity V in some reference frame. According to the special theory of relativity, the relativistic mass and the rest mass are related by Lorentz' time-dilation factor. Whenever V approaches the velocity of light itself, the relativistic mass – the rest mass divided by β – becomes infinite. It follows that to accelerate an object to a velocity equal to or greater than c would require an infinite expenditure of energy.

Einstein concluded that c represents a universal limiting velocity, a physical "speed limit." Nothing with finite mass can travel faster than light. As bizarre as the implications of relativity seemed to slow-moving humans, they describe perfectly the observed effects for fast-moving particles. In 1901, predating relativity, experiments to accelerate electrons to very high speeds observed an initially inexplicable dependence of mass upon velocity, which violated Newtonian mechanics. When, following

1905, such experiments were repeated at relativistic velocities (those approaching the velocity of light), the results were found in close agreement with the predictions of relativistic mass increase.[10]

How then could a photon, a particle of light, travel at light's velocity? There was only one explanation: photons possess no rest mass. A photon, when it ceases moving, ceases to exist. But although it possesses no rest mass, paradoxically a photon does carry momentum and energy while moving. After all, the sun's light warms us; so photons carry energy. They also carry momentum, which can be inferred from a Nichols radiometer, an instrument designed to detect and measure "radiation (light) pressure." A minuscule beast of great power but no mass, the photon is a remarkable creature.

The unique nature of the photon led Einstein to formulate another revolutionary *Gedanken* experiment. He considered the conservation of mass, momentum, and energy when a photon is exchanged between two spatially separated points. From the peculiar "mass-less" nature of the photon at rest, Einstein concluded that mass and energy are interchangeable. Energy is liberated mass; mass is congealed energy. In his own words:

> The most important result of a general character to which the special theory of relativity has led is concerned with the conception of mass. Before the advent of relativity, physics recognized two conservation laws of fundamental importance, namely, the law of conservation of energy and the law of conservation of mass; these two fundamental laws appeared to be quite independent of each other. By means of the theory of relativity they have been united into one law.[11]

$E = mc^2$. Something in the formula captivates. Recognized the world over, it appeals to the educated and to the scientifically illiterate and lays bare a long-held secret of nature. If the energy contained in the tiniest grain of sand were liberated completely, the result would be stupendous because the square of c is astronomically large. Why, Einstein asked rhetorically, did the prodigious quantities of energy imprisoned in matter go unnoticed for so long?

His homespun answer: "It is as though a man who is fabulously rich should never spend or give away a cent; no one could tell how rich he was."[12] By only one activity observable on Earth at the time did nature betray the wealth of matter: the spontaneous disintegration of radioactive particles. In 1907, Einstein speculated that radioactive processes more energetic than the decay of radium might exist. In 1917, confirming Einstein's suspicions, Ernest Rutherford split atoms at Cambridge University's prestigious Cavendish Laboratory. The first successful alchemist, Rutherford had transmuted one element into another, freeing vast quantities of energy in the process. The equivalence of mass and energy was no longer in doubt, though the survival of the human species was.

Seven

Einstein's Happiest Thought

> I am now exclusively occupied with the problem of gravitation, and hope, with the help of a local mathematician friend, to overcome all the difficulties. One thing is certain, however, that never in my life have I been quite so tormented. A great respect for mathematics has been instilled within me, the subtler aspects of which, in my stupidity, I regarded until now as pure luxury.
>
> **– Albert Einstein**

Newton's theory of gravitation was doomed from the outset, its creator beset by grave doubts. In public, Newton projected assurance. "Hypotheses non fingo," he was fond of saying in Latin: "I don't deal in hypotheses." But in private, he brooded:

> It is inconceivable that inanimate brute Matter should without the Mediation of something else which is not material, operate upon, and affect other Matter without mutual Contact. Gravity . . . is . . . so great an Absurdity that I believe no Man who has in philosophical Matters a competent Facility of thinking can ever fall into it.[1]

In Einstein's modern turn of phrase, what haunted Newton was "spooky action at a distance." How can two celestial bodies affect one another instantly at great distance if nothing fills the intervening void?

The answer had to wait for science's "New Testament." If the three books of Newton's *Principia Mathematica* comprised the scientific Old Testament, then Einstein's papers on special and general relativity comprised the first two books of the New. Book One, in particular, resumed the investigation of gravitation where Newton had quit.

Others might have written Book One. Not so with Book Two. General relativity demanded what Einstein uniquely possessed: "the most amazing combination of philosophical penetration, physical intuition, and mathematical skill."[2] By nearly universal consent, the general theory of relativity represents the greatest individual achievement ever in synthetic scientific thought. Like a rope of great strength, general relativity braids together three strands: the gravitational theory of Newton; the non-Euclidean geometry of Bernhard Riemann; and the special theory of relativity of Einstein himself. To Einstein's friend and fellow physicist Max Born, general relativity was simultaneously a "great work of art" and "the greatest feat of human thinking about nature."

"No one who has really grasped it can escape the magic of this [new] theory,"[3] Einstein effused at the birth of his new theory. Far from dry and lifeless, general relativity brims with magic and mystery: black holes, time warps, wormholes, and the origins of the universe. But foremost it is a generalized theory of gravitation that lay to rest Newton's concerns about "spooky action at a distance."

Child's Play

Magazine cover photos from 1916 reveal an uncharacteristically gaunt and weary Einstein. A letter from Einstein to fellow physicist Arnold Sommerfeld revealed why. "Compared with this problem, the original [special theory of] relativity is child's play."[4] Crafting the general theory of relativity consumed Einstein for ten exhausting years, during which he endured divorce, world war, and near-fatal illness.

In 1906, however, Einstein remained blissfully unaware of the grueling effort that lay before him. Still languishing as an underemployed twenty-seven-year-old patent examiner, Einstein feared his creative days had already passed. The previous year he had launched four papers that were soon to overturn physics. But reaction was initially slow, and additional golden eggs were not immediately forthcoming. The limbo was agonizing. Mercifully, before despair could root, a simple idea rescued him: "the happiest thought" of his life.

The universe of special relativity is highly idealized. All frames of reference are inertial; that is, all motion must be uninteresting – in a straight line and at a constant velocity. Moreover, gravitational attraction is disallowed. Special relativity has little relevance to the fighter pilot in a tight turn, to the thrill-seeker on the newest roller coaster, or to the child on a merry-go-round, whose paths are not lines. As Einstein struggled to extend special relativity to the universe of common experience, a sudden realization seized him, one that we who grew up during the space age take for granted: one is weightless during free fall. As he put it, "The gravitational field has only a relative existence . . . because for an observer falling freely from the roof of a house there exists . . . no gravitational field."[5]

Galileo, too, had pondered the issue of free fall and, countermanding Aristotle, had arrived at the realization that bodies of different masses fall to earth at the same rate. That is, the acceleration of a falling object toward a celestial body is independent of the object's mass. While on the moon during the Apollo 15 mission, astronaut David Scott playfully demonstrated this fact. To a live television audience, Scott simultaneously released a geology hammer and a falcon feather from chest height. In the absence of an atmosphere, the feather and the hammer hit the lunar soil at the same instant, vindicating Galileo.

Scott's experiment exposed an astounding "coincidence." In physics, the notion of mass arises in two seemingly different contexts. Einstein marveled, "Mass is defined by the resistance that a body opposes to its acceleration [inertial mass]. It is also measured by the weight of the body [gravitational mass]. That these two radically different definitions lead to the same value for the mass of the body is . . . an astonishing fact."[6]

Einstein refers, of course, to the remarkable fact that the gravitational mass of an object and its inertial mass, measured by different techniques, always turn out identical. Five years following the emergence of general relativity, the Hungarian experimental physicist Roland von Eötvös developed the ingenious torsion balance, which effectively magnified the difference between inertial mass and gravitational mass, should any difference exist. He found the two measures to be equal to within a few parts in one billion. In 1960, a team led by Robert Dicke of Princeton University refined Eötvös' method to significantly reduce the experimental error. Modern experiments confirm the equality of gravitational and inertial masses to at least 12 significant digits.[7] The issue is not that the two measures are exactly equal. The issue is *why*, because they measure different things.

The equivalence of gravitational and inertial mass was recognized long before Einstein but went largely unappreciated. Indeed, the equivalence accounts for Galileo's observations that objects of different masses fall at the same rate *in vacuo* or that the period of a pendulum depends only on its length and not upon the mass of its bob. With characteristic clarity, Galileo reported:

> I took two balls, one of lead and one of cork, . . . and suspended them from two equal thin strings. . . . Pulling each ball aside from the vertical, I released them at the same instant, and they, falling along the circumferences of the circles having the strings as radii, passed thru the vertical and returned along the same path. This free oscillation . . . showed clearly that the heavy body kept time with the light body so well that neither in a hundred oscillations nor a thousand will the former anticipate the latter by even an instant, so perfectly do they keep step.[8]

The ordinary person and ordinary physicist alike accept this fact as just that, without a second thought. In contrast, Einstein, like a child, paused to flip over the stone to see what hid beneath. Why? Why are gravitational and inertial masses equal?

He devised a hallmark *Gedanken* experiment. Imagine a room-sized "trunk," he asks in *Relativity*, his popular treatise. A rope attached to its lid suspends the trunk, which hides an "observer" whose job is to infer from his experience what type of motion the trunk is undergoing. To update the analogy, let's consider an elevator car with a single passenger, insulated from the sights and sounds of the outside world with only his or her remaining sensory clues to make inferences.

We'll consider four distinct cases and try to anticipate the passenger's sensory state for each. First (Case I) suppose that the elevator car and its passenger are in fact located in deep space, far from significant gravitational fields, and either motionless or drifting along at constant velocity. The passenger will experience the weightlessness familiar to astronauts en route to the moon. In equal and opposite reaction to the slightest of forces, the passenger glides about the car, careening off the walls. If she wishes to stabilize herself, she must secure her feet to the floor with Velcro attachments, or constantly hold onto an anchor point. A ball released from the passenger's hand floats freely, motionless relative to the occupant who carefully set it free.

For Case II, the elevator car, in which an occupant sleeps, plummets from the top floor of the Empire State Building. By an unfortunate combination of failures, the cable snaps and the emergency brake fails. The hapless occupant awakens just as the car begins its downward plunge. Without external clues, she floats freely about the car, reveling in weightlessness, mercifully oblivious to the "gravity" of her predicament, unaware that she experiences Case II rather than Case I. We conclude, as did Einstein, that there is no easy way for the occupant of a sealed "trunk" to distinguish between Cases I and II. That is, free fall abolishes gravity.

> For an observer falling freely from the roof of a house there exists – at least in his immediate surroundings – no gravitational field. Indeed if the observer drops some bodies then these remain to him in a state of rest or uniform motion . . . The observer therefore has the right to interpret his state as "at rest."[9]

Now consider two additional cases. In Case III, the elevator car, initially floating freely in deep space, is subjected to the uniform "upward" acceleration of rocket thrusters from below. What sensation does the occupant experience? In response to the upward motion of the floor, he must either crumple to the floor or stiffen his legs to oppose the upward force. If the upward acceleration remains constant, say, at 32 feet-per-second per second (feet per second squared), the occupant (presumed to be an earthling) will sense that he is pulled downward by a force exactly that of gravity at the earth's surface. What happens to the ball in the compartment? Initially it floated freely. But during acceleration, the floor of the compartment rushes upward to meet the ball, which appears to the observer to fall to the floor as if under the influence of gravity. Indeed, without external visual or auditory clues, the occupant is unable to distinguish whether the car is at rest on earth (Case IV) or accelerating in free space. Cases III and IV are compared in Fig. 12.

Free fall abolishes gravity, Einstein concluded, and acceleration creates it. Gravitational mass exactly equals inertial mass because gravitational at-

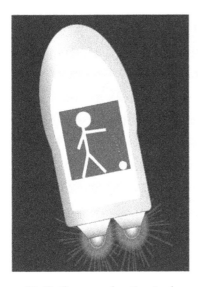

Figure 12: Uniform acceleration in deep space (Case III, left) is equivalent to gravitation on, say, Earth (Case IV, right). (A. Crockett, illustrator.)

traction and uniform acceleration are indistinguishable. Like mischievous twins, gravity and acceleration masquerade as one another. Is there a way to infer which twin is which? With very sensitive instruments, the observer in Einstein's elevator could indeed infer whether he was suspended in the gravitational field of a heavenly body or uniformly accelerating in deep space. On earth, for example, the gravitational lines of force, which all point toward the earth's center, are not quite parallel. Consequently, the occupant of an earthbound elevator experiences a minuscule horizontal "pinching" force termed the *tidal force* because similar gravitational pinching of the earth by the moon induces the tides.

This then is the crux of the principle of equivalence, the first strand of general relativity: wherever we desire, we may interchange the words *gravitation* and (uniform) *acceleration*.

The Shortest Distance between Two Points

Ask any man or woman on the street, "What path gives the shortest distance between two points?" and with virtual certainty, he or she will reply "A straight line." Ask any bright-eyed and well-educated college student, and he or she will respond similarly. However, ask the physicist or professional mathematician, and with wrinkled brow, he or she will mumble something about the question being ill-posed.

Properly phrased, the question should ask: What is the path of shortest distance between two points *in a plane*? For, if the underlying surface is not planar, then the shortest path cannot be linear. Consider for example transatlantic flights, which follow the arcs of "great circles" whose radii are that of the earth. When projected onto flat maps, such routes appear indirect, when in fact they are optimally short, given that the aircraft is constrained to skim the surface of a globe, like a fruit fly in the fuzz on a peach. The paths of shortest distance along curvilinear surfaces are termed *geodesics*. In a plane, geodesics are straight lines; on a sphere, they are circular arcs (Fig. 12). On a more complicated surface – for example, a saddle – geodesics have complex paths.

Ask the same man or woman on the street, "What is the sum of the interior angles of a triangle?" and most likely he or she will reply, "180 degrees." Again, the question is ill-posed. The familiar answer is correct only if the triangle is constructed in a plane. If, however, a "triangle" is constructed from three great circular arcs on the surface of a sphere, the sum of the interior angles is greater than 180 degrees. For a small triangle so constructed, the curvature of the underlying surface is virtually unde-tectable and the angles sum to just over 180 degrees. On the other hand, if two of the triangle's vertices lie along the equator, and the third lies at the North Pole, then the sum of its three angles can greatly exceed 180 degrees (Fig. 13).

Other cherished geometric notions also fail whenever the underlying geometry is not planar. For example, the magical number π represents the ratio of the circumference of a circle to its diameter, *provided each lies in the same plane*. On the earth's spherical surface, however, that ratio lies between 2 and π. For a small circle – one whose diameter is tiny relative to the earth's – the ratio is very nearly π. In contrast, for the equatorial circle, the ratio of circumference to "diameter" (also a great circular arc) is 2.

The notions of straight lines and planes are so deeply ingrained in the Western psyche that we commonly mistake them for physical reality rather than mathematical abstractions. One conceptualizes a plane by a sheet of glass, for example. But glass has finite thickness; planes don't. Make the glass very thin then. But thin glass sags under its own weight, no longer truly planar.

Equally vexed is the notion of a straight line. How, in practice, does one construct a straight line? By using a ruler or yardstick. But who made the ruler straight? And how? Under an electron microscope, a ruler's edge appears as jagged as the coast of England. Modern construction demands straight lines and square corners, so the bricklayer levels bricks with a chalked snap-line. But were he or she to attempt to extend that chalk line over a distance of miles, it would sag under its own weight and oscillate in the breeze. How then are straight roads constructed across miles of prairie? For long distances, surveyors use a transit, a glorified telescope. And now we have the glimmer of the answer to our question. A "straight" line, as we know it, is the line of sight; that is, the path followed by a beam of light. There is no other arbiter of "straightness."

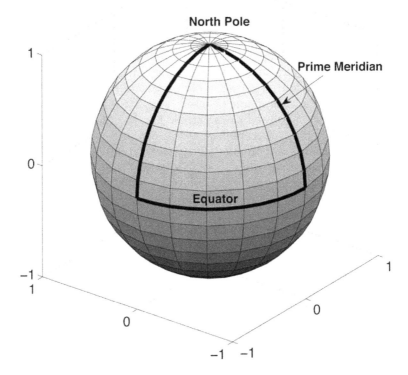

Figure 13: Interior angles of large "triangles" on a spherical surface may sum to more than 180 degrees – for example, to 270 degrees for the highlighted "triangle." (Author.)

If light beams are straight, then the underlying geometry of nature is that of Euclid. If, perchance, light rays bend when subject to gravity, as Newton conjectured, then nature's geometry is fundamentally non-Euclidean. The mathematics of non-Euclidean geometry – general relativity's second strand – tormented Einstein.

Special Relativity

A central tenet of special relativity – general relativity's third strand – is the constancy of c in all inertial frames of reference. Thus special relativity intrudes into the domain of general relativity literally as a beam of light. What happens to light, space, and time when the frame of reference accelerates, no longer inertial?

Mimicking the master himself, we consider a common analogy to illustrate non-inertial motion. Consider a line drive to shortstop. For this somewhat implausible thought experiment, presume the baseball is hit so hard that it follows a nearly perfect straight-line trajectory despite the earth's gravitation. The shortstop runs forward to meet the ball, hoping to shave milliseconds off his reaction time to complete a double play at second. How does the path of the ball appear to the moving shortstop? It depends. If he runs along a straight line, oblique to the path of the ball, and at a constant speed, then the path of the ball also appears linear. In fact, he could run along any straight path – toward or away from the ball – at any constant velocity – large or small –and, to the shortstop, the ball's flight would appear as straight as an arrow. On the other hand, should he accelerate toward the ball, its path would appear to bend slightly toward him. In fact, because motion is relative, it would not be incorrect to say that the shortstop's acceleration "bent" the path of the ball.

Now replace the baseball with a photon of light traveling at 300,000 kilometers per second. We return to the frame of reference provided by the thought experiment of Einstein's elevator, this time, however, with increased sensitivity to the well-being of its occupant. Recognizing that sensory deprivation over long periods of time is harmful, we install a window in the compartment through which the traveler may view the outside world and perhaps even communicate with other distant travelers.

Let's modernize the analogy though. Our passenger travels aboard a spacecraft in deep space. Another spacefarer, separated by a great distance, sends the first spacefarer a message by means of laser light. What path do the photons follow from the vantage of the first traveler? Again, it depends. If both sending and receiving spacefarers travel at constant velocities, a photon racing between them will appear to each to follow a linear path. If, on the other hand, say, the receiver accelerates modestly relative to the sender, the beam will appear ever so slightly deflected. However, could our spacefarer withstand tremendous acceleration, the laser beam would appear to bend significantly.

The principle of equivalence permits us to globally edit the paragraphs above, replacing every occurrence of acceleration with gravitation. Experience tells us that baseballs subject to gravitation follow curved trajectories. Similarly, the previous thought experiment forces us to acknowledge that gravitation deflects light as well. In the words of Einstein, "In general, rays of light are propagated curvilinearly in gravitational fields. . . . Hence, the idea of a straight line loses its meaning."[10] But if beams of light are the only practical mechanisms for generating "straight" lines in space, and if they themselves are bent by gravity like the bricklayer's sagging chalk line, then we are forced to conclude the underlying geometry of the universe is inherently non-Euclidean. Spacetime is curvilinear. Straight lines and flat planes are idealizations found nowhere in the universe.

Once general relativity undermined Euclidean geometry as inherent in nature, it quickly dispatched other time-honored notions. In particular, general relativity cast a second "wrinkle in time." In yet another *Gedanken* experiment – the rotating disk experiment – Einstein deduced that gravitation alters the flow of time. To modernize Einstein's analogy, let's find a children's playground.

Any casual observer of children concludes that there exist two types of people in the world: those who have no fear and those who do. Imagine then a large merry-go-round with two riders, one timid and the other intrepid. Each wears a watch and takes his or her place on the merry-go-round. The careful rider prefers to be near the hub. The fearless rider holds on at the rim and leans yet further out to experience maximum centrifugal acceleration as the angular velocity of the merry-go-round increases to dizzying levels. After a while, each rider glances at his or her watch, the careful rider eager to end his agony, the fearless one wishing her ride could last forever. Their watches, synchronized at the beginning, no longer agree. The watch of the outer rider, who, near the whirling rim, has traveled at the faster rate, experiences relativistic time dilation. Her watch therefore has run slightly slower than that of the careful rider, who has traveled slower by virtue of his position near the hub. But the rider whose watch has run slower has also experienced the greatest centrifugal acceleration.

By the principle of equivalence, we may substitute *gravitation* for *acceleration* in the scenario above. If clocks run more slowly when subjected to acceleration, then they must likewise run more slowly when subjected to gravitation. Thus, gravity not only bends light rays, it slows the passage of time.

General relativity is so named because it generalizes special relativity to permit non-inertial frames of reference, such as the rotating merry-go-round in the previous analogy. One by one, Einstein deduced all the following from the bedrock principle of equivalence:

1. The concept of a straight line loses its meaning in a gravitational field.

2. Rays of light propagate curvilinearly in a gravitational field.

3. The speed of light is *not* constant in a gravitational field.

4. Clocks run more slowly in a gravitational field.

5. Euclidean geometry does *not* hold in a gravitational field.

Even the casual student of relativity cannot escape the feeling that, like Alice in Wonderland, he or she too has stepped through the looking glass. If spacetime is non-Euclidean, what does it look like? It is precisely this question that presented Einstein with his most formidable obstacle. Were if not for a fortuitous synchronicity, the general theory of relativity would have foundered on the shoals of mathematics.

The Fabric of the Universe

Mathematicians long suspected a weakness in the foundations of Euclidean geometry. The likely culprit was Euclid's fifth postulate. All the rich constructs of Euclidean geometry rested atop just five assumed truths – postulates – that originated with the ancient Greek geometer Euclid. Of these, the first four seemed more or less self-evident to the geometers of yore:[11]

1. A straight line may be drawn between any two points.

2. A line segment may be extended indefinitely in either or both directions.

3. A circle of any radius may be drawn with any point as center.

4. All right angles are equal.

The fifth postulate, which may be expressed in a number of logically equivalent forms, is less intuitively obvious. John Playfair, a Scottish mathematician, formulated the most popular version of the fifth postulate, know as Playfair's axiom, which states: given a line and a point not on the line, there exists *exactly one* line through the point parallel to the original line. From these five simple postulates follows the familiar and immensely useful geometry of Euclid. At the end of the eighteenth century, the German philosopher Immanuel Kant declared Euclid's geometry to be essential truth.[12] To Kant, Euclidean geometry was hard-wired in the human brain. Kant was right about a lot, but he was no mathematician.

"Euclid's fifth," or the "parallel postulate" as it became known, attracted special scrutiny. A number of mathematicians attempted to deduce Euclid's fifth from the first four, which, if successful, would have promoted the fifth postulate to the status of a theorem. All such attempts failed, and by 1868 mathematicians conclusively established the independence of the fifth postulate from the other four.

Unconvinced by Kant's authoritative assertion as to the Euclidean nature of reality was German mathematician Karl Friedrich Gauss, perhaps the brightest comet ever to light the mathematical sky. Convinced that the geometry of space, like mechanics, must be tested by experiment rather than postulated as theory, Gauss dabbled with non-Euclidean geometry. Confessed Gauss to a friend, "Indeed I have from time to time in jest expressed the desire that Euclidean geometry would not be correct."[13]

Early in the nineteenth century, several powerful mathematicians, two allied with Gauss, found a way to establish once and for all the status of Euclid's fifth. Direct assaults on the fifth postulate having failed, mathematicians turned to indirect assaults, exploiting a mathematical technique known as proof by contradiction. Suppose that one accepts an alternate fifth postulate that negates Euclid's fifth. Two possibilities then exist. If a logical contradiction is uncovered, then the original fifth postulate must be essential and sacrosanct to *any* system of geometry. On the other hand, if one obtains a self-consistent geometric system that differs from Euclid's, then the fifth postulate is nonessential, and fundamentally different fifth postulates prove equally viable.

Two alternate pathways negate Playfair's axiom: by assuming there exist no lines through a point parallel to a given line, or that there exist many lines parallel to the given line. Surprisingly, both alternate fifth postulates resulted in successful and self-consistent non-Euclidean geometries. Russian mathematician Nikolai Lobachevsky developed the first. Lobachevskian geometry, also known as hyperbolic geometry, follows logically from the latter assumption: multiple parallel lines exist. The former assumption – no parallel lines – leads to elliptic geometry. Developed by Gauss' student Bernhard Riemann, elliptic geometry is known also as Riemannian geometry.

In the 1840s, the University of Göttingen, Germany, marked the epicenter of the mathematical world. There, the legendary Gauss held a distinguished professorship. In 1843, at age nineteen, the brilliant but shy Riemann matriculated at Göttingen to study theology in preparation for the ministry. Soon disenchanted with the direction his father had charted for him, Riemann turned his attention to his first love: mathematics. To his dismay, in the Teutonic fashion of the time, the god-like Gauss was utterly unapproachable to beginning students.

Disillusioned, Riemann transferred to Berlin to study under kindlier but lesser lights: Gustav Dirichlet and Carl Jacobi. Two years later, Riemann returned to Göttingen, earning his doctorate in 1851 under Gauss, who bestowed rare praise upon his protégé for his "gloriously fertile originality."[14]

Riemann then began a dehumanizing climb up the German academic ladder. The first rung was *Privatdozent*, an unpaid lectureship. Hope of attaining the next rung rested upon a trial lecture before Göttingen's entire faculty. According to custom, Riemann submitted three titles, from which the department head chose one, normally the first by tradition. As his third topic, Riemann hastily included the foundations of geometry, upon which Gauss had been ruminating for nearly sixty years. His interest piqued, Gauss picked the third topic, for which Riemann was wholly unprepared. Abandoning all other interests for two months, Riemann delved feverishly into preparing the lecture.

After several postponements, in June of 1854 Riemann delivered the long-awaited lecture. "On the Hypotheses which Lie at the Foundation of Geometry" secured Riemann's place in mathematical history. "The result," according to no less a light than Einstein, "was one of the great classical masterpieces of mathematics, and probably the most important scientific lecture ever given."[15] In a lecture of vast sweep, Riemann generalized Gauss' early geometric work to develop a framework so broad that it encompassed all known systems of geometry, Euclidean and non-Euclidean.

To do so required the calculus of tensors, complicated mathematical objects that, like vectors, obey certain laws under coordinate transformations. With consideration toward his audience, Riemann omitted the laborious mathematical details and cast his talk in nontechnical language. The aging Gauss, seldom impressed, shone with enthusiasm at the end of Riemann's magnificent treatise. In 1859, following the deaths of Gauss and his successor Dirichlet, Riemann ascended to Gauss' chair at Göttingen, the highest throne of the mathematical world at the time.

But Riemann's years of debilitating poverty had broken his health. He succumbed to tuberculosis at just thirty-nine years of age, an incalculable loss to the mathematical and scientific worlds. Still, he had fathered Riemannian geometry and tensor calculus, the tools that would rescue Einstein from his mathematical woes just sixty years later. Thanks to Riemann, Einstein successfully completed the general theory of relativity, after "years of anxious searching in the dark, with their intense longing, their alternations of confidence and exhaustion and the final emergence into light."[16]

General Relativity

In March 1916, *Annalen Der Physik* published the momentous paper *"Die Grundlage der allgemeinen Relativitätstheorie"* ("The Foundation of the General Theory of Relativity"). Like a fairy-tale marriage, general relativity was a perfect union of mathematics and physics, as evidenced by Einstein's deceptively simple-looking field equations:

$$E_{ij} = \frac{8\pi G}{c^4} T_{ij}$$

Of these magnificent equations – the result of eight years of arduous mathematical labor – Einstein's future colleague Banesh Hoffman said:

> . . . if written out in full instead of in the compact tensor notation, [they] would fill a book with intricate symbols (in one form, millions of them). And yet there is something about them that is intensely beautiful and almost miraculous. Their power and their utter naturalness in both form and content give them an indescribable beauty.[17]

One aspect of their beauty appeals even to the uninitiated. The constant of proportionality on the right side of the equation, namely $8\pi G/c^4$, thrills. It combines the most fundamental constant of mathematics, π, with two of the three most fundamental physical constants: G, the Newtonian gravitational constant, and c, the velocity of light *in vacuo*. When such a constellation of quantities arises, one cannot help but sense, as did Einstein, that one has glimpsed the mind of God.

In the curvilinear geometry of general relativity, geodesics replace the straight lines of familiar Euclidean geometry. Einstein summarized the situation as follows: "Riemann's geometry of an n-dimensional space bears the same relation to Euclidean geometry on an n-dimensional space as the general geometry of a curved surface bears to the geometry of a plane."[18] This relationship was illustrated earlier in Fig. 13, which compared geodesics on flat and curved surfaces. Recall that in flat geometry the geodesics trace straight lines, whereas on a sphere geodesics follow great circles. Remarkably, the geodesics of relativity's four-dimensional spacetime are exactly the paths traced by a beam of light. In general relativity as in special relativity, light plays a leading role.

Einstein's field equations above relate two mathematical quantities: the Einstein curvature tensor E and the energy-momentum tensor T. The former expresses mathematically the departure (or deviation) of spacetime geodesics from the familiar straight lines of Euclidean space. The latter is related to the strength of the gravitational field. Succinctly, the equation states that the curvature of spacetime depends upon the gravitational intensity. Gravity distorts spacetime by bending light rays, a phenomenon euphemistically called *gravitational lensing*. In the absence of gravitation, the right-hand side of the equation vanishes, and spacetime is Euclidean.

In its full glory, general relativity is difficult even for physicists to comprehend and next to impossible for the layperson. A possibly factual anecdote relates an encounter between a journalist and Sir Arthur Eddington, fresh on the heels of the success of the Sobral expedition that first confirmed Einstein's predictions about gravitational lensing. Asks the journalist: "Is it true that only three people in the world understand relativity?" Quips Eddington (at least in the mythology of physics), "Who's the third?"

The mathematical objects E_{ij} and T_{ij} in the equation above are tensors, square arrays each comprised of sixteen components. Because relativity entangles space and time, the subscripts (indices) i and j are row and column counters that each run over the number of spacetime dimensions; that is, from one to four. By uniting space and time, relativity also teases humans to think and to visualize in four dimensions, for which we are poorly equipped.

Einstein, like most mortals, found it difficult to visualize four-dimensional spacetime and relied upon common analogies for an intuitive grasp of relativity. A two-dimensional analog attributed to Eddington captures the essence of spacetime distortion by gravitational lensing. Consider a game of billiards played on a regulation table with balls all of the same size and weight. Barring a collision, the presence of one or several balls at the far end of the table does not affect the path, say, of the cue ball in regulation play.

Now consider a billiard game for which both the table and the balls are nonregulation. Specifically, the table's surface is pliant, like a thick pad of foam rubber. Furthermore, the balls differ dramatically in weight. Some, steel ball bearings the size of billiard balls, are very heavy; others, of

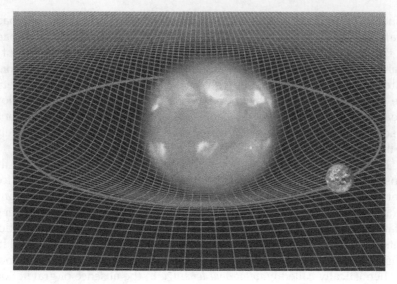

Figure 14: A planetary trajectory along the geodesics of spacetime, which are warped significantly in the vicinity of the sun. (Gravity Probe B Image and Media Archive, Stanford University.)

hollow plastic, are light. Near a very heavy ball, the table's spongy surface dimples deeply. If a lightweight ball passes the vicinity of the dimple, even without directly colliding with the heavy ball, its path is deflected. The deeper the dimple, the greater the deflection. Although no motive force acts upon the lightweight ball after the cue stick strikes it, the ball nevertheless follows a curved path.

And so it was that general relativity cleared up the issue that had troubled both Newton and Einstein: gravity's apparent "spooky action at a distance." In Newtonian mechanics, objects not subject to outside forces coast along forever in straight lines. The closed orbits of planets are, therefore, unnatural and attributed to gravitational forces that nudge the planets off their natural trajectories. By contrast, in relativity planets coast naturally along geodesics. Geodesics may be straight or curved, depending upon the distribution of mass (Fig. 14). Massive bodies distort the surrounding spacetime fabric, bending the geodesics, as in our two-dimensional billiard-table analogy. The heavier the body, the greater the departure of nearby geodesics from linear. Thus, in curved spacetime, "planets move in their orbits around the sun by simply coasting along geodesics instead of being pulled into curved paths by a mysterious force of gravity whose nature no one has ever understood."[19]

General relativity in a nutshell: "Space tells matter how to move, and matter tells space how to curve."[20] There you have it.

Einstein's cosmos abides magic. Spacetime is not what Newton and Kant supposed: the Euclidean backdrop to events, like the stage of a play. Spacetime is participatory. The stage and the players interact; the spider and its web are partners. The spider's motions jostle and distort the web, even as the web's fabric makes possible the spider's locomotion. General relativity constructs a web-like cosmos. Space, time, and all events that occur in spacetime are integrally interconnected. Elementary particles, planets, stars, humans, and creatures: all perturb the spacetime fabric, sending ripples throughout the cosmic pond.

Eight

Lights All Askew

> How can it be that mathematics, . . . after all a product of human
> thought . . . independent of experience, is so admirably appropriate
> to the objects of reality?
>
> – **Albert Einstein**

In the early 1920s, a debate that had raged in astronomy was settled
by an exquisite marriage of experiment and theory. The experimental apparatus was the hundred-inch telescope at the Mt. Wilson Observatory
in California, and the experimenter was astronomer Edwin Hubble. The
unwitting theorist was Albert Einstein.

When Einstein and wife Elsa visited Hubble at Mt. Wilson in 1931, Elsa
inquired as to the function of the huge apparatus. Told that it was a telescope for studying the structure of the universe, she teased, "Well, well!
My husband does that on the back of an old envelope."[1] In her homespun
way, Elsa had put her finger squarely upon her husband's unique genius.
Using only the laboratory of the mind, Einstein could ferret out the deepest secrets of nature, and one such secret was how the universe began.

Island Universes

Until the twentieth century, the Milky Way comprised the entire known
universe, although some suspected otherwise. At issue was the nature of
nebulae, from the Latin *nebulosus*, meaning "misty" or "foggy." Patches
of the night sky that appear cloudy, nebulae had intrigued astronomers
for centuries. Of the few visible to the naked eye, the Andromeda Nebula
predominates.

Some believed nebulae to be luminous masses of interstellar dust particles and gases. Others suspected nebulae to be "island universes" like
the Milky Way, so distant that their individual stars could not be resolved
with existing telescopes. Nebulae, it turns out, are both. But that's getting
ahead of the story.

An early proponent of the latter school of thought was Immanuel Kant,
the German philosopher. In 1750, as a twenty-six-year-old student, Kant

127

dabbled in natural philosophy, especially astronomy. Aware from observations and avid reading that nebulae manifest in several forms – disks,
spirals, or ellipses – Kant made an intellectual leap. On the correct presumption that the Milky Way is a disk-like aggregation of stars, it stood
to reason that other similarly constructed galaxies, depending upon the
viewing angle, could manifest with any of the aforementioned appearances. Kant concluded that nebulae are "systems of many stars" lying "at
immense distances."[2]

The competing theory was the "nebular hypothesis," which originated
in 1734 with the Swedish inventor, scientist, and mystic Emanuel Sweden-
borg. It maintained that spiral nebulae are whirlpools of gas that condense
to form stars and planets. The nebular hypothesis gained credence in 1796
by the analytical work of the French mathematical powerhouse Pierre Simon de Laplace (1749-1827). Using mathematics, Laplace showed that,
under the influence of gravity, spinning gaseous clouds tend to flatten into
rings, which further accrete into planets or stars.[3] The nebular hypothesis
fell temporarily from favor when mathematical difficulties with Laplace's
analysis came to light.

Laplace's mathematics suffered primarily from insufficient computational power. Modern computer simulations mimic stellar formation and
link the two theories. Gaseous nebulae are the incubators of stars, the "primordial soup" from which stars are born. Mature stars, however, aggregate into clusters, the island universes we know as galaxies. Both theories
are correct but represent different stages in the life cycles of galaxies.

To summarize, galactic nebulae are clouds of luminous gas and dust
particles within our own galaxy, the stuff of stars not yet born. Extragalac-
tic nebulae, on the other hand, are clusters of billions of stars outside our
Milky Way, which appear gaseous only because of their extreme remoteness.

The issue of nebulae raised passions among astronomers as late as
1920, when optical telescopes attained sufficient power to resolve extra-
galactic nebulae into constituent stars. Princeton-trained Harlow Shapley of the Mt. Wilson Observatory, whose work with cepheid variable
stars had made him a legend,[4] strongly supported the nebular hypothesis. Cepheids, by the way, pulse in their brightness, waxing and waning
in periodic cycles. The period of the luminosity cycle – from one day to
one hundred days – relates directly to the star's intrinsic brightness. Prior
to the discovery of cepheids, determining stellar distances was immensely
problematic. Astronomers relied upon apparent brightness to indicate distance. But apparent brightness depends upon many factors besides distance, the size, type, and age of the star among them. Cepheids came
to the rescue. The precise relationship between the period of a cepheid
variable and its brightness was first noticed by Henrietta Swan Leavitt,
one of a host of lowly "computers" hired by Harvard University to perform tedious computations on mechanical calculators.[5] From countless

hours of observing cepheids within the Magellanic Clouds, Leavitt teased out a relationship between the luminosity and the period that had eluded those less intimate with cepheids. Knowing a cepheid's period allowed astronomers to calculate its *absolute* luminosity, independent of distance. Cepheids therefore afforded astronomers a "standard candle" by which to calibrate distance measurements.

Mt. Wilson's Shapley, although brilliant, remained adamant that the Milky Way was unique. On the other side of the debate, Heber Curtis of California's Lick Observatory adhered to the island-universe theory. On April 26, 1920, the National Academy of Sciences got into the fray by hosting a debate between Shapley and Curtis. Although Curtis was judged the victor,[6] no hard data yet existed to back him up. The times, however, were changing.

Shapley's colleague at Mt.Wilson – and ultimately his nemesis – was the dashing and vain Edwin Hubble. Hubble had been a Rhodes scholar, a boxer, and a lawyer prior to turning to astronomy. Using Mt. Wilson's magnificent telescope, Hubble set about categorizing and classifying nebulae. To the dismay of Shapley, Hubble found most nebulae to resolve into "dense swarms of images in no way different from ordinary stars,"[7] which strongly supported the island-universe theory.

Still, Shapley, who had departed Mt. Wilson to direct Harvard College Observatory, remained skeptical. In 1924, however, Hubble made a momentous discovery. His one-sentence report to Shapley dispatched Shapley with his own sword and resolved the debate once and for all. "You will be interested to hear that I have found a cepheid variable in the Andromeda Nebula."[8] The presence of a cepheid variable star in Andromeda incontrovertibly established that the apparent "cloud" was in fact a swarm of stars. And on this news, the known universe exploded in size.

During the two decades following Hubble's discovery, he and others meticulously pushed back the outer boundaries of the observable universe, adding ever more distant galaxies to those named and catalogued. Since the 1990s, a NASA space telescope named to honor Hubble has continued what he started, probing to the very edge of the universe. Copernicus' sky held a mere 4,000 stars, those visible to the naked eye. In the five centuries since, the observable universe has grown incomprehensibly. Astronomers now count some 100 billion "island universes," each averaging 100 billion stars.

While Hubble, the experimentalist, painstakingly mapped the heavens night after night, taking photographic time exposures, some over multiple nights, theorists also coaxed cosmological secrets from Einstein's field equations. A theory of gravitation, general relativity applies to massive objects or collections of objects such as stars or galaxies. The bolder of the theorists applied relativity to the whole of the universe.

In 1919, immediately on the heels of Eddington's successful confirmation of general relativity, Einstein unleashed his theory on the cosmos, giving birth to a new discipline: cosmological relativity. Confronted by its unexpected implications, however, Einstein uncharacteristically lost confidence. The man who had overthrown Newton fudged his field equations with a term he called the "cosmological constant." His motivation? An assumption that the physical universe is static. The sole purpose of the arbitrary cosmological term was to ensure that the mathematical model kept the universe at a constant size, because Einstein's unadorned equations demanded that it either grow or shrink.

Others had more confidence in Einstein. First to notice that the cosmological equations of relativity admit nonstatic solutions was Willem de Sitter, a Dutch physicist.[9] Soon thereafter, Russian mathematician Alexander Friedmann turned his formidable talents to the study of general relativity. In 1922 Friedmann published cosmological solutions to Einstein's field equations without the suspect cosmological term and obtained, depending upon certain reasonable assumptions, two different classes of results, neither of which was static.[10] Friedmann's work, now the basis of modern cosmology, suggested that the universe must either expand indefinitely or expand and then contract, like cosmic breathing, exhalation following inhalation. Another who seized upon the novel idea of an expanding universe was Georges Lemaître (1894-1966), a little-known mathematician, Belgian priest, and former student of Eddington.[11] Unaware of Friedmann's earlier work, in 1927 Lemaître independently derived a nonstatic solution to Einstein's field equations. As a Jesuit, Lemaître found motivation in the religious implications of his scientific work. Convinced that the universe is in an expansion phase, Lemaître pondered how the universe might have appeared in cosmic antiquity.

Imagine that one could, with a God's-eye view, make a motion picture of the evolution of this nonstatic universe. Playing the movie forward would show the universe to expand as time progresses. Lemaître then performed a neat conceptual trick. Mentally running the time sequence in reverse, he envisioned a universe that collapses in upon itself as one peers ever more deeply into the past. The reversed-time thought experiment suggested a startling possibility to which Lemaître quickened: here was hard scientific evidence for a moment of creation.

Because he had no audience, Lemaître's ideas initially went nowhere. Lacking confidence, Lemaître published in obscure journals. Still, the timid priest summoned the courage to write Einstein of his findings. Einstein was uncharacteristically blunt, "Your calculations are correct, but your physics is abominable."[12] Fortunately, Lemaître persisted, for experimental evidence was beginning to back him up.

The Big Bang

As Hubble and others catalogued galaxies,[13] a remarkable trend emerged. In the mean, distant galaxies appear to recede from us. Startlingly, the more distant the galaxy, the faster its recession velocity. Hubble based his findings on a Doppler shift in the color of the light that reaches the earth from remote galaxies. Humans perceive the frequency of light as color in a spectrum that ranges from red at the low-frequency end to blue at the high. In the same way that sound from a police siren changes in pitch depending upon whether the vehicle is speeding toward or away from us, so light emitted from a distant star changes frequency depending upon the star's velocity relative to our own. On average, light from distant galaxies is consistently red-shifted; hence, distant galaxies recede from the Milky Way.

Due to a magical feature of nature, the frequency shift of distant starlight can be measured precisely. Each of the nearly 120 elements that make up the objects of the physical universe has a unique fingerprint in the light spectrum it emits when heated or burned. Passing light from a chemical source through a prism separates the light into its component colors (frequencies), revealing the fingerprint. Unique to each element, a series of dark, narrow bands – known as Fraunhofer lines – manifests at precise locations in the frequency spectrum.

Stars shine primarily by fusing hydrogen in a thermonuclear furnace to create helium. Because young stars are comprised primarily of hydrogen, the red shift in the Fraunhofer lines of hydrogen affords a precise gauge of a young star's recession velocity. Setting the stage for Hubble, the Lowell Observatory's Vesto Slipher had calculated that the relatively nearby Andromeda Galaxy races away from us at the speed of 300 kilometers per second. By 1929, Hubble and a colleague had calculated recession velocities as high as 15 percent of the velocity of light for the most distant galaxies then observable.[14]

Hubble's genius was to plot recession velocity against distance for each of the galaxies catalogued. Although he could determine a galaxy's recession velocity accurately by its Doppler shift, galactic distances were subject to considerable uncertainty, unless a cepheid variable could be located within the galaxy. However, all factors taken into consideration, the trend was unmistakable: more distant galaxies recede from us more quickly. Hubble expressed the relationship between recession velocity and distance as a simple linear law that now bears his name: $V = H \times d$. Here V is the recession velocity of a given galaxy, d is its estimated distance, and H is the Hubble constant, which quantifies, astoundingly, the expansion rate of the universe (Fig. 15). It is sometimes mistakenly presumed that, if galaxies in every direction appear to race away from earth, the earth occupies some sort of privileged position in the cosmos. Simple analogies dispel the misconception. The classic example in three space dimensions is that of raisin bread, with the raisins representing galaxies. As the bread bakes, the dough rises, and the distance between any pair of raisins increases. No raisin's point of view is privileged in any regard. Similarly,

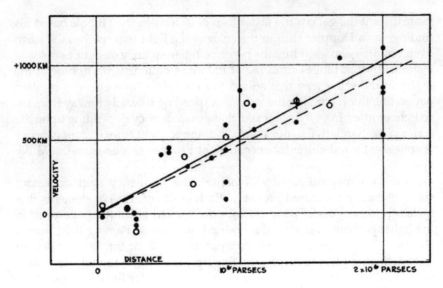

Figure 15: Edwin Hubble's original correlation of estimated distance of 46 galaxies versus their recession velocities determined by red shift. The Hubble constant H is the slope of the line that best fits the data. (This item is reproduced by permission of the Huntington Library, San Marino, California.)

the frame of reference of the Milky Way Galaxy is in no way privileged. Nor is that of the earth.

For all its simplicity, Hubble's law encapsulates a stunning fact: Ptolemy, Aristotle, Copernicus, and Einstein all erred. The universe is dynamic. Moreover, Einstein's field equations predict that the cosmos must be dynamic, as De Sitter, Friedmann, and Lemaître calculated.

The once iconoclastic Einstein stubbornly resisted this new development. In 1930, during his first visit to the United States, Einstein scheduled a meeting with Edwin Hubble at Mt. Wilson. Upon hearing of Einstein's imminent visit with Hubble, Lemaître seized the opportunity and arranged for the three men – the astronomer, the physicist, and the priest – to meet casually in order to discuss the state of the universe.

The meeting had a profound impact on the recalcitrant Einstein. During dinner, Hubble, with his data, and Lemaître, with his calculations, began to chip away at Einstein's preconceptions. As the evidence mounted, and Einstein's reluctance dissolved, the tension in the room mounted. Near the end of the meal, in a moment of triumph for Lemaître and epiphany for Einstein, the latter rose to his feet and exclaimed, "Yes! It is the most beautiful thing I have ever seen."[15] Thereafter Einstein referred to the cosmological constant as the "biggest blunder" of his life.[16]

A year after his triumphant meeting with Hubble and Einstein, Lemaître proposed that the universe originated from a dense "primeval atom."[17] Waxing poetic in a scientific journal, he speculated that the current universe is the "ashes and smoke of bright but very rapid fireworks."[18] Whereas Einstein, once convinced of the scientific validity of the "primeval atom," found the theory beautiful, others found it anathema.

Cambridge astrophysicist Fred Hoyle (1915-2001), an ardent atheist, vehemently objected to Lemaître's theory for its creationist overtones. Hoyle, who by sheer intellect had clawed his way from England's working-class textile towns to the hallowed halls of Cambridge University,[19] retained the combativeness that had made possible his ascendancy. With relish, he took to the airwaves to debunk Lemaître's cosmology. During a broadcast in March 1949 of the BBC radio program *The Nature of Things*, Hoyle coined the term "Big Bang," initially intended to ridicule Lemaître's "primeval atom." The name stuck.

Hoyle had legitimate scientific reasons for skepticism. From the accepted value of the Hubble constant at the time, Big Bang cosmologists estimated the age of the universe at two billion years. In contradiction, radiological methods of dating the earth placed its age at double that of the universe, an inconsistency of the first order. In 1948 Hoyle and colleagues Herman Bondi and Thomas Gold had proposed an alternate theory: the "steady-state" model.[20]

Steady-state theory attempted to account for the apparent expansion of the universe – which the theory's promoters accepted – without the dramatic creation event inherent in the Big Bang. The theory required, however, that new matter be created continuously to fill the growing gaps between receding galaxies. The proponents argued that continuous generation of matter was no more suspect than Lemaître's spontaneous creation of a universe out of nothing. By analogy, the universe was like a river that appears static even when it is constantly replenished by tributaries.

Despite zealous attempts to refute Big Bang cosmology, Hoyle inadvertently did much to bolster the theory. Hoyle's group of cosmologists (having worked together on radar during the Second World War) turned their analytical talents to examining how stellar furnaces create the elements, particularly the heavier ones. By the 1950s, following the development of atomic and thermonuclear weaponry, the physics of nuclear fusion was well understood. Stable by nature, fusion reactions require enormous temperatures and prove difficult to initiate. (This is why H-bombs require A-bombs as triggers.) Once initiated, however, fusion liberates stupendous quantities of energy as the strong nuclear force forges new bonds in atomic nuclei.

How a star dies depends upon its mass. A main-sequence star like our sun evolves into a red giant once its hydrogen fuel depletes, leaving behind vast quantities of helium. As the star's outer envelope expands, its core contracts gravitationally, eventually initiating a new fusion reaction that transmutes helium into carbon. What happens next depends upon size. A star with mass less than four times that of the sun ends as a white dwarf. More massive stars end more spectacularly. Those on the order of eight solar masses erupt into supernovae. Supernovae are cataclysmic, the most energetic events in the universe save for the Big Bang itself. In its lifespan of a few weeks, a supernova liberates more energy than the sun will emit in its multi-billion-year lifetime. Supernovae, which briefly outshine their host galaxy, are visible to the naked eye. Both Galileo and Kepler witnessed a spectacular supernova in 1604, which was later named for Kepler. During such explosions, successive fusion reactions sequentially produce all the heavier elements, ending with iron. Iron, being the most stable of all elements, cannot fuel fusion, and there the process stops. The remnant of a supernova is an inconceivably dense neutron star, a teaspoon full of which weighs, it is estimated, five billion metric tons.

Elements heavier than iron are forged by the bombardment of existing elements by free neutrons. Two heavy-element-forging processes exist, known pedantically as s and r, for slow and rapid, respectively. Process s occurs in the envelope of red giants. Here, free neutrons remain relatively rare, and their capture one at a time by the atoms of existing elements results in a slow alchemy over eons that terminates in the creation of bismuth and lead.[21] By contrast, the intense neutron flux during a supernova gives rise to the r process, the virtually instant formation of an entire smorgasbord of heavy elements and their isotopes, most of which are radioactive. The radioactive elements in turn furnish ticking clocks, whose decay over time, measured in their half-lives, provides the basis of radiochronology, a means of dating the earth and its constituents, the moon, and meteorites.

Hoyle, Bondi, and Gold did yeomen's work for physics and cosmology by developing a theory for the generation of the chemical elements. The $64,000 question that Hoyle and his company could not answer, however, concerned the prevalence of the light elements hydrogen and helium in the early universe. Where did all the hydrogen that fuels the stars come from? And why is so much helium scattered about interstellar space? To Hoyle's mortification, subsequent analyses showed that the lightest elements could only originate from the Big Bang itself.

In 1948, Ralph Alpher, a graduate student at George Washington University, elucidated the mathematical details of the Big Bang. His principal coauthor and thesis advisor was the Russian-born physicist George Gamow, a student of Alexander Friedmann. A prankster, Gamow included as the paper's second author his friend, the eminent physicist Hans Bethe, although Bethe had done no work whatsoever on the manuscript. Gamow then listed the authors in the order Alpher, Bethe, and Gamow, a cosmic pun on the first three letters of the Greek alphabet.[22] Alpher, who had done much of the work, was not amused. Adding insult to injury, the paper appeared on April 1, April Fools' Day. To the end of his life in 2007, Alpher resented Gamow's lapse into frivolity, which masked the significance of their work. Only recently has Alpher received due credit as the "forgotten father of the Big Bang." No doubt, however, the paper was groundbreaking. It predicted the creation of hydrogen and helium in the ratio of 10 to 1 during the primordial explosion, the exact ratio that astronomers observe today.[23]

As a last gasp attempt to preserve his steady-state theory, Hoyle surmised that if there had been a Big Bang, remnants of the colossal explosion would persist in the form of "fossil" radiation. Moreover, because we remain imbedded in the "fireball" of the conjectured primeval explosion, its faint wisps should be detectable from any direction in space. A few months following publication of the notorious Alpher-Bethe-Gamow paper, Alpher and Robert Hermann, the latter of Johns Hopkins University's Applied Physics Laboratory, calculated that the afterglow of the Big Bang should manifest as a blanket of nearly uniform microwave radiation across the entire sky – at a temperature of just 5 degrees Kelvin, close to absolute zero. Perhaps thinking it impossible to observe such low-level radiation, radio astronomers were slow to pick up the scent. But by the early 1960s, the hunt was on for the telltale "ashes" of the Big Bang.

As the theoretical work of Hoyle, Gamow, Alpher, and others shored up Big Bang cosmology, other pieces of the cosmological creation puzzle fell into place. As methods to measure stellar distances improved, the Hubble constant was continually refined. Today's accepted value of the constant suggests that the universe is just under 14 billion years old, a value now compatible with the earth's estimated 4.6 billion years. However, Big Bang cosmology received its biggest boost from an unlikely source: the theory of black holes.

Black Holes

On November 10, 1919, the *New York Times* headlined a momentous scientific development: "Lights All Askew in the Heavens: Men of Science More or Less Agog." News raced around the globe on the heels of Eddington's successful expedition to observe the gravitational lensing of the sun during a solar eclipse. As predicted by general relativity, gravitational lensing involves the bending of rays of light by massive objects such as stars. As early as the 1700s, on the basis of Newton's corpuscular (particle) theory of light, scientists had toyed with the notion that gravity could bend light rays. The question remained: is there a limit to how severely the path of light can be bent?

To free itself of the earth's gravitational tug, Apollo 11 needed a velocity of 25,000 miles per hour. More generally, the "escape velocity" of a celestial body depends upon its density. For a neutron star, for example, which packs enormous mass inside a sphere of small radius, the escape velocity is monstrously high. Could there exist celestial objects whose escape velocity exceeded the velocity of light?

In 1783, in a brilliant synthesis of Newton's corpuscular theory of light and his theory of gravitation, British natural philosopher John Michell surmised the existence of "dark stars." His reasoning was elegantly simple. If a star of given mass is sufficiently small – that is, if all its mass is concentrated within a "critical circumference"[24] – then the escape velocity of a light particle would be the speed of light itself. For a star the mass of our sun, Michell calculated the critical circumference at 18.5 kilometers.[25] If all a star's mass resides inside a sphere of circumference smaller than the critical value, light particles cannot overcome the tug of gravity and fall back to the surface like baseballs thrown upward with insufficient velocity. Such a star would be invisible: that is, it would be a dark star.[26] Perhaps the universe was full of dark stars. Who could know?

Michell presented his findings to the British Royal Society on November 27, 1783, to enthusiastic reception. Michell was eclipsed when, thirteen years later, the mathematical titan Laplace published his famous popular account of the universe: Le systeme du monde. It contained Laplace's own theory of dark stars and no mention of Michell.[27] By the third edition of Du monde in 1808, Laplace had abandoned all references to dark stars, and for good reason.[28] In the years intervening the first and third editions, views on the nature of light had undergone drastic revision. In 1801, Thomas Young's famous double-slit experiments with light demonstrated interference patterns when a beam of light, split in two, recombines. Interference phenomena are properties of waves, not particles. When science temporarily abandoned Newton's corpuscular theory of light in favor of the new wave theory, dark stars fell from favor.

If Einstein overturned most of Newtonian physics, he redeemed Newton on the subject of light. The first of four papers Einstein published in his anni mirabilis concerned the photoelectric effect. Einstein reasoned that the color-dependent properties of photovoltaic cells are explicable only if light behaves as a stream of particles, later termed photons. Moreover, according to Einstein, the momentum of a photon – the wallop it packs upon colliding with the photovoltaic cell – depends upon its color. Newton's corpuscles, it seemed, were back, and so, by implication, were dark stars.

The 1915 issue of the Proceedings of the Prussian Academy of Sciences dropped another Einsteinian bombshell: the general theory of relativity. The First World War was raging. Karl Schwarzschild, a distinguished astrophysicist, had been called to serve the German army on the Russian front.[29] For diversion from life at the front, Schwarzschild studied Einstein's article and used it to theorize about stars. When mathematical analyses are too daunting, the theorist often resorts to simplifying assumptions. For starters, Schwarzschild assumed that stars are perfectly spherical. In truth, stars – like planets – rotate, giving them a somewhat oblate shape. Still, the simplification led quickly to a beautiful and tantalizing result.

Four months after Schwarzschild mailed his findings to Einstein, Schwarzschild was dead, having succumbed to disease in the deplorable conditions of warfare. The sad task of reporting Schwarzschild's untimely death to the Prussian Academy fell to Einstein, who had presented an earlier paper to the Academy on Schwarzschild's behalf.

A brilliant meteor rather than an enduring star, Schwarzschild nevertheless left a lasting legacy. His work established the first rigorous theoretical basis for understanding dark stars, now known as "black holes." In the close proximity of a massive star, spacetime warps significantly. If the star is compact enough, spacetime distorts so radically that strange things happen to the spacetime fabric. Time ceases to exist, and light no longer propagates through space.

Schwarzschild's work corroborated the earlier work of Michell and Laplace, albeit by an entirely different mechanism: general relativity. In the classical theories of light and gravity embraced by Michell and Laplace, gravitational attraction slows light; sufficient gravity stops it altogether. In contrast, from the vantage point of relativity, the speed of light *in vacuo* remains constant. It is time and space that are relative. By slowing time, gravitation alters frequency, shifting the frequency of light toward the lower end of the spectrum. Sufficient gravity shifts the frequency to zero, at which point light vanishes. To every star of given mass, there corresponds a critical radius, now termed the Schwarzschild radius. If the star's mass is concentrated within a sphere of that limiting radius, the star is dark; it is a black hole.

The world's leading authorities on relativity in the 1920s and 1930s were Einstein and Eddington. Neither giant warmed to the notion of black holes. Black holes might be invisible, but to Einstein they had "smell," and they just didn't "smell right."[30] It took a young Indian superstar to convince the world of physicists that black holes were worthy of serious consideration.

The year was 1928 when Arnold Sommerfeld, one of Germany's leading theoretical physicists, arrived in Madras, India. Upon learning of Sommerfeld's imminent visit, Subramanyan Chandrasekhar, who had been reading Sommerfeld's book *Atomic Structure and Spectral Lines*, arranged a face-to-face meeting. It took some gall. Chandrasekhar was a mere seventeen years old at the time, but he was extraordinarily gifted in mathematics and physics.

During their meeting, Sommerfeld quickly deflated the lad: "The physics you have been studying is a thing of the past. Physics has all changed since my textbook was written."[31] To restore the crestfallen young man, Sommerfeld kindly gave him galley proofs of a yet unpublished paper regarding a new development in quantum mechanics: so-called Fermi-Dirac statistics for large collections of electrons. The encounter with Sommerfeld was the "single most important event" in Chandrasekhar's life.[32] Picking up a new scent, the young upstart followed the quantum trail like a bloodhound. It led to a seminal paper by British physicist R. H. Fowler and from there to Eddington's classic *The Internal Composition of Stars*, where Chandrasekhar encountered an intriguing mystery concerning white dwarf stars. White dwarfs are abundant. After all, our close stellar neighbor Sirius B is a white dwarf. White dwarfs are also extraordinarily dense, and therein lay the mystery of Eddington's paradox. No known physical mechanism accounted for a white dwarf's resistance to gravitational collapse.

Recognizing Chandrasekhar's unique gifts, the government of India created a scholarship for the young man to further his education in England. In 1930, with a fresh baccalaureate degree and a head full of quantum mechanics, astrophysics, and questions, Chandrasekhar boarded the steamer *SS Pilsna* and packed off to Cambridge University to study under Fowler and Eddington. His burning interest: how stars die. Unlike Eddington, Fowler had applied the new quantum physics to stars. The new approach changed everything.

The eighteen days en route at sea were seminal. In a *tour de force* that combined quantum mechanics and relativity, Chandrasekhar came to a startling conclusion: the maximum mass possible for a white-dwarf star is 1.4 times that of our sun. No known mechanism in all of physics could repulse the gravitational collapse of a white dwarf whose mass exceeded 1.4 solar masses, a number now referred to as the Chandrasekhar limit. What fate then awaits large white dwarfs? They continue to collapse to a pinpoint in space – a *singularity* – to become, in today's lingo, black holes.[33] The fertility of those days would culminate in a Nobel Prize, but only after more than half a century had elapsed.

The reception that greeted Chandra (as his friends called him) in England was not what he expected. Astronomers, on the whole, are cautious types. Although Fowler was personally cordial, he remained skeptical of Chandra's startling result and unwilling to forward Chandra's paper to the Royal Society. After a respectful wait, Chandra bypassed his mentor and published the paper in the respected *Astrophysical Journal*.

Chandra completed his degree at Cambridge in 1933, having been nurtured by a star-studded cast of characters: Fowler, Eddington, quantum physicist Paul Dirac, and – during an interlude in Sweden – Niels Bohr. Following graduation, Chandra landed a prestigious fellowship at Cambridge University's Trinity College and delved deeper into the fate of massive stars. Despite genuine affection for Chandra, Eddington, like Einstein, vehemently opposed the concept of black holes, and at a meeting of the Royal Academy of Sciences in 1935 publicly humiliated Chandra. Traumatized by the ferocity of Eddington's attacks, Chandrasekhar fled England for an academic post at the University of Chicago and abandoned astrophysics for a quarter century.

By 1956, a number of luminaries had tackled many questions involving dying stars: Swiss astronomer Fritz Zwicky, with his overbearing personality; Russian physicist Lev Landau, who narrowly escaped Stalin's purges; Robert Oppenheimer, who directed the Manhattan Project; and American theoretical physicist John Archibald Wheeler, a designer of the H-bomb. Of these, Wheeler in particular grasped the import of such investigations: "Of all the implications of general relativity for the structure and evolution of the universe, this question of the fate of great masses of matter is one of the most challenging."[34]

The collective attention of brilliant minds produced continual refinements to the theory, tweaks of the numbers, and more questions. But ultimately nature yielded. Chandra's original thesis was correct: massive stars suffer wildly peculiar fates.

It was Wheeler who coined the term "black hole."[35] He was given to meditating – often while relaxing in the bathtub or in bed – on the names physicists ought to assign to characterize new phenomena. In 1969, Wheeler hit upon just the right appellation for the strange phenomenon physicists had lately been probing: black holes. Theoretical evidence was mounting that dying stars of sufficient mass collapse without rebound. But to what, pray tell, did they collapse? Most of the supporting analysis involved the collapse of "ideal" stars – those with no rotation and perfect spherical symmetry. Would the developing theory of black holes hold also for real stars, with all their imperfections?

In 1952, as a mathematics and sciences undergraduate at University College London, Roger Penrose chanced upon one of Fred Hoyle's aforementioned addresses on cosmology. Penrose was both enamored by Hoyle's lively presentation and puzzled by it. No shrinking violet, Penrose booked a train to Cambridge to visit his brother and to meet one-on-one with Dennis Sciama, an eminent astrophysicist well versed in Hoyle's steady-state theory. Presumably, Sciama would quickly set him straight. Instead, it happened the other way round. On the back of a napkin, with a simple spacetime diagram, the student convinced the professor that Hoyle had to be wrong.[36] Shortly thereafter, Penrose began doctoral studies at Cambridge University under the direction of Sciama, whom Penrose had duly impressed.

By 1964, Penrose was back in London, this time as a professor of physics at Birkbeck College and deeply engrossed in the study of quasars. While walking back to his office with a friend, Ivor Robinson, a flash of insight hit Penrose like a bolt from the blue.

> My conversation with Robinson stopped momentarily as we crossed a side road, and resumed again at the other side. Evidently, during those few moments an idea occurred to me, but then the ensuing conversation blotted it from my mind! Later in the day, after Robinson had left, I returned to my office. I remember having an odd feeling of elation that I could not account for.[37]

After some effort, Penrose retrieved the inspiration, which had to do with stellar implosion. After a few weeks of polishing, the stray thought emerged as Penrose's first singularity theorem. Published in *Physical Review Letters*, the theorem required that "every black hole must have a singularity inside itself."[38] Its power rested in the sweep of its application. Any sufficiently large imploding star – ideal or nonideal, symmetric, or asymmetric, rotating or nonrotating – was destined to collapse to a black hole. Black holes were no longer just theoretically possible. Penrose's mathematics proved they were likely.

Meanwhile, Sciama had taken on another promising graduate student, whose name was Stephen Hawking. Hawking's studies got off to a rocky start. First, he had trouble settling on a topic. And then, in 1963, he received a devastating medical diagnosis: ALS (Lou Gehrig's disease), a motor neuron disorder. With a damning prognosis, Hawking lost interest in physics and in life. There was, however, a slim glimmer of hope. Hawking's variant of ALS progressed slowly. He would eventually lose all muscle control, but the diminishments would proceed gradually; he would have time to adapt. With the clock ticking, Hawking plunged back into physics.[39]

With no time to waste, Hawking suggested a thesis topic to his mentor: the mathematics of the Big Bang. The idea was not entirely new. Estonian physicist Yakov Zel'dovich, in a moment of piercing insight similar to Penrose's, had envisioned a promising approach. The collapse of a star into a black hole and the origin of the universe from the Big Bang are similar but time-reversed. That is, suppose one could record by video camera the death throes of a giant star as it collapsed into a black hole. Then, playing the video in reverse, the star would appear to emerge in time from a single point in space: Lemaître's "primeval atom" in microcosm. Hawking's genius was to suggest that, by reversing the direction of time, he could appropriate Penrose's rigorous mathematics of black holes to study the Big Bang.

Like the black holes in the previous analyses of Schwarzschild and Chandrasekhar, the Big Bang suffered from a symmetry problem. The universe, as we observe it today, is lumpy. It consists mostly of empty space in which matter is more or less randomly distributed. If the Big Bang originated from a perfect mathematical singularity with exact symmetry, then the ensuing explosion would likely retain symmetry forever. Matter would be evenly distributed in every direction. Some asymmetry – some "symmetry breaking" event – was necessary to seed the galactic lumpiness now observed. Adapting Penrose's mathematics, Hawking proved his own singularity theorem: under reasonable assumptions, symmetry would naturally break in the first few nanoseconds following the Big Bang.

By 1970 Hawking required a walker; by 1972 a motorized wheelchair. By 1975 he could no longer feed himself, and in 1985 breathing problems forced a tracheostomy. His voice, already failing, was completely lost.[40] Hawking now communicates by computer-synthesized speech, laboriously controlled by a muscle in his cheek, virtually the only mobility he has remaining. But as Hawking's body atrophied, his mental powers seemed to increase commensurately, by which he frequently out-thought and out-intuited his robust colleagues. Stephen Hawking, who has beaten ALS for some five decades, was for three of those decades (1979-2009) Lucasian professor of mathematics at Cambridge University, the chair once occupied by Isaac Newton. Thanks largely to Hawking, the Big Bang is widely accepted as the scientific account of the origin of the universe.

Faint Wisps

By the early 1960s, the physics of the Big Bang was relatively well understood and the search was on for experimental verification in the form of "fossil" microwave radiation, the faint afterglow of the primordial fireball. The lead contenders in the race, which almost certainly held a Nobel Prize, were Princeton University's Bob Dicke and his cadre of graduate students. Dicke, a brilliant theoretician, had proposed a theory of gravity that had briefly competed with Einstein's general relativity.

Meanwhile, in nearby Holmdel, New Jersey, two engineers at Bell Laboratories were having a heck of a time with Bell's new horn antenna. The large radio telescope had been designed to detect faint radio waves reflected from orbiting echo-balloon satellites. For a year, Arno Penzias and Robert Wilson had labored to locate the source of a persistent and annoying hiss, two orders of magnitude larger than any anticipated background noise. The pair carefully checked and rechecked the electronics. Finding no problems whatsoever with the electronics, they forcibly relocated family of pigeons that had nested in the telescope. Unfortunately, they were homing pigeons. Upon their return, the family was sacrificed in the cause of science. But the hiss remained. Moreover, it remained regardless of the direction in which the antenna pointed.

In 1965, unable to locate the source of the hiss, Wilson called Princeton University's physics department for help. The call was put through to Bob Dicke during a lunch break with graduate students. Within a few minutes, Dicke surmised that the hiss plaguing Penzias and Wilson was none other than the sought-after remnants of the Big Bang. David Wilkinson, a twenty-eight-year-old graduate student at the time, recalled verbatim Dicke's words upon his hanging up, "Well, boys, we've been scooped."[41]

A short thirty-five-mile drive from Princeton to Holmdel confirmed Dicke's worst suspicions: the Bell Lab antenna registered microwave radiation at 3 degrees Kelvin, close to the value predicted in 1948 by Gamow and Alpher. Wilson and Penzias had been eavesdropping on a "whisper from the moment of creation."[42] In a stroke, a half-century of cosmological debate was settled; the universe had revealed its violent origins.

Scientist-priest Georges Lemaître learned of the discovery of the "ashes and smoke" of his "primeval atom" just before his death in 1966. In 1978, thirteen years after their fateful call to Bob Dicke, Penzias and Wilson received the Nobel Prize in physics "for their discovery of cosmic microwave background radiation."

The discovery of the cosmic background radiation was the *coup de grace* by which Big Bang cosmology ultimately conquered Hoyle's steady-state theory. Lingering questions remained, however. Hawking's mathematics had shown that broken symmetry was possible shortly following the Big Bang. However, anisotropy was essential for the formation of the universe as we know it. (The technical term anisotropy refers to slight irregularities that depend upon the direction in which one observes.) Galaxy formation must be seeded. "Ripples" in the Big Bang's otherwise uniform distribution of matter and energy provided the seeds.

Finding small ripples in a sea of noisy cosmic data was problematic in the extreme, especially from the laboratory of the earth. To eliminate as much atmospheric noise as possible, measurements were made initially from high-altitude balloons and later from high-flying U2 spy planes converted for NASA use. Mapping the sky by such techniques was productive but laboriously slow.

To circumvent these difficulties, NASA Goddard Spaceflight Center designed COBE – the COsmic Background Explorer. Launched from Vandenberg Air Force Base atop a Delta rocket on November 18, 1989, the 5,000-pound COBE satellite succeeded far beyond expectations. By mid-June of 1990, COBE had collected background radiation for the entire heavens, data that allowed astrophysicists to map the distribution of matter and energy when the universe was very young.

COBE's major findings were twofold. First, "The cosmic microwave background (CMB) spectrum is that of a nearly perfect blackbody with a temperature of 2.725 ± 0.002 K." The finding fit hand-in-glove with the predictions of hot Big Bang theory. Second, "The CMB was found to have intrinsic 'anisotropy'. . . at a level of a part in 100,000." Through a process still poorly understood, these early structures have evolved over billions of years "into galaxies, galaxy clusters, and the large-scale structure that we see in the universe today."[43]

Big Bang cosmology is now the accepted scientific creation story, backed by hard evidence and sound theory. It is a legacy that Copernicus could not have dreamed of. Cataclysms in the heavens – supernovae, black holes, and the Big Bang – all resisted initially by astronomers, broke the back of Aristotle's 2,000-year hold on astronomy. The cosmos is not, as Aristotle fashioned it, cozy, immutable, and perfected.[44] Creation is not a *fait accompli*. It is dynamic, sometimes violent, and still very much in progress.

Epilogue

If seeing is believing, then black holes, by definition, remain objects of faith. Nevertheless, circumstantial evidence for their existence is overwhelming. Advances in radio astronomy now permit astrophysicists to peer into the hearts of distant galaxies, and what they see is spine-tingling. At the center of a garden-variety galaxy, stars race around an *apparent* gravitational source. These stars behave like mere planets, orbiting tightly on short gravitational leashes, trapped in a monstrous gravitational field. Their orbits are elliptical, in accordance with Kepler's first law. The location of the gravitational source can be pinpointed precisely at the coincident foci of overlapping ellipses. But the gravitational source, unlike the orbiting stars, remains invisible because the source is a black hole. At the heart of every healthy galaxy resides a black hole. And, as we shall see, in a yet stranger twist, black holes may be necessary for conditions that support life.

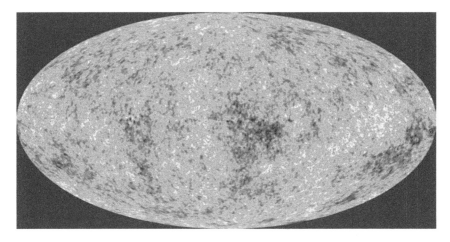

Figure 16: Composite image of the heavens showing cosmic microwave temperature fluctuations from seven-year WMAP data over a full sky. The average temperature is 2.725 Kelvin; gray tones represent deviations from average temperature of about ±200 microKelvin. (NASA Goddard Spaceflight Center.)

At the age of sixty-five, Subramanyan Chandrasekhar returned to his scientific first love: astrophysics. The year was 1975. Eight years later, in 1983, he completed *The Mathematical Theory of Black Holes*, a scientific classic praised in words usually reserved for fiction: "splendorous, joyful, and immensely ornate."[45] Chandrasekhar died in 1995 at the age of 84. Four years later, the space shuttle *Columbia* deployed *Chandra*, an orbiting x-ray observatory named in honor of the Indian astrophysicist and naturalized American citizen who first opened our eyes to invisible stars.

In 1993, COBE retired after completing its mission. For mapping the cosmic microwave background radiation, COBE pioneers John Mather of NASA Goddard and George Smoot of UC Berkeley were jointly awarded the 2006 Nobel Prize in physics.

In 2001, NASA launched the Wilkinson Microwave Anisotropy Probe (WMAP), COBE's successor, named for David Wilkinson, Bob Dicke's former student who followed Dicke's footsteps to become a Princeton University professor of experimental physics. Wilkinson died of cancer in 2002 at the age of sixty-seven. Figure 16, a composite of seven years of WMAP data, displays the entire "celestial sphere." It reveals the telltale wisps of the Big Bang as a sea of microwave background radiation at the dawn of creation. Gentle ripples in that sea, indicated by variations in color, are the seeds of our current galaxies, the anisotropies predicted by the mathematics of Stephen Hawking.

PART II

The Second "Copernican" Revolution: Biology

Figure 17: A boulder in the embrace of a tree, Grayson Highlands State Park, Virginia. The image is symbolic of the intimate relationship between the physical and biological worlds and also of the tenacity of life on Earth. (Don Carroll.)

Nine

The Entangled Bank

No object in nature, whether flower, or bird, or insect of any kind, could avoid his loving recognition.

— **Dr. Edward Eickstead Lane, of Darwin**

On February 20, 1835, while on shore near Concepción, Chile, Charles Darwin experienced firsthand a devastating earthquake. Prolonged and intense, the quake destroyed Concepción's great cathedral and laid waste to innumerable nearby villages. Darwin recorded his immediate impressions in his journal.

> I happened to be on shore, and was lying down in the wood to rest myself. It came on suddenly, and lasted two minutes, but the time appeared much longer. . . . There was no difficulty in standing upright, but the motion made me almost giddy: it was something like the movement of a vessel in a little cross-ripple, or still more like that felt by a person skating over thin ice, which bends under the weight of his body.[1]

Darwin's matter-of-fact description belies the disquieting emotional impact the event had upon him. "A bad earthquake at once destroys our oldest associations: the earth, the very emblem of solidity, has moved beneath our feet . . . [while] one second of time has created in the mind a strange idea of insecurity, which hours of reflection would not have produced."

In the epilogue to the *Voyage of the Beagle*, Darwin reflected upon the earthquake at Concepción in words almost identical to those he had written years before. One senses that the earthquake had become a metaphor for Darwin's own inner upheaval, which figuratively rocked the foundations of Christianity. In 1859, he reluctantly published *On the Origin of Species*, scandalizing Victorian England and sending scientific shock waves around the world. "Seldom in the history of English prose," writes David Quammen in the introduction to the *Illustrated Edition of Origin*, "has such a dangerous, disruptive, consequential book been so modest and affable in tone."[2]

Today, the progenitor of the incendiary theory of evolution is revered by scientists, reviled by the religiously fundamental, and underappreciated by nearly everyone else. How did such a shy, gentle, and reluctant revolutionary as Charles Robert Darwin unleash a revolution every bit as potent as that of Copernicus?

One glimpses Darwin's soul in an 1881 portrait that hangs in the British National Portrait Gallery. John Collier's striking likeness of Darwin, painted just two years before his death, is nearly life-sized. The viewer stares directly into Darwin's unblinking, slightly downcast eyes, in whose depths are found all manner of qualities and emotions: intelligence, gentleness, sadness, wisdom, and world-weariness. In those eyes one feels the lingering grief that Darwin never quite shook in the loss of three children. In those eyes, one appreciates the gentleness that rendered Darwin incapable of defending himself, and one encounters a man grown weary of life's vicissitudes. With his dark traveling cape and flowing beard, Darwin seems more a biblical prophet than an iconoclastic scientist. How strange that he would seem a religious figure, given the widespread revulsion to Darwin's "blasphemous" theory within large segments of Christendom and Islam.

The Ne'er Do Well

Both of Darwin's grandfathers were "lunaticks." The colorful moniker was ascribed to members of the Lunar Society. Second in importance only to the Royal Society as a watering hole for scientists, inventors, and natural philosophers during the latter half of the eighteenth century, the Birmingham-based society acquired its name from its tradition of meeting monthly on the Monday nearest the full moon so that the reveling and inebriated members might find their ways home at night. The society boasted among its many illustrious members James Watt, who perfected the steam engine, and Joseph Priestley, Unitarian minister and discoverer of the element oxygen. Numbered also among these guiding lights of the Industrial Revolution were Charles' paternal grandfather, Erasmus Darwin, and his maternal grandfather, Josiah Wedgwood, who had acquired fortune and fame as the manufacturer of fine china that bears his name.

Of Darwin's grandfathers, Erasmus was the more colorful. Corpulent and jovial, a freethinking libertine and composer of erotic verse, Erasmus enjoyed a lucrative medical practice in Lincolnshire. Like Priestly, most "lunaticks" were Unitarians, save for Erasmus, an atheist who considered "Unitarianism . . . a feather bed to catch a falling Christian."[3]

Robert Waring Darwin, the youngest son of Erasmus, followed in the practice of medicine and in 1796 married Susannah Wedgwood, daughter of the Wedgwood China magnate. The Darwins settled comfortably at "The Mount," their rural home in Shrewsbury. There, Charles Robert Darwin was born on February 12, 1809, the fifth child of Robert and Susannah. Three sisters and an older brother preceded him.

As political liberals, "lunaticks" abhorred slavery and sympathized with both the French and the American revolutions. As industrialists, they believed in private property, the entrepreneurial spirit, and reducing the power of the aristocracy. And as Unitarians, they chafed at the rigid authority of the Anglican Church. All such views were unwelcome to Victorian traditionalists, typically both Anglican and aristocratic. Backlash to the unconventional views of the "lunaticks" culminated in the Priestley riots of 1781, in which the houses of Birmingham's elite were looted and burned, starting with Priestley's.

In the tradition of their societal circle, the Robert Darwins were Unitarians. To maintain an air of respectability, however, the Darwins christened Charles as an Anglican. Nevertheless, Susannah discreetly took her children to Unitarian services. When Charles was eight, his mother contracted a tumor and died after suffering greatly, and his upbringing fell to his older sisters. With his father aloof, his sisters overbearing, and his brother away at school, Charles sought refuge in nature, filling many a solitary hour by walking, collecting, hunting, and fishing.

Charles' pastimes, however, earned him a father's enmity: "You care for nothing but shooting, dogs, and rat-catching, and you will be a disgrace to yourself and all your family."[4] Robert Darwin expected a son to carry on the family tradition of medicine. He boarded Charles at nearby Shrewsbury School, where older brother Erasmus was already studying. No scholar, Charles found every excuse to come home and indulge his passions. Not to be thwarted, in 1825 Robert Darwin enrolled Charles, then sixteen, at the University of Edinburgh to study medicine where Charles' grandfather, father, and uncle had all attended.

Charles lacked the constitution for medical practice: the sight of blood nauseated him and the trauma of witnessing surgery on a child before the advent of anesthesia haunted him for years. To his father's ire, Charles abandoned medicine. Worse, so mind-numbing were the geology lectures at Edinburgh that he briefly gave up completely on a scientific career.[5] His vocational possibilities were quickly diminishing.

As a last resort, Charles entered Cambridge University to prepare for the clergy. Resigned to a modest profession, Charles hoped that a country parsonage might afford ample time for dabbling in hobbies. By his own admission, however, Charles squandered the three years at Cambridge as completely as those at Edinburgh and Shrewsbury. "I got into the sporting set, including some dissipated, low-minded young men . . . more intent on having a good time than determined to pursue their academic studies."[6] As soon as classes ended, singing, drinking, and card playing commenced. Yet in old age, Darwin confessed with a wistful twinkle, "I know I ought to feel ashamed of days and evenings thus spent, but as some of my friends were very pleasant, and we were all in the highest spirits, I cannot help looking back to these times with much pleasure."[7]

Membership in the sporting set had its perks. While continuing to indulge his passion for hunting, Charles unexpectedly discovered a new interest: beetle collecting. Seeing his name under a rare specimen in an illustrated volume of British insects, Charles was elated, later recalling, "No poet ever felt more delighted at seeing his first poem in print than I."[8] Among his beetle-collecting companions at Cambridge were men destined to become a leading archeologist, a great agriculturist, a railway magnate, and a member of parliament, prompting Darwin to recall with a chuckle, "It seems therefore that a taste for collecting beetles is some indication of future success in life!"[9]

The new hobby had the fortuitous side effect of convincing Charles to regard his studies more seriously, particularly natural history and geology. Pivotally, this led to a friendship with the Reverend John Stevens Henslow, professor of botany at Cambridge. A likeable fellow, Henslow enjoyed the company of enthusiastic students while on his fieldwork and frequently invited them afterwards to his home for discussions. Darwin quickly became Henslow's favorite, earning him the reputation among students and dons alike as "the man who walks with Henslow."[10]

The Voyage of the *Beagle*

On August 29, 1831, Darwin's fortunes improved dramatically. Two letters from Cambridge professors awaited him upon his return from a geological tour of North Wales. Each offered him opportunity to accompany a round-the-world scientific expedition on the *HMS Beagle*. Henslow himself had been asked to assume the position, but, on the strong objections of his wife, he declined, naming Darwin as his stand-in. In heartfelt gratitude, Darwin acknowledged in the preface of *Voyage of the Beagle* his debt to Henslow, "who . . . rendered me every assistance which the kindest friend could offer."

Thrilled by the *Beagle* offer, Charles was devastated when his father denounced the notion as a total waste of time. Despondent, Charles rode to his uncle Josiah's estate to seek solace in shooting. To Charles' surprise, Uncle Josiah exuded enthusiasm about the opportunity and intervened immediately on Charles' behalf, countering all of Dr. Darwin's arguments against the trip. With utmost respect for Josiah as a man of good sense, Charles' father relented.

Against the odds, Charles Darwin – a haphazard student and ill qualified for the position – landed it nonetheless. The three months spent waiting to sail were among "the most miserable that I ever spent."[11] During this period Darwin first experienced signs of a mysterious illness that would torment him for the remainder of his life. Among the symptoms – almost certainly anxiety related – were heart palpitations and a rash. Mysteriously, when the long-awaited day arrived, the symptoms abruptly vanished.

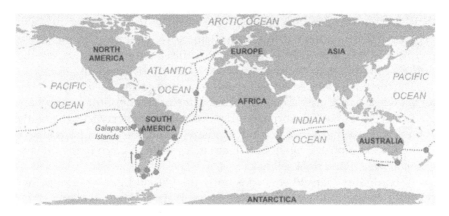

Figure 18: Route of the *H.M.S. Beagle*, 1831-1836. (Oxford University Museum of Natural History.)

The *Beagle* set sail on Dec. 27, 1831, and returned on Oct. 2, 1836, after a round-the-world voyage of nearly five years (Fig. 18). Throughout the long voyage, Darwin faithfully kept a journal. *The Voyage of the Beagle*, published originally under the ponderous title *Journal of Researches into the Natural History and Geology of the Countries Visited during the Voyage round the World of H.M.S. Beagle*, commences with this delightful passage:

> After having been twice driven back by heavy south-western gales, Her Majesty's ship *Beagle*, a ten-gun brig, under the command of Captain Fitz Roy, [Royal Navy], sailed from Devonport on the 27th of December 1831. . . . On the 6th of January we reached Tenerife, but were prevented landing, by fears of our bringing the cholera: the next morning we saw the sun rise behind the rugged outline of the Grand Canary Island, and suddenly illumine the Peak of Tenerife, whilst the lower parts were veiled in fleecy clouds. This was the first of many delightful days never to be forgotten.[12]

At the time, it was not uncommon for the emotional isolation of authority to send ships' captains into madness on long voyages. Darwin's official role aboard the *Beagle* was as a "supernumerary" passenger and cabin companion for Captain FitzRoy. A troubled man whose life eventually ended by his own hand, FitzRoy initially rejected Darwin, objecting to the shape of the young man's nose. Despite sometimes violent arguments en route, the two men developed a friendship and remained amicable following the voyage. The official naturalist, the ship's surgeon, Robert McKormick, was confined to the ship by duties, so Darwin, who enjoyed greater freedom, served as de facto naturalist.[13]

The *Beagle*'s human cargo included an unlikely contingent: three Fuegian Indians whom FitzRoy had "kidnapped" on a previous voyage of the *Beagle* in retribution for the loss of a boat that the Fuegians, notorious for theft, had stolen. Having educated the captives in England at his own expense, FitzRoy intended to return them, now "civilized," to their own people and to drop off Mr. Matthews, a missionary accompanying the *Beagle*'s expedition, to complete the conversion of the Fuegians.

The three Fuegians – two males and one young female – had been conferred whimsical Christian names: York Minster, Jemmy Button, and Fuegia Basket. The crew's favorite, Jemmy Button, was a merry fellow, greatly sympathetic to those in any kind of distress. Frequently suppressing a smile, Button repeated again and again in a plaintive voice to the oft seasick Darwin, "Poor, poor fellow!"[14]

Aboard ship, Darwin was constantly ill. On land, however, whether pushing through the claustrophobic rain forests of Brazil, walking upon the windswept plains of Patagonia, or scrambling over the volcanic rock of the Galapagos Islands, Darwin was in his element. In a passage written from a Brazilian rainforest early in the voyage, the young man betrays a wandering naturalist's lofty enthusiasm:

> A most paradoxical mixture of sound and silence pervades the shady parts of the wood. The noise from the insects is so loud, that it may be heard even in a vessel anchored several hundred yards from the shore; yet within the recesses of the forest a universal silence appears to reign. To a person fond of natural history, such a day as this brings with it a deeper pleasure than he can ever hope to experience again.[15]

Beauty sometimes caught Darwin unawares, overwhelming the emotions. For example, when he emerged abruptly from a noble forest to a breathtaking vista near Rio de Janeiro, he recalled, "At this elevation that landscape attains its most brilliant tint; and every form, every shade so completely surpasses in magnificence all that the European has ever beheld in his own country, that he knows not how to express his feelings."[16]

Keenly observant, Darwin chronicled in detail everything new to him, often waxing poetic: geological strata; creatures of the rainforest; the Indians of Tierra del Fuego; the wretched mistreatment of Indians and slaves in South America; and the tortoises, lizards, and finches of the Galapagos Islands. Nothing escaped his attention, not even a mortal battle between a hapless spider and a determined wasp that "commenced as regular a hunt as ever hound did after fox."[17]

In another delightful passage following the ship's departure from Rio, Darwin describes rare spectacles of a different type, mostly of the open ocean:

> One dark night we were surrounded by numerous seals and penguins, which made such strange noises, that the officer on watch reported he could hear the cattle bellowing on shore. On a second night we witnessed a splendid scene of natural fireworks; the masthead and yard-arm-ends shone with St. Elmo's light; and the form of the vane could almost be traced,

as if it had been rubbed with phosphorus. The sea was so highly luminous, that the tracks of the penguins were marked by a fiery wake.[18]

Darwin's daily entries more often than not concerned geology, in part because as a bon-voyage gift Henslow had bequeathed Darwin a copy of Charles Lyell's *Principles of Geology*, admonishing him to ignore the "blasphemous" parts. An early proponent of *uniformitarianism*, Lyell espoused the unorthodox view that the geology of the earth is shaped gradually by the accumulation over eons of small changes, like the slow accretion of sediment. Henslow, on the other hand, espoused *catastrophism*, the view that the earth had been shaped by sudden, brief, and catastrophic events. Separating catastrophism and uniformitarianism lay the chasm of time.

Of the two theories, catastrophism was far more compatible with the prevailing religious orthodoxy: young-earth creationism, as inferred from *Genesis*. By counting the "begats," biblical scholars derived the number of human generations descended from Adam and Eve to estimate the age of the earth at approximately 6,000 years. The number was widely accepted and agreed coincidentally with the emergence of written language. Two scientific giants of the seventeenth century – Johannes Kepler and Isaac Newton – independently estimated the age of the earth from biblical accounts, each arriving at a creation date of roughly 4000 BCE. Irish bishop James Ussher went them one better by calculating the very day of creation as "the 23rd day of October, in the year . . . 4004 B.C."[19] Young-earth geology was to time what the cosmology of Aristotle was to space: confining. Christian orthodoxy demanded a "cozy" universe: small, young, and complete.

By the late eighteenth century, England's burgeoning canal industry had begun laying bare the earth's geological layers and dredging up a plethora of "damned facts," among them fossilized seabeds on mountaintops and periodic wholesale extinctions of species. To Lyell, catastrophism was "a dogma . . . calculated to foster indolence, and to blunt the keen edge of curiosity."[20] Catastrophism is inherently static. Between catastrophes, stasis reigns. Thus, catastrophism compresses massive change into a short time by fracturing the flow of time. Uniformitarianism, on the other hand, is inherently dynamic and therefore continuous. But for the uniformitarian view to be viable, the earth must be exceedingly old. Lyell's "blasphemous" view: "The age of the earth must be reckoned not in thousands but in millions of years."[21]

When the twenty-two-year-old Darwin embarked upon the *Beagle*, other than being Unitarian, he was conventionally religious, a creationist and a catastrophist who accepted more-or-less unquestioningly the biblical chronology. Soon, however, Darwin's observations and deductions corroborated Lyell, not Henslow. Accordingly, the second edition of *Voyage*, published in 1845, bears a dedication to Lyell, later to become one of Darwin's two closest friends.

One of Darwin's earliest speculations about geological time occurred during the *Beagle*'s survey of Brazil. Along Brazil's entire 2,000-mile coast, granite formations predominated. Forged by crystallization under enormous heat and pressure, granite requires some covering layer, a lid for the pressure cooker. Darwin pondered: "Was this effect produced beneath the depths of a profound ocean? Or did a covering of strata formerly extend over it, which has since been removed? Can we believe that any power, acting for a time short of infinity, could have denuded the granite over so many thousand square leagues?"[22]

By late August 1832, Lyell's geology had begun to exert a steady tug on Darwin's biology. It was then that the *Beagle* made its first port of call at Bahia Blanca in the northern reaches of Patagonia. There, along the beach at Punta Alta, Darwin made a momentous find: "The remains of these nine great quadrupeds and many detached bones were found embedded on the beach, within the space of about 200 yards square. It is a remarkable circumstance that so many different species should be found together; and it proves how numerous in kind the ancient inhabitants of this country must have been."[23]

The quadrupeds, all mammalian species whose remains lay in the stratified gravel and mud of a shallow bank at the water's edge, varied wildly in type from *Scelidotherium* – a giant ground sloth – to an archaic giant armadillo. All nine species were extinct. Yet, in the same gravel strata, Darwin found twenty-three species of shells, of which most were either current or recent varieties of testacea. By the well-preserved state of the mammalian remains, Darwin concluded that the giant quadrupeds and testacea had coexisted, but only the latter remained among the earth's present inhabitants. The finding corroborated Lyell's assertion that "the longevity of the species in the mammalia is, upon the whole, inferior to that of the testacea."[24] Mammals, it seemed, came and went. More rudimentary life forms endured. More importantly, geology informed biology.

Although many frequently associate the theory of evolution with the Galapagos Islands, the seeds of those ideas were planted much earlier in the voyage, particularly in windswept Patagonia.

What a history of geological changes does the simply-constructed coast of Patagonia reveal! . . . [T]he land, from the Rio Plata to Tierra del Fuego, a distance of 1200 miles, has been raised in mass (and in Patagonia to a height of between 300 and 400 feet), within the period of the now existing sea-shells. The old and weathered shells left on the surface of the upraised plain still partially retain their colours. The uprising movement has been interrupted by at least eight long periods of rest, during which the sea ate deeply back into the land, forming at successive levels the long lines of cliffs . . . like steps . . .[25]

Patagonia's vast plains laid bare two countervailing geological processes: uplift and erosion, both fingerprints of uniformitarianism. Remnants of their gradual tug-of-war, the majestic step-like plains extended laterally for hundreds and hundreds of miles. Whereas the lowest of these plains rose a modest ninety feet above sea level, the highest had been up-lifted to a striking 950 feet, the full height of which Darwin ascended on foot. At an intermediate level of 300 to 400 feet above sea level, he found fossil shells identical to those of existing species, concluding that the en-tire mass of Patagonia had been elevated within the relatively recent past. Shells found at Santa Cruz and Port St. Julian further revealed the intrigu-ing geological history of Patagonia. By reading a contemporary authority, Darwin knew that the extinct marine species could have lived only in wa-ter of 40 to 250 feet in depth. Yet they lay buried by more than 800 feet of sea-deposited sediment. Darwin came to the only viable conclusion: "the bed of the sea, on which these shells once lived, must have sunk down-wards several hundred feet, to allow the accumulation of the superincum-bent strata."

At Patagonia's Port St. Julian, Darwin collected additional mammalian fossils, among them the *Macrauchenia patachonica*, a camel-like quadruped. From the strata in which the fossil lay entombed, Darwin inferred that the curious *Macrauchenia* was a relatively recent dweller of the region, albeit extinct, and puzzled "how a large quadruped could so lately have sub-sisted, in lat. $49°$ 15', on these wretched gravel plains with their stunted vegetation."[26] The answer came partially in the realization that its living relative, the guanaco, akin to the llama, adapted well to its equally sterile and forbidding environment. Each creature seemed somehow suited to its particular domain.

Near Bahia Blanca, Argentina, Darwin had earlier unearthed a treasure trove of extinct mammals, including several varieties of giant sloths and a *Toxodon platensis*, which Darwin considered one of the strangest animals ever beheld. It rivaled the elephant in size and chewed like an herbivore, but "judging from the position of its eyes, ears, and nostrils, it was proba-bly aquatic, like the Dugong and Manatee, to which it is also allied. How wonderfully are the different Orders, at the present time so well separated, blended together in different points of the structure of the Toxodon!"[27]

In today's jargon, we would recognize the *Toxodon* as a "missing link." Once common, *Toxodons* disappeared just 16,500 years ago. The prevalence of arrowheads at such fossil sites suggested that the *Toxodon* may have been hunted to extinction. In all, Darwin collected three *Toxodon* skulls, including one he purchased for eighteen pence from a farmer near Monte-video. The local boys had used it for target practice, flinging stones to see if they could dislodge the giant's teeth. Of the three skulls, one eventually found its way into the hands of the author and his students during visits to the British Natural History Museum between 2003 and 2011.[28]

A Mind in Ferment

"What a book a Devil's Chaplain might write on the clumsy, wasteful, blundering low & horridly cruel works of Nature,"[29] Darwin confided to a private notebook many years following the *Beagle*'s voyage. In particular, what could be more appalling than the wholesale extinction of species? Extinction was also a tremendous problem for creationists. What sort of God would take the care to create individual species and then selectively destroy them willy-nilly?

While in Port St. Julian in Patagonia, Darwin took a breather to ponder the findings of his contemporaries Lund and Clausen, who had stumbled onto an extraordinary collection of archaic remains in Brazilian caves. Of the thirty-two genera of mammals currently living in the region, Lund and Clausen had unearthed extinct species of all but four. The find included archaic "ant-eaters, armadillos, tapirs, peccaries, guanacos, opossums . . . and monkeys" among numerous other mammals. Sensing a link between past and present, Darwin noted in his journal, "This wonderful relationship in the same continent between the dead and the living, will, I do not doubt, hereafter throw more light on the appearance of organic beings on our earth, and their disappearance from it, than any other class of facts."[30]

"Certainly, no fact in the long history of the world is so startling as the wide and repeated exterminations of its inhabitants."[31] With increasing perplexity, Darwin pondered the fate of the great creatures that have vanished from the earth. "Formerly the [South] American continent must have swarmed with great monsters: now we find mere pigmies, compared with the antecedent allied races."[32] Over and over Darwin mulled the question: what mechanisms have extinguished so many species and entire genera? At first he appealed to catastrophism for the answer. "The mind at first is irresistibly hurried into the belief of some great catastrophe; but thus to destroy animals, both large and small, in Southern Patagonia, in Brazil, on the Cordillera of Peru, in North America up to Bering's Straits, we must shake the entire framework of the globe."[33] Gradually he concluded that catastrophism was an unlikely explanation, but no viable alternative presented itself. Turning to uniformitarian culprits, he considered changes in climate and the failure of food sources but was left scratching his head.

The first faint outline of "descent through modification" emerged as Darwin continued to wrestle with the issue of extinction. Each animal, he reasoned, is subject to natural checks that prevent too rapid an increase in population. We humans are not yet privy to nature's delicate balancing mechanisms; slight differences in climate, the availability of food, or the prevalence of enemies, subtle causes beyond human ken, might just "determine whether a given species shall be abundant or scanty in numbers."[34]

Darwin began to appreciate the sensitive interplay of conditions that make life suitable or unsuitable for a given species at a given place and a given time. Small changes, unnoticeable on the scale of a human lifetime, when accumulated over time, could tip the balance for or against a particular type of creature. Uniformitarianism might well apply to biology as well as geology.

Just as Darwin's philosophical moorings were slipping, he experienced the great quake at Concepción. The quake wrought many changes, physically and intellectually. The ground moved beneath his feet metaphorically as well as literally. For pages, he recounted in his journal the quake's devastating consequences.

As is common, a tidal wave accompanied the quake. Cresting at nearly thirty feet, the wave obliterated seventy coastal villages.

> It is impossible to convey the mingled feelings I experienced. . . . It is a bitter and humiliating thing to see works, which have cost man so much time and labour, overthrown in one minute; yet compassion for the inhabitants was almost instantly banished, by the surprise in seeing a state of things produced in a moment of time, which one was accustomed to attribute to a succession of ages.[35]

The most dramatic effect of the quake was its permanent elevation of the landmass. At the Bay of Concepción, the land shifted upward three feet from its previous height above sea level. Thirty miles distant, on the island of S. Maria, FitzRoy "found beds of putrid mussel-shells still adhering to the rocks, ten feet above high-water mark: the inhabitants had formerly dived at lower-water spring-tides for these shells."[36] It did not escape Darwin's notice that similar shells were to be found near Valparaiso at a height of 1,300 feet above sea level. Once again, Darwin rightly concluded, "[I]t is hardly possible to doubt that this great elevation has been effected by successive small uprisings, such as that which accompanied or caused the earthquake of this year, and likewise by an insensibly slow rise, which is certainly in progress on some parts of this coast."[37]

In a sense, the earthquake revealed the hands of both catastrophism and uniformitarianism. It all depended upon the perspective of time. To living creatures, with short life spans, the earthquake was catastrophic. To the earth herself, the lingering consequence of the quake was a relatively small uplift in a long chain of such events. Each event, however, left a geological clue, and through the sheer abundance of those clues, Darwin began to fathom the antiquity of the earth. Gradual processes acting over countless eons had molded the earth like a potter patiently working her clay. Moreover, those processes were ongoing.

By far the most profound shift in Darwin's thinking during the five-year voyage concerned the flow of time. As Copernicus liberated the universe from its narrow Ptolemaic spatial confines, so Lyell and Darwin liberated the earth from its narrow temporal confines. First, however, Darwin had to liberate himself from young-earth thinking. It happened most likely in Chile. Scholars have long puzzled why Darwin was not particularly productive as a naturalist while in Chile. Darwin aficionados now speculate that Chile was for Darwin a time of ferment,[38] the subject of which may have been time itself.

A land of majestic mountains, Chile thunders with cataracts. Unlike the rivers Darwin had known, Chile's mountain torrents, the color of mud, race down precipitous inclines, churning the earth and transporting small boulders all the way to the sea. In the thrall of such hydraulic power, one could not help but ponder the cumulative effects of the incessant churning. Of one impressive cataract, the Maypu, Darwin wrote with particular reverence:

> [T]he noise from the stones, as they rattled one over another, was most distinctly audible even from a distance . . . night and day . . . along the whole course of the torrent. The sound spoke eloquently to the geologist; the thousands and thousands of stones which, striking against each other, made the one dull uniform sound, were all hurrying in one direction. It was like thinking on time, where the minute that now glides past is irrevocable. So was it with these stones; the ocean is their eternity, and each note of that wild music told of one more step towards their destiny.[39]

One cannot help but be struck by the poetry of Darwin's journal entries about time. As the passage continues, the reader experiences vicariously Darwin's mental transition from human time to "deep time."

> As often as I have seen beds of mud, sand, and shingle, accumulated to the thickness of many thousand feet, I have felt inclined to exclaim that causes, such as the present rivers and the present beaches, could never have ground down and produced such masses. But, on the other hand, when listening to the rattling noise of these torrents, and calling to mind that whole races of animals have passed away from the face of the Earth . . . I have thought to myself, can any mountains, any continent, withstand such waste?[40]

The Scottish visionary and early uniformitarian James Hutton (1726-1797) once wrote: "The ruins of an older world are visible in the present structure of our planet."[41] By the time the *Beagle* neared the end of her voyage, Darwin could read the history of the earth as well as any geologist, and the stories she told were spellbinding.

At Villavicencio in the Andes, Darwin stumbled upon a petrified forest at an elevation of 7,000 feet, where thirty to forty silicified trunks protruded a few feet above a bare slope. "It required little geological practice to interpret the marvelous story which this scene at once unfolded; though I confess I was at first so much astonished that I could scarcely believe the plainest evidence," the now-consummate geologist recorded.[42] The trees had once "waved their branches on the shores of the Atlantic," when the great ocean lapped at the foot of the Andes. Since that time, the gradual uplift of the continent had driven back the ocean some 700 miles. The volcanic soil from which the trees sprung had subsequently been covered by ocean and layered over with sedimentary rock. Submarine lava then covered the sediment to a depth of 1,000 feet. Like a layer cake, alternate layers of "molten stone and aqueous deposits" repeated five times. "Again the subterranean forces exerted themselves," Darwin marveled, "and I now beheld the bed of that ocean, forming a chain of mountains more than seven thousand feet in height."[43]

Nature, it seemed, was never content to rest upon her laurels. Ever restless, her opposing forces continued the work of resculpting and recreating. What had once been lush forest was "utterly irreclaimable" desert to Darwin's eyes, such a wasteland that even the hearty lichen could find no opportunity to adhere to the stone hulks of the former trees.

But perhaps the most stunning insight revealed by the stone forest was a sense of the relativity of time. "Vast, and scarcely comprehensible as such changes must ever appear," Darwin observed, "yet they have all occurred within a period, recent when compared with the history of the Cordillera; and the Cordillera itself is absolutely modern as compared with many of the fossiliferous strata of Europe and America."[44]

Time, therefore, is like the ocean covering the earth. Its depths are great. And yet some oceanic chasms, like the Marianas Trench, are unfathomably deeper than others. It is not uncommon when swimming in the vast unknown of very deep waters to experience a feeling of disorientation almost like vertigo. As Darwin's observations carried him further from the young-earth shore and over the abyss of deep time, an analogous sense of mental disorientation may have prevailed.

If the physical earth is subject to inexorable change, so too might her inhabitants be. Once Darwin had steeled himself mentally to entertain Lyell's heresy – geological uniformitarianism – biological dynamism was the next logical step. With new eyes, Darwin began to see evidence of natural creativity everywhere he turned. In the strata of Patagonia, in the Galapagos and Cocos islands, and in the coral reefs of the Dangerous Archipelago, Darwin found incontrovertible evidence of biological change: "crabs that could crack coconuts, dogs that could catch fish, and fish that lived off coral."[45] But nowhere was biological adaptation subtler than in the Galapagos Archipelago.

The Galapagos

The *Beagle* entered the Galapagos Islands on September 15, 1835. Austere and sterile in comparison with the *Beagle*'s previous ports, the Galapagos afforded Darwin the perfect laboratory. The timing of the visit – near the end of the voyage – and the lack of floral and geological diversity was, in hindsight, fortuitous, for it focused Darwin's attention sharply on the few prevailing species of fauna: tortoises, lizards, finches, and a few variations of mice and rats, likely the descendents of stowaways aboard ships.

Of volcanic origin and isolated 500 miles off the coast of Ecuador, the Galapagos Islands are microcosms, natural biological laboratories. In such laboratories, to so astute an observer as Darwin, the mechanism of speciation, that "mystery of mysteries," began to reveal itself. In the following passage from October 8, 1835, he uses the word aboriginal to mean any "native" creature.

> Considering the small size of these islands, we feel the more astonished at the number of their aboriginal beings, and at their confined range. Seeing every height crowned with its crater, and the boundaries of most of the lava-streams still distinct, we are led to believe that within a period geologically recent the unbroken ocean was here spread out. Hence, both in space and time, we seem to be brought somewhat near to that great fact – that mystery of mysteries – the first appearance of new beings on this Earth.[46]

In this land of reptiles, Darwin felt like a time traveler, and indeed he was. "These huge reptiles, surrounded by the black lava, the leafless shrubs, and large cacti, seemed to my fancy like some antediluvian animals."[47] Describing a playful encounter with a giant tortoise, stone deaf according to the natives, Darwin appears more a curious boy than a keen naturalist. He delighted in surprising the "great monsters," which, at the instant of his passing, would draw in their legs and heads and drop straightway to the ground with a resounding hiss. At this point, Darwin would mount the shell, "and then giving a few raps on the hinder part of their shells, they would rise up and walk away" carrying Darwin, the tortoise *gaucho*.[48]

The other dominant reptilian inhabitants of the Galapagos Islands were two related species of iguana that exist nowhere else on Earth and in fact inhabit only the central four Islands of the Galapagos Archipelago. Darwin referred to the creatures as his "imps of darkness." Both herbivorous, one species was terrestrial, the other aquatic. Although similar in most respects to the terrestrial species, the aquatic variant sported a flattened tail and webbed feet. Both species were tame, having no natural predators. Birds perched contentedly upon the backs of the terrestrial lizards, which inhabited burrows, presumably for protection only from extreme temperatures. Darwin delighted in watching them dig, favoring the legs on one side of the body, and when tiring, switching to the rested legs on the other side. As playful with lizards as with tortoises, he recounted bemused satisfaction in the puzzled reaction of a half-buried lizard when unceremoniously removed from his burrow by a tug on its tail.

Of all the biological novelties of the Galapagos, none was more striking than the vast number of finches, whose beaks came in a stunning array of sizes and shapes. With some imagination, one might surmise that the beak of each variety was specially adapted to that variety's dominant food source. In all, Darwin counted twenty-six types of finch, all peculiar to the Galapagos, with one exception, a lark-like marsh dweller also common to North America. But for their blatantly varied beaks, the birds looked extraordinarily similar.

> The most curious fact is the perfect gradation in the size of the beaks in the different species of Geospiza, from one as large as that of a hawfinch to that of a chafinch, and . . . even to that of a warbler. . . . There are no less than six species with insensibly graduated beaks. The beak of Cactornis is somewhat like that of a starling, and that of the fourth sub-group, Camarhynchus, is slightly parrot-shaped.[49]

Darwin's Galapagos journal then continues with a vague but stunning speculation, tentative thoughts that would ultimately coalesce into *Origin of Species*. "Seeing this gradation and diversity of structure in one small, intimately related group of birds, one might really fancy that from an original paucity of birds in this archipelago, one species had been taken and modified for different ends."[50]

But Darwin was not yet an evolutionist. Still firmly under the spell of creationism and its proponents, Darwin collected finch specimens two by two, as if he were Noah loading the Ark with male and female representatives of each and every species, species presumed to have been immutable since the creator had formed them 6,000 years ago.[51] Having collected type specimens of all the varieties scattered about the Galapagos Islands, Darwin committed a *faux pas* that no true evolutionist would commit. Not recognizing the importance of correlating each specimen with its location, he dumped the specimens from two islands into a single bag, a bungle he would soon regret.

Nine months later, while criss-crossing the Pacific as the *Beagle* beat an erratic course home, Darwin's attention trailed back to the tortoises and finches of the Galapagos. From the natives, he had learned that the tortoises of the various islands of the archipelago differed ever so slightly, each possessing attributes unique to the residents of the particular island. So, too, the mockingbirds. And the finches. Having finches commingled in one bag, rather than identified by island, put Darwin at a distinct disadvantage in trying to correlate beak attributes with habitat. Still, patterns began to emerge. The beak of each finch seemed finely tuned to its primary food source. Darwin must have felt chills down his spine when first entertaining the "dangerous" thought that would haunt him for the remainder of his life: that each species was uniquely suited to the environment of its particular island. "If there is the slightest foundation for these remarks," he confided to himself in his notebook, "the zoology of Archipelagoes will be well worth examining; for such facts would undermine the stability of species."[52] The explosive question was: were the multifarious types of finch just variations of a few species, or was each a species unto itself? Official classifications would have to wait until the *Beagle* settled snugly back in England.

As astute an observer of human nature as of flora, fauna, and geology, Darwin found the gradations of *Homo sapiens* a source of continual curiosity and, sometimes, sheer amazement. While going ashore at Wollaston Island, Darwin's landing party encountered six Fuegian Indians in a canoe, "the most abject and miserable creatures I anywhere beheld."[53] Despite the harsh climate of Tierra del Fuego, the Fuegians were completely naked. The sight of a newborn suckling at his mother's naked breast in a sleeting rain appalled Darwin. More revolting was the practice of cannibalism against warring tribes and, in extreme circumstances, within their own circle. When hunger pressed, the Fuegians occasionally sacrificed the oldest woman in their midst. When asked why they did not first kill their dogs, Jemmy Button replied matter-of-factly, "Doggies catch otters, old women no."[54]

When reintroduced to his countrymen at Tierra del Fuego, Jemmy Button, who could no longer speak his native tongue, expressed shame at the Fuegians' uncivilized ways. Nevertheless, both York Minster and Jemmy Button soon reverted to their natural state, Jemmy's return hastened by his finding a wife. FitzRoy's experiment in "elevating the savage" thus ended in dismal failure, and Mr. Matthews, in a change of heart, abandoned his plans of mission work.

At the other end of the spectrum of aboriginal man were the Tahitians, a noble race. Tahiti, lush in vegetation and abounding in tropical fruit, afforded a welcome respite on the long journey home in 1836.

Upon entering Tahiti's Matavai Bay, the *Beagle* was immediately surrounded by canoes. Nothing pleased Darwin about Tahiti so much as its people. Physically beautiful, with athletic and well-proportioned bodies, the Tahitians exhibited a combination of gentleness and intelligence that Darwin had not previously encountered in indigenous cultures. In some ways they seemed superior to civilized Western man, noted Darwin. "A white man bathing by the side of a Tahitian was like a plant bleached by the gardener's art compared with a fine dark green one growing vigorously in the open fields."[55]

Homeward Bound

On the final leg of the voyage, Darwin began to reflect upon his experience *in toto*. Despite having spent as much time as possible on land, Darwin's dominant impression was of the sea. "It is necessary to sail over this great ocean to comprehend its immensity," he penned. Otherwise, "we do not rightly judge how infinitely small the proportion of dry land is to water of this vast expanse."[56]

The first four years of the *Beagle*'s voyage were an extraordinary adventure for its young naturalist. By the final year of the journey, though,

Darwin was more than eager to return to England. His journal entry of Christmas Day, 1835, written from New Zealand, betrays his weariness. "In a few more days the fourth year of our absence from England will be completed. Our first Christmas Day was spent at Plymouth, the second at St. Martin's Cove near Cape Horn; the third at Port Desire in Patagonia; the fourth at anchor in a wild harbour in the peninsula of Tres Montes, this fifth here, and the next, I trust in Providence, will be in England."[57]

Despite growing fatigue, Darwin made the most of the return. A stop in the Cocos (Keeling) Islands proved one of the more seminal experiences. There, among other things, Darwin observed yet another marvelous adaptation: "a huge land-crab . . . furnished by nature with the means to open and feed on this most useful production [a coconut]."[58] But it was the atoll itself that captured his attention and turned his mind from biology back to geology.

> The shallow, clear, and still water of the lagoon, resting in its greater part on white sand, is, when illumined by a vertical sun, of the most vivid green. This brilliant expanse, several miles in width, is on all sides divided, either by a line of snow-white breakers from the dark heaving waters of the ocean, or from the blue vault of heaven by the strips of land, crowned by the level tops of the cocoa-nut trees. As a white cloud here and there affords a pleasing contrast with the azure sky, so in the lagoon bands of living coral darken the emerald green water.[59]

These words, penned with the sensibility of an artist, belie the assiduous and ever-vigilant scientist. From observations and soundings in the Cocos during the *Beagle*'s homeward leg, and earlier in the Dangerous Archipelago, Darwin completed the *coup de grace* in his emergence as a geologist: a theory of coral formations.

The fragility of coral atolls impressed Darwin, in stark contrast to the "all-powerful and never-tiring waves of that great sea, miscalled the Pacific."[60] Careful reading of geology texts had convinced Darwin that atolls were built upon the remnants of volcanoes. If volcanoes could spew forth out of the sea, given sufficient time, they could also collapse back into the sea. Atolls, he surmised were wreaths of coral attached to the flanks of volcanoes just below the waterline. As the volcanic cone sinks, new coral deposits atop the old. Eventually, when the cone has completely submerged, only the frail reef remains. Because coral cannot survive "at a depth greater than twenty to thirty fathoms,"[61] the submersion of the cone must be gradual, otherwise slow-growing coral would die when carried too quickly beyond the range of nourishing sunlight. Here was another triumph for the uniformitarian view.

In the final dated entry of his journals, Darwin concluded the momentous voyage with the simplest of sentiments, words that underplay its cumulative effect upon him: "On the last day of August we anchored for the second time at Porto Praya in the Cape de Verd archipelago; thence we proceeded to the Azores, where we stayed six days. On the 2nd of October we made the shores of England; and at Falmouth I left the *Beagle*, having lived on board the good little vessel nearly five years."[62]

Not quite content to let the journal rest there, Darwin attached an epilogue. It was perhaps his way of processing the emotional power of experiences that had altered him irrevocably and would soon change the world. In the final pages, he reflected upon the pros and cons of the whole endeavor. Chief among the pros were indelible images of places visited.

> Among the scenes which are deeply impressed on my mind, none exceed in sublimity the primeval forests undefaced by the hand of man; whether those of Brazil, where the powers of Life are predominant, or those of Tierra del Fuego, where Death and Decay prevail. Both are temples filled with the varied productions of the God of Nature – no one can stand in these solitudes unmoved, and not feel that there is more in man than the mere breath of his body.[63]

It came as a surprise to Darwin that, of all nature's temples, Patagonia was to him most sacred. "In calling up images of the past, I find that the plains of Patagonia frequently cross before my eyes; yet these plains are pronounced by all wretched and useless."[64]

Yet on further reflection, he understood the appeal of Patagonia, as "partly owing to the free scope given to the imagination."[65] Sometimes the senses are shocked by unfamiliarity into perceiving what they would not otherwise have detected. Anyone who has, for the first time, gazed upon the vast unearthly expanse of Monument Valley has likely experienced the ineffable. So too did the Apollo astronauts who stood upon the surface of the moon – barren, forbidding, and yet possessed of a stark, otherworldly beauty. Among the other spectacles that Darwin beheld were

> the Southern Cross, the cloud of Magellan, and the other constellations of the southern hemisphere – the waterspout – the glacier leading its blue stream of ice, overhanging the sea in a bold precipice – a lagoon – island raised by the reef-building corals – an active volcano – and the overwhelming effects of a violent earthquake.[66]

Not all Darwin's experiences were beautiful or positive. The atrocities he witnessed by slaveholders against men, women, and children haunted him for years. As a man of integrity, Darwin could not escape his own complicity, by virtue of his nationality, in these heinous crimes of man against man:

> And these deeds are done and palliated by men who profess to love their neighbours as themselves, who believe in God, and pray that His Will be done on Earth! It makes one's blood boil, yet heart tremble, to think that we Englishmen and our American descendants, with their boastful cry of liberty, have been and are so guilty.[67]

On balance, the unexpected delights of unfamiliar places, the freedom from the encumbrances of civilization, and the sublime experience of a wild and largely pristine Earth far outweighed for Darwin the travails of the journey. In a concluding paragraph he writes wistfully: "I do not doubt that every traveller must remember the glowing sense of happiness which he experienced when he first breathed in a foreign clime where the civilised man had seldom or never trod."[68]

A Triumphant Return

Two days after the intrepid *Beagle* rested at last in Falmouth on October 4, 1836, Darwin returned to the Mount in Shrewsbury, having slept the previous night in a hotel so as not to disturb his family at a late hour. Seasick for five long years, Darwin never returned to sea.

He returned to some notoriety, thanks to the abundance of correspondence and the number of specimens forwarded to Henslow at Cambridge. The two years immediately following the *Beagle*'s return were immensely active and productive. Through Lyell's influence and his own carefully honed powers of geological observation, Darwin was elected a fellow of the Royal Geological Society and appointed its secretary. His first order of business: classification of the vast collection of specimens in Henslow's safekeeping. Publication of his journals nipped close on the heels of his first priority. The instant success of *Voyage of the Beagle* upon its 1839 publication surprised no one more than its author.

Darwin wasted little time in relegating his collections to the Zoological Society of London, soon abuzz with talk of Darwin's treasure trove of unusual creatures. Ornithologist John Gould immediately set about classifying Darwin's finches. To be near Gould, Darwin rented an apartment in London's Bloomsbury district. Excitement mounted as Gould neared completion. When the news finally broke, it made the London papers. Of the various finches collected by Darwin in the Galapagos, eleven were species found nowhere else in the world. As Darwin had suspected, "all these species, marooned in their lonely archipelago, had diverged from their ancestral stocks and then gone right on diverging." They had, in fact, "broken the species barrier."[69]

The year 1839 was pivotal in another way. At the age of thirty, Charles Darwin married Emma Wedgwood, his first cousin. Charles and Emma, one year his senior, had been companions since childhood, their mutual affection long nurtured by Charles' frequent hunting excursions at his Uncle Josiah's estate "Maer." As he was inclined to do in weighty decisions, Darwin debated the pros and cons of marriage, even to the point of drawing up a ledger. Under the "Not Marry" column: the loss of personal freedom; the responsibility, expense, and anxiety of children; and perhaps most significantly for a determined scientist and adventurer, the loss of time. "How should I manage all my business if I were obliged to go walking every day with my wife. Eheu! I never should know French, or see the Continent, or go to America, or go up in a Balloon."[70] Fortunately for Darwin and the world, petty concerns were overruled, and the "Marry" side of the balance sheet won handily. He concluded:

> My God, it is intolerable to think of spending one's whole life, like a neuter bee, working, working, and nothing after all. No, no, won't do. Imagine living all one's day solitarily in a smoky dirty London house. Only picture to yourself a nice soft wife on a sofa with a good fire, and books and music perhaps. Marry – Marry – Marry. Q.E.D.[71]

Shortly after their marriage, Emma joined Charles at the house he had let on Upper Gower Street near University College. Harkening back to splendors observed in South American rainforests, Charles dubbed their London abode the "Macaw Cottage" because the eclectic furniture of the newlyweds combined the macaw's colors "in hideous discord."[72] The Darwins immersed themselves in the affairs of high Bloomsbury society. Neighbors included publishers, surgeons, lawyers, and Emma's brother Hensleigh. At the end of December 1839, Emma gave birth to her first child, William Erasmus, later known as Willy. To his delight, Charles took at once to fatherhood, enraptured by his "little animalcule of a son."[73] There also, on March 2, 1841, following a difficult labor, Emma gave birth at their Gower Street residence to a second child, Annie.

The Reluctant Revolutionary

During a visit to Maer in November 1840, Charles fell ill, beset by a host of unpleasant symptoms: dizziness, gastrointestinal upset, nausea, visual black spots, and skin disorders. "Maer's disease," as Darwin was to term the syndrome, plagued him for the duration of his life. Episodes of ill health were particularly acute and frequent during the Darwins' nearly four years on Gower Street. What energy he could muster he devoted to publishing his theory on coral. "This book, though a small one, cost me twenty months of hard work,"[74] he lamented. The theory, a triumph of deductive reasoning, was especially notable in that Darwin had worked it out in full before ever laying eyes upon a coral reef. Firsthand experience with coral atolls while aboard the *Beagle* served merely to validate the theory. The work was well received, particularly by the brilliant, kindly, and sympathetic geologist Charles Lyell, whose book had companioned Darwin on his voyage. Having met through the Geological Society, the men developed a fast and lifelong friendship.

Darwin had already begun to sense that he was ill suited to the pace, noise, dirt, and demanding social obligations of London. Illness, however, made London intolerable. The excitement of social engagements frequently brought on "violent shivering and vomiting attacks."[75] Periods of relative equanimity during lengthy visits to Maer convinced the Darwins that country life better suited them, and in 1842, Charles, Emma, and the two children moved permanently to the village of Downe in Kent.

In Downe, just sixteen miles from London, Charles and Emma purchased a rambling, two-story house. Down House – as they called it, dropping the "e" – was, in today's lingo, a fixer-upper. Located a quarter mile from the village, Down House was ideal in location and potential: "a good, ugly house on 18 acres, situated on a chalk flat, 560 feet above sea, [whose] chief merit is its extreme rurality."[76] Of its relative isolation, a friend of Darwin quipped that the house could be reached only by mule-track. In some disrepair, the home was quickly restored to the serviceability and comfort necessary for a rapidly growing family, and its gardens provided both food and opportunities for horticultural experimentation.

At Down House, Darwin lived for forty years, shielded "from every avoidable annoyance"[77] by his devoted wife. Charming after its restoration, Down House was soon abustle with servants and children. Of the latter, Charles and Emma had ten. The odds of the day, however, virtually guaranteed parental heartbreak. On average, one in five children died in their first year, and another would not live to adulthood.

While in South America, Darwin had climbed numerous peaks with youthful zest. In the year the Darwins moved to Downe, Charles again attempted a climbing excursion, this time in Wales. Although the Welsh countryside afforded pure delight to any naturalist, Maer's disease had exacted a toll, prematurely robbing Darwin of his vigor. Never again was Darwin "strong enough to climb mountains or to take long walks such as are necessary for geological work."[78]

Mystery shrouds the nature of Darwin's illness. While on the *Beagle*'s voyage, he was bitten by a *Benchuca* beetle, which is known to cause lingering symptoms. However, he manifested peculiar symptoms prior to the voyage as well. Abnormally sensitive, Darwin was something of a hypochondriac. Most likely his malady stemmed both from physical causes and neurosis, the latter originating from anxiety related to his potentially explosive ideas. Prior to their marriage, Emma had written an aunt, "He is the most open transparent man I ever saw and every word expresses his real thoughts."[79] Such a man had little stomach for controversy and was ill-equipped to defend himself against criticism. Were his ideas to take root, Darwin would become a lightning rod. That knowledge, conscious or unconscious, took a physical toll. Fortunately, he had Emma for comfort. Charles, who never felt secure in her absence, probably would not have long survived bachelorhood. A friend observed of the couple, "the perfect nurse had married the perfect patient."[80]

Life in the Darwin home settled into two spheres of activity: managing a lively household where children could thrive under minimal supervision and establishing an environment conducive to Charles' scientific work. For indulgent parents, the two spheres cohabited remarkably well

and sometimes intersected. It was not uncommon for a sick child to occupy the daybed in Charles' study while he read or wrote.

Maer's disease was a double-edged sword. Although illness interrupted and complicated Charles' scientific work, it also necessitated that
he develop a routine to manage his symptoms. He became a creature of
habit. The grounds at Down House featured the "sand-walk," a triangular path that led from the house, past the gardens, and then left along the
woods before tightly looping back to rejoin itself near the gardens. Thrice
daily – morning, noon, and afternoon – rain or shine, Darwin paced the
sand-walk as he contemplated an unruly jumble of observations and ideas
that demanded order. Frequently accompanied by Annie, Darwin the
kindly father ambled along lost in thought while his effervescent daughter pranced before him. A game of billiards with his butler, Parslow, also
afforded occasional respite from obsessive rumination.

One conceptual question gnawed at Darwin, the missing link to his
theory: what process governs natural selection? In 1838, Darwin stumbled onto Reverend Thomas Malthus' essay on population, and in a flash,
Darwin had his answer. Although Malthus' essay focused on economics
rather than biology, his thought contained a germinal idea: "The power
of population is indefinitely greater than the power in the Earth to produce subsistence for man."[81] Darwin had failed to fully appreciate nature's profligacy. Almost universally among species, far more offspring
are produced than can possibly survive. Hence, the lot of all living things
is an intense struggle to survive against the odds. "Struggle for survival"
was the key breakthrough for which Darwin had Malthus to thank, and
by 1842, Darwin had sketched, in pencil, a thirty-five-page draft of *On the
Origin of Species*.

To this point, Darwin had confided the theory to no one. In order to
continue, he needed a confessor. In 1843, Darwin met Joseph Hooker, a
young botanist whom he had approached for help with the arduous task of
classifying the vast array of plant specimens harvested from South America and the Galapagos. Hooker would soon succeed his father William as
director of the unrivaled Royal Botanic Gardens at Kew. A brilliant naturalist in his own right, Hooker became Darwin's confidant, gentle critic,
and devoted friend. That deep friendship, which both men cherished and
cultivated, lasted a lifetime.

By 1844, Darwin's draft had grown to a whopping 230 pages, in ink,
and he was ready to sea-test the theory in the safest waters he could find.
Tentatively, he wrote to Hooker: "I am almost convinced (quite contrary
to opinion I started with) that species are not (it is like confessing a mur-
der) immutable."[82] The confession is remarkable, for two reasons. First, it
demonstrates how deeply Darwin's own ideas unnerved him. Second, it
reveals how deeply he feared the rejection that radical ideas might elicit.
Fortunately, Hooker, whose objective feedback was just the medicine Darwin needed, considered his friend neither murderer nor blasphemer.

Darwin had few epiphanies. Mostly the theory of evolution emerged clue by stubborn clue from a long, hard slog through mounds of observational data and years of cogitation. There were, however, a few "Eureka" moments, like that following the reading of Malthus. Much latter, after the theory of evolution was well along, Darwin had a similar epiphany: "I can remember the very spot in the road, whilst in my carriage, when to my joy the solution occurred to me, and this long after I came to Down."[83]

At issue was another nagging question: Why are there so many species? Why so many species of finch in the remote, barren, and largely inhospitable Galapagos Archipelago? Or why such an abundance of species of the order *Coleoptera*? There were so many that a British biologist wryly accused the creator "of an inordinate fondness for beetles." Forty percent of the earth's insects are beetles!

The answer came as "the principle of divergence." Incipient species – that is, biological varieties on the cusp of speciation – will be driven apart by natural selection. Darwin's metaphor was branching in the tree of life. When times are hard, competition for resources is intense. In the natural world or the laboratory, "no two species eating the same food in identical ways can peaceably co-exist in the same test tubes, on the same rocks, or on the same islands without driving one or the other to extinction."[84] Organisms reduce the pressure of competition by specialization, by finding and occupying an open niche.

Imagine for example that in the early days of human civilization every male sought livelihood as a cobbler. The quality of shoes would vary wildly, and maladroit cobblers would likely starve to death. To avoid disaster, some would become millers instead, or loggers, or blacksmiths. One avoids disaster, and the larger society is better served, when one finds what one does best and concentrates on that.

Regarding the Galapagos finches, thirteen species occupy four genera. The two tree-dwelling genera are distinguished primarily by diet: insectivorous or vegetarian. The finches of a third genus act and look like warblers. The most populous genus, *Geospiza*, contains six species of ground finches, which look so much alike that, during mating, even the birds have difficulty telling one another apart. The ground finches differ appreciably only in their beaks, which are finely tuned to crack seeds of specific size, shape, and hardness. For example, cactus finches feed on the cactus itself, extracting seeds from the fruit. Others depend upon scavenging seeds that have fallen to the ground. On the lower, less lush islands, which present only a single niche for the ground feeders, either *Geospiza fuliginosa* or *G. difficilis* can be found – never both. On the larger islands, both can be found, but topographically separated, one species feeding at the base of the island, the other near the top.[85] Wherever there exists a niche, some species will evolve to fit it to near perfection. And wherever two varieties or species compete for a single niche, the less well adapted one will surely perish from the earth.

With the theory of natural selection well in hand and given a green light by Hooker to proceed with publication, Darwin did the unthinkable: he procrastinated. He shelved the theory indefinitely to take up, of all things, the study of barnacles, confiding to a friend that his new interest "will put off my species book for a rather long period."[86] First, though, he took an unusual precaution. Darwin deposited the manuscript and a sum of £400 for safekeeping and drafted a letter to Emma, dated July 5, 1844. In it he instructed her to publish the manuscript posthumously should he die prematurely. His two friends, Lyell and Hooker, were suggested as editors.

Several factors contributed to Darwin's incongruous shift of focus. He had simply not acquired the intestinal fortitude necessary to withstand the firestorm of controversy that *Origin* would generate. At the core of the issue lay inevitable conflict between traditional religious views of creation and his revolutionary uniformitarian views, which rendered the deity's role in creation much murkier. Darwin's internal conflict was further exacerbated by Emma's conventional religious beliefs. At all costs, he wished to spare her anguish and avoid drawing her into controversy. Finally, Darwin was not one who could bully detractors into submission. His chief recourse was to accumulate such a mountain of facts that no thinking person could deny the weight of evidence. One senses that, at times, Darwin even longed for death as an escape from the vilification he was sure to face.[87]

Eight years later, Darwin completed his exhaustive taxonomy of the lowly barnacle, which he published in five volumes that ran collectively to well over 1,000 pages. The work is a testament to Darwin's unsurpassed powers of observation, his humility, and his deeply felt dread about going public with his major work. But the diversion took its toll. "I hate a Barnacle as no man ever did before, not even a Sailor in a slow-sailing ship," Darwin confessed in a letter to a friend.[88]

And then a series of family tragedies threatened Darwin's undoing, unraveling the rags of his Christian faith and ultimately driving him back to his major theme.

In 1848, Dr. Robert Darwin died, an unbeliever. Neither Charles nor his brother Erasmus could countenance a god who would punish their father for his unbelief, as they had been taught to believe.[89] And then, in the summer of 1850, Annie fell ill. At first her symptoms were mild and indistinct, but the trend was alarming. Despite many doctors and therapies, by Easter 1851 Annie's health had turned precarious.

Emma, tending to the other children, remained at Down House while Charles tenderly cared for Annie in the sanitarium at Malvern, where the family had sent her following the failure of conventional remedies. Annie received Dr. Gully's "water treatment," an act of desperation. From Annie's bedside, Charles wrote to Emma daily, sometimes hourly, to report Annie's state. Despite her distress, Annie never complained, appreciative of every small kindness her father could render. Despite all that love, devotion, and medical care could offer, Annie slipped the bonds of earth at the age of ten. Dr. Gully had diagnosed Annie with "smart bilious gastric fever," but more likely the child succumbed to a form of tuberculosis.

Darwin's devastation at Annie's loss was bottomless. Whereas Emma could find solace in her religious beliefs, Charles' faith teetered on the brink. He processed his grief in the ways truest to himself: by observation and the written word, hoping by the latter to fix Annie forever in his memory. "Our poor child, Annie, was born in Gower St on March 2nd 1841 and expired at Malvern at Midday on the 23rd of April 1851. I write these few pages as I think in after years, if we live, the impression now put down will recall more vividly her chief characteristics." Chief among those characteristics was a radiance that endeared her to all.

> From whatever point I look back at her, the main feature in her disposition which at once arises before me is her buoyant joyousness, tempered by two other characteristics, namely her sensitiveness, which might easily have been overlooked by a stranger, and her strong affection. Her joyousness and animal spirits radiated from her whole countenance and rendered every moment elastic and full of life and vigor. . . . Her eyes sparkled; she often smiled; her step was elastic and firm; she held herself upright, and often threw her head a little backwards, as if she defied the world in her joyousness. . . . All her movements were vigorous, active and usually graceful; when going round the Sand-walk with me, although I walked fast, yet she often used to go before pirouetting in the most elegant way, her dear face bright all the time, with the sweetest smiles.[90]

Darwin's tender and loving tribute ran for more than four pages and concluded: "We have lost the joy of the household, and the solace of our old age: she must have known how we loved her; oh that she could now know how deeply, how tenderly we do still and shall ever love her dear joyous face. Blessings on her. April 30, 1851." From this blow he never fully recovered.

The deepest pain that life can inflict is the loss of a child. This inconsolable loss Darwin shared with other great men of science, among them Kepler and Galileo. And like each of them, Darwin sought solace in the concentration of scientific work, which afforded the only respite, however brief, from heartbreak. In a letter later written to his son Francis after the death of Francis' wife during childbirth, Darwin offered up his poor palliative for grief, "The only chance of forgetting for short times your dreadful loss [lies in] the habit of close mental attention."[91]

Beneath the Waves

Those who would denigrate Darwin sometimes assume that atheism drove Darwin to formulate the theory of evolution. Nothing is further from the truth. Darwin experienced a "slow drift from Christianity" during which his religious beliefs gradually evolved toward agnosticism, a transformation driven equally by personal tragedies and scientific revelation.

While a university student at Cambridge, Darwin had attended Christ College, the home of John Milton and the Reverend William Paley. The latter's books – *Evidences of Christianity* and *Natural Theology*, in which faith and science coexisted harmoniously – had great effect on Darwin. Time and again he found inspiration in their logic. Reading them gave him "as much delight as did Euclid."[92]

On the voyage of the *Beagle*, Darwin took a Bible, and a copy of Milton's *Paradise Lost* accompanied him on all his inland forays. Both speak eloquently and powerfully of special creation and the perfection of God's handiwork in that creative process. What is perfect has no need of change. Darwin had no cause to doubt either Milton or the deity. He was then a conventionally religious Christian, although not particularly devout. By his own admission, Darwin "never gave up on Christianity until the age of 42."[93] Why then? It was in his forty-second year that Annie died.

In *Natural Theology*, Paley presented an argument for design that is still invoked today: the analogy of the watchmaker. If, while crossing a heath, with its abundance of nature, one were to stumble upon a watch – beautiful but unnatural – one would be forced to recognize the hand of an intelligent watchmaker. Perhaps Paley's watchmaker analogy contributed to Darwin's keen appreciation for nature's designs and his unsurpassed powers of observation.

But Darwin's own observations ultimately revealed the flaw in Paley's argument – it was incomplete. Stuck in the present, Paley overlooked the watch's very purpose: to document the *flow* of time.

Paley's watch is itself the result of evolutionary processes. Today's quartz chronometer has many antecedents, the analogs of biological ancestors. First came the *gnomon*, then the sundial, the water clock, and the pendulum clock. Then Thomas Harrison's marine chronometers, the Bulova Accutron of the 1970s with its miniature tuning fork, and finally the quartz-crystal digital timepiece with its extraordinary precision. The watch of today is as different from the first crude timekeeper as is the first patch of light-sensitive tissue is to the fully evolved eye. There may be a watchmaker, but he is never content to rest on his laurels.

In sum, it was the suffering of innocents, his repugnance at the idea of eternal punishment for nonbelievers, and his scientific skepticism about special creation that pushed Darwin away from belief in a beneficent God. His belief simply sank beneath the waves, like the central volcano of a coral atoll.

Almost Scooped

Following Annie's death, Darwin buried himself in writing his "big species book," whose theory was now solid, and he steeled himself for the inevitable controversy to ensue. Still, he hesitated to publish.

On June 18, 1858, Darwin received a jolt. Arriving in the form of a letter, it forced his reluctant hand. From Alfred Russel Wallace, a young naturalist in Malaya with whom he had communicated several times, the letter outlined a theory of speciation uncannily similar to Darwin's own. Wallace had independently hit upon "natural selection" as the mechanism for "survival of the fittest."

Wallace's epiphany occurred during a night of fever in the Spice Islands at the conclusion of four years in the Amazon basin.[94] Darwin was shocked by the resemblance of their ideas. "I never saw a more striking coincidence; if Wallace had my MS. sketch written out in 1842, he could not have made a better short abstract!"[95] At a loss for how to respond, Darwin contemplated ceding credit for the theory to Wallace, thereby voiding nearly three decades of scientific labor. For advice, he wrote Lyell and Hooker. While awaiting their responses, he was dealt yet another devastating blow. The family's two-year-old son, Charles Waring Darwin, contracted scarlet fever and died five days later.[96] Despondent and unable to respond adequately to Hooker's advice, Darwin deferred to his two friends to resolve the matter of intellectual property in whatever way they deemed best.

Lyell and Hooker worked out a Solomonic compromise by which proxies of Darwin and Wallace would present in tandem their papers on natural selection at a meeting of the Linnaean Society. Darwin's first inclination? "I was first unwilling to consent, as I thought Mr. Wallace might consider my doing so unjustifiable."[97] Fortunately, Darwin relented to the combined wisdom of Lyell and Hooker. The papers were read on July 1, 1858, causing little stir.[98] Neither Wallace nor Darwin was present. The only published response came from a Professor Haughton of Dublin, "whose verdict was that all that was new [in the papers] was false, and all that was true was old."[99] It was, as we might say today, a big yawn.

Nevertheless, the close call freed Darwin of inhibitions and impressed upon him the urgency of publication. After thirteen months of feverish work, Darwin published his phenomenal work: *On the Origin of Species by Means of Natural Selection*. Upon completion of the 490-page tome, which Darwin considered a mere abstract, the exhausted author wrote a friend, "So much for my abominable volume, which has cost me so much in labour that I almost hate it." But by the time the book appeared in print two months later, Darwin's mood had lifted. "I am infinitely pleased & proud," he praised his publisher, "at the appearance of my child."[100]

All copies of the first edition of *On the Origin of Species* sold out on the day of publication. Later editions dropped "On the," and the book became famous by the foreshortened title *Origin of Species*. Darwin had for decades feared the criticism publication would unleash, yet when it finally materialized, he was unprepared and plunged into anxiety and depression. Fortunately, by then Darwin had strong champions willing to fight the battles he was ill constituted to fight.

The theory of descent with modification split the world into two camps: strong proponents and equally vociferous opponents; few were noncommittal. In the former camp were his friends Lyell and Hooker, and a recent convert, the pugnacious educator Thomas H. Huxley, who was "prepared to go to the stake" in the defense of Darwin's work. Also effusive in his praise was Asa Gray, Harvard University's preeminent botanist. To Hooker, Gray wrote glowingly, "[*Origin*] is done in masterly manner. It might well have taken twenty years to produce it. It is crammed full of the most interesting matter – thoroughly digested – well expressed – close, cogent, and taken as a system makes out a better case than I had supposed possible."[101]

In a letter to Darwin, Gray raved that he had never learned as much from a book as from *Origin*. But Gray also reported that Harvard colleague Louis Agassiz was much annoyed by the book. It grated also on Adam Sedgwick, Cambridge's professor of geology and Darwin's dear friend, for whom *Origin* gave "more pain than pleasure." Disturbed that Darwin had abandoned God's will as the primary causal agent, Sedgwick admitted that Darwin's conclusions "greatly shocked [his] moral taste."[102] Huxley, on the other hand, bolstered Darwin against the "considerable abuse" in store for him. To this encouragement he added sincerely the "lasting gratitude of all thoughtful men."[103]

The two decades of procrastination had been far from fallow. On the contrary, they had yielded Darwin opportunity to anticipate the numerous objections that the theory was certain to elicit. It is testimony both to Darwin's confidence in his ideas and to his integrity that he himself originated most of the theory's objections, including many that find voice today. He devoted two meaty chapters of *Origin* to elucidating bona fide concerns and countering them as completely as possible. Gray marveled at his candor and lightly chastised him for its excess, for in Chapters VII and VIII – entitled "Difficulties of the Theory" and "Miscellaneous Objections," respectively – Darwin had handed his would-be assassins their weapons on a silver platter.

First and foremost among the "difficulties," and still routinely evoked by detractors, is the issue of so-called "missing links." In Darwin's words: "[W]hy if species have descended from other species by fine degradations, do we not everywhere see innumerable transitional forms?"[104] In modern terms we might phrase the question as follows: why is speciation discrete (i.e., punctuated) rather than continuous? In Darwin's more picturesque prose: "Why is not all Nature in confusion, instead of species being, as we see them well defined?"

The answer to the question has three parts, of which Darwin nailed the first two. Regarding the third part, he was completely ignorant through no fault of his own.

First, today's detractors claim that the fossil record contains too few "missing links." Darwin correctly countered that the fossil record was incomplete. It was certainly true in his day, and it remains true today. However, since Darwin's time, numerous transitional forms have been unearthed, and the number of missing links continues to dwindle. Darwin himself was not overly concerned by the absence of transitional creatures. "Opponents will say – 'show them [to] me.' I will answer yes, if you will show me every step between bulldog and greyhound."[105]

Second, a deeper and more compelling reason for the relative paucity of transitional forms is that such forms are in essence biology's failures. Missing links are those species not as well adapted as their kin. Hence, in the struggle for survival, they are first to perish. To Darwin "extinction and natural selection go hand in hand."[106] We see relatively few transitional forms in the fossil record because these forms did not inhabit the earth for long or in great numbers.

Third, Darwin was ignorant of genetics. We do not see a continuous spectrum of species because the mechanism of heredity, of which Darwin was unaware, is discrete. Variation is the result of genetic mutation, and mutation stems from a quantum jump in the genetic molecule. (See subsequent chapters.) Thus, by analogy, variations are distinct steps on a biological stairway rather than positions on a biological ramp.

A second objection Darwin posed as follows: "Why, again, do whole groups of allied species appear . . . to have come in suddenly on the successive geological stages?"[107] Darwin countered that the geological record, like the fossil record, is incomplete. In his day, for example, the theory of plate tectonics, which explains continental drift, had not yet even been anticipated. Today we find in the geologic record close connections between catastrophic geological events and speciation. For example, following the meteoric impact that precipitated the extinction of dinosaurs, mammals proliferated and flourished because biological niches previously filled by reptiles opened to mammalia. In truth, catastrophism and uniformitarianism work in concert. Following catastrophes, which are relatively rare, life responds by adaptive radiations in a profusion of new species. Between cataclysms, life finds near-equilibrium states, and evolution lolligags along.

A favorite foil of creationists is the eye, an organ that seems too exquisitely engineered and too specialized to have evolved by successive stages. Darwin anticipated this objection, too. "Can we believe," he posited, "that natural selection could produce, on the one hand, an organ of trifling importance, such as the tail of a giraffe, which serves as a fly-flapper, and, on the other hand, an organ so wonderful as the eye?"[108] Lyell encouraged him to leave out this particular excess of candor, which he believed to be a stumbling block to the reader. The eye is indeed marvelously complex, but modern biology can now trace its development through successive adaptations. Although Darwin did not explain the evolution of the eye *per se*, he nevertheless offered quite plausible arguments backed up by considerable evidence for equally perplexing adaptations, specifically, for the gradual migration of an eye from one side to the opposite in a bottom-dwelling flounder.[109]

The gravest objection to evolution originated with the scientific luminary Sir William Thomson (Lord Kelvin). Kelvin's estimates of the earth's age were far from sufficient for evolutionary processes to effect significant change. Lord Kelvin's objection was particularly onerous to Darwin, who had no good answer for it. Neither did he try to sweep it under the rug. He simply acknowledged the problem and asserted that we do not yet know enough "of the universe and of the interior of our globe to speculate with safety on its past duration."[110] His simple statement was prophetic. Cosmology and chemistry have since vindicated Darwin and revealed Kelvin's oversight. But it was not until the discovery of radioactivity in the late nineteenth century, and its use in radiochronology in the twentieth, that the scientists credibly established the necessary antiquity of the earth.

Following the publication of *Origin*, Darwin returned for a time to safe topics. In 1862, while the Civil War raged in the United States, he wrote about orchids with devotion similar to what he had earlier lavished on barnacles. In 1864 he published "The Movements and Habits of Climbing Plants" in the journal of the Linnaean Society. In 1868, he attempted an explanation of heredity in a book entitled *The Variation of Plants and Animals under Domestication*. While Darwin studiously avoided the limelight and further controversy, his stature steadily grew, as the waves generated by the *Origin* reached shores increasingly distant. He received numerous accolades: from the British Royal Society, the Royal Society of Edinburgh, and the Berlin Academy, among others. However, none meant as much as did an honorary doctorate in 1877 from his alma mater, Cambridge University.

Following the reprieve from controversy, Darwin once again turned his attention to a "big" idea. In 1871, after three years of labor, much of it lost to illness, Darwin published his second *magnum opus*: *The Descent of Man*. *Descent* is organized into two major themes, the first of which deals with the origins of humanity and the second with natural selection in general, and sexual selection in particular. In *Descent* Darwin establishes that the human race is no different in kind from the higher animals; it differs only in degree. To Darwin's surprise and immense relief, *Descent* generated relatively little controversy. The battle had largely been won. By the 1870s, in Britain at least, dissent in the scientific establishment, the public, and the Church had largely succumbed to the logic of Darwin's arguments and the sheer weight of his facts.

Having made his point and survived, Darwin turned forever from controversial topics, and not surprisingly, his health improved. In 1872, he published *Expressions of the Emotions in Man and Animals*, a book that drew in part on experiences as commonplace as trips to the zoo and parenthood, but seen through the eyes of the most uncommon of observers. Originally intending the topic as a chapter in *The Descent of Man*, Darwin soon realized that it stood on its own merits and demanded a separate treatise.

For the topics of his final books, he looked no further afield than the familiar grounds of Down House. In a greenhouse adjacent to the gardens, Charles and son Frank raised *drosera*, a species of carnivorous plant, the subject of *Insectivorous Plants*, published in 1875. In 1880, with Frank's assistance, Darwin published the *Power of Movement in Plants*. "It has always pleased me," he recalled in his autobiography, "to exalt plants in the scale of organized beings."[111]

No work of creation, regardless of how seemingly insignificant, escaped his attention or loving admiration. Darwin's final work, published in 1881, focused on the lowly earthworm, whose contributions to the soil and the web of life the great man championed with enthusiasm.

Darwin's mentor and dear friend, Charles Lyell, died in 1875; his elder brother, Erasmus, died in 1881. Afterwards, in a letter to Wallace, Darwin confided, "Life has become very wearisome to me."[112] After suffering a heart attack in early 1882 and recovering sufficiently to tread once again his beloved sand-walk, Charles Darwin collapsed on April 15 and died four days later on April 19, 1882. In an interlude of calm and alertness, Charles whispered to Emma, "I am not in the least afraid of death."[113] Wallace – Darwin's generous and noble competitor, friend, and co-conspirator in the theory of evolution – served as pallbearer.

The British have their faults, but a failure to recognize greatness is not among them. Charles Robert Darwin lies buried in Westminster Abbey among the greatest and noblest of the species. Nearby lies his good friend Charles Lyell, both not far from the resting place of Isaac Newton.

The *Origin of Species* concludes with perhaps the most famous passage of any scientific work:

> It is interesting to contemplate an entangled bank, clothed with many plants of many kinds, with birds singing on the bushes, with various insects flitting about, and with worms crawling through the damp earth, and to reflect that these elaborately constructed forms, so different from each other, and dependent on each other in so complex a manner, have all been produced by laws acting around us. These laws . . . being Growth with Reproduction . . . Inheritance . . . Variability . . . Struggle for Life . . . Natural Selection . . . Divergence of Character and . . . Extinction. Thus, from the war of Nature, from famine and death, the most exalted object which we are capable of conceiving, namely, the production of the higher animals, directly follows. There is grandeur in this view of life, with its several powers, having been originally breathed into a few forms or into one;

and that, whilst this planet has gone cycling on according to the fixed law of gravity, from so simple a beginning endless forms most beautiful and most wonderful have been, and are being, evolved.[114]

Hardly the words of a blasphemer.

Epilogue

In his autobiography, Charles Darwin gratefully acknowledged that "not having to earn his bread" had afforded him the luxury of unencumbered pursuit of science. Even his ill health had a silver lining, by "saving him from the distractions of society and amusement." Surprised by his worldwide influence, Darwin humbly credited his success to four qualities: a love of science, "unbounded patience," "industry in observing," and common sense.[115] As always, his observations, even of himself, were dead on, if incomplete. To those four must surely be added a fifth: uncommon integrity.

The year 2009 marked the bicentennial of Darwin's birth, and simultaneously the sesquicentennial of the publication of *Origin of Species*. Around the world, Darwin was fondly remembered and his contributions to human understanding celebrated in countless magazines, newspapers, and presentations. Still, controversy regarding the theory of evolution rages on, particularly in the United States, where, for decades, polls consistently show nearly half of all Americans agreeing with the statement: "God created human beings pretty much in the present form at one time within the last 10,000 years or so."[116]

At bottom, the controversy boils down to a dichotomy: Is creation a done deal or an unfolding process? If the latter is difficult to swallow, it may help to know that Darwinism was difficult also for Darwin. For the entirety of his post-voyage life, Darwin suffered "Maer's disease," the mysterious malady that was at least partly psychosomatic. His longsuffering is testament both to his hypochondria and to his courage. Few persons are willing to endure what Darwin suffered for the sake of an unpopular idea whose time has come.

Fundamentalism offers the conflicted a stark choice: either/or, meaning versus no-meaning, faith versus science. The courageous alternative is both/and: synthesis. Let one example suffice. Since 2004, when the Clergy Letter Project was founded to counter the misperception that modern science and religious faith are incompatible, 12,000 members of the American clergy have signed letters embracing science in general and evolution in particular. Each February, mainstream congregations in the United States and around the globe gather to participate in Evolution Sunday, a celebration of the life and legacy of Charles Darwin. By February 2009, at presumably the midpoint of the Second Copernican Revolution, those courageous congregations numbered more than 1,000.[117]

Ten

The Time Bomb

The creation is never over. It had a beginning but it has no ending.
Creation is always busy making new scenes, new things, and new
Worlds.

– **Immanuel Kant**

Since its publication in 1859, *Origin of Species* has never gone out of
print – or out of controversy. Fearing that its sequel, *Descent of Man* (1871),
would reignite the passions that *Origin* incited, Darwin predicted in a letter to Wallace, "I shall soon be viewed as the most despicable of men . . .
the most arrogant, odious beast that ever lived."[1]

Fortunately for Darwin, times had changed since Bruno and Galileo,
when the rack and flames regularly dispatched heretics to the nether-
world. Instead, those who espoused heresies were simply humiliated and
satirized in cartoons, such as that which appeared in *The Hornet* in 1871
(Fig. 19).

One hundred fifty years after the publication of *Origin*, Darwin's "dangerous" ideas still incite passions, and skirmishes still flare in the pro-
tracted war between the defenders and detractors of evolution. A case
in point: in 2005, in its ruling in the landmark case *Kitzmiller v. Dover Area
School District*, the U.S. Third District Court castigated proponents of "intelligent design," the modern guise of creationism, for violating separation
of church and state.

The noxious atmosphere that continues to poison the reconciliation of
faith and reason formed just months after the publication of *Origin* in a
heated public debate between Anglican archbishop Samuel Wilberforce
and educator Thomas Huxley, evolution's powerful advocate.

"How extremely stupid of me not to have thought of that," Huxley
lamented upon completing his initial reading of *Origin*. Nevertheless, eager to champion the cause, he wrote Darwin, "I am sharpening my beak
and claws in readiness."[2]

In the most infamous exchange of the debate, Wilberforce turned toward Huxley to ask smugly, "Was it through his grandfather or his grandmother that [Huxley] claimed his descent from a monkey?"[3] Jumping to
his feet, Huxley then verbally dismembered Wilberforce with the rejoin-

Figure 19: Caricature of Charles Darwin published in *The Hornet* in 1871. (UCL Library Services, Special Collections.)

der: "A man has no reason to be ashamed of having an ape for his grand-father. If there were an ancestor whom I should feel shame in recalling it would rather be a man [Wilberforce] . . . who . . . plunges into scientific questions with which he has no real acquaintance, only to obscure them by aimless rhetoric."[4]

The debate then descended into chaos as the audience erupted in laughter. When the dust settled, it was generally conceded that Huxley had routed Wilberforce. But there are no winners when battle lines are drawn and articles of war declared.

Present controversies over the teaching of evolution, genetic modification, embryonic stem-cell research, and cloning are but echoes of the "second blow to human narcissism" levied by the mild-mannered Darwin.

In the din engendered by evolution, its central tenet often goes unheard: "When all the foolish wind and wit that [the theory] raised had blown away, the living world was different because it was seen to be a world in movement."[5] Indeed, the Greek philosopher Heraclitus had unwittingly struck the gong of evolution twenty-five centuries ago when he declared, "There is nothing permanent but change."

The Devil's Chaplain

Darwin did not coin the term *evolution*, nor was he first to propose the notion of variation of species. In *Zoonomia* (1801), Erasmus Darwin, Charles' grandfather, wrote with grandiose flourish:

> [I]n the great length of time since the earth began to exist, perhaps millions of ages before the commencement of the history of mankind would it be too bold to imagine that all warm-blooded animals have arisen from one living filament, which THE GREAT FIRST CAUSE [imbued] with animality, with the power of acquiring new . . . propensities . . . and of delivering down these improvements by generation to its posterity, world without end.[6]

Similarly, in *Philosophie Zoologique* (1809), French naturalist Jean-Baptiste Lamarck had laid out a theory of *transmutation*, by which species transform continuously in the general direction of greater biological complexity. Furthermore, Lamarck audaciously proposed that creation is ongoing, its manifold species driven by some innate tendency toward advancement. Considered heretical on religious grounds, Lamarck's views earned him the vilification of contemporaries. When his theory that acquired traits are heritable was discredited, Lamarck's star fell precipitously, and he died in obscurity and poverty. But still, Lamarck's legacy endures. He coined the term *biology*, contributed substantially to the new science, and earned Darwin's everlasting respect for recognizing the influence of environment upon heredity.[7]

"Nothing is so powerful," wrote French novelist Victor Hugo, "as an idea whose time has come." Such was the case with the theory of evolution, which sprouted simultaneously in disparate locations. As Darwin honed his ideas while visiting the Galapagos Archipelago, so Wallace honed his in the Malay Archipelago. Recalling the night in the Spice Islands in 1858 when he had conceived the notion of *survival of the fittest* while delirious with malarial fever, Wallace wrote: "Then I at once saw, that the ever present variability of all living things would furnish the material from which, by the mere weeding out of those less adapted to the actual conditions, the fittest alone would continue the race. There suddenly flashed upon me the idea of the survival of the fittest."[8]

It was as if Wallace had read Darwin's mind. The resemblance of thought and phraseology was uncanny. In perhaps an even more remarkable passage from Wallace, he not only appropriated Darwin's term *natural selection* but introduced sentience into the evolutionary landscape, foreshadowing, along with Lamarck, Teilhard de Chardin's twentieth-century notion of complexity-consciousness: "Natural selection could only have endowed savage man with a brain a few degrees superior to that of the ape, whereas he actually possesses one very little inferior to that of a philosopher. With our advent there had to come into existence a being in whom that subtle force we term 'mind' became of far more importance than mere bodily structure."[9]

But at the end of the day, it was Darwin who best articulated the theory and who planted the flag of Victorian England upon the highest mountain of observational evidence. To the victor belong the spoils – and the heat.

Scientific theories, like nature herself, are works in progress. Nevertheless, the theory of evolution sits upon an Everest of corroborating evidence. Some evidence Darwin accumulated during his lifetime; much was pieced together thereafter. The theory, like a barstool, rests solidly upon three sturdy legs, each elegant tenet verifiable: (1) variation, (2) struggle for survival, and (3) natural selection.

Variation means simply that not all individuals of a species are identical; nature does not use a cookie cutter. At least two different mechanisms contribute to variation. Variation most commonly arises from the intermixing of the parents' genetic material during sexual reproduction. Occasionally, a new variant arises spontaneously through mutation.

Mutations are small, random, heritable changes that originate in the genetic material and are thereby passed down to progeny. In this recognition, Darwin distinguished himself from Lamarck, who incorrectly believed that acquired traits could also be passed down to future generations. For example, although I may work out at the gym faithfully to build massive biceps, my child will not inherit large biceps because working out does not alter the genetic code.

In addition to being heritable, mutations occur randomly. Some mutations are favorable to the organism, some neutral, and others unfavorable. Rarely is a mutation so detrimental that the organism dies immediately. More commonly, whether a mutation is favorable or unfavorable depends upon the organism's natural environment. Albinism in a hare, for example, might afford camouflage in the arctic but not in a temperate zone.

Finally, most mutations are small, involving a single site in the genomic sequence encoded by the organism's DNA or RNA. Depending upon the site, the mutation may affect a single genetic trait – pigmentation, size, bone density, strength, armament, or mobility, for example – or multiple traits. Mutations may be induced by outside influences – radiation, mutagenic chemicals, or viruses – or they may occur naturally from replication errors during meiosis. Whatever their origins, however, mutations are spontaneous and discontinuous.

Nature is nothing if not profligate. How many cones fall each season from a single pine? How many nuts do the whirls of a pinecone contain? And of this total, how many pine nuts mature into trees? For every apple that flowers into an apple tree, hundreds of others feed the worms and enrich the soil. Why are millions of sperm necessary to fertilize a single egg? Of the roe laid by a sturgeon, what fraction matures to the point of repeating the process, and what fraction ends as caviar for other fish or the lucky human? One cannot study nature without being in awe of her extravagance, embarrassed by her wastefulness, or both. Why such excess? The answer fell to Darwin upon reading Malthus' essay on population. In *Origin*, Darwin explains:

> As many more individuals of each species are born than can possibly survive; and as consequently, there is a frequently recurring struggle for existence, it follows that any being, if it vary however slightly in any manner profitable to itself, under the complex and sometimes varying conditions of life, will have a better chance of surviving, and thus be naturally selected. From the strong principle of inheritance, any selected variety will tend to propagate its new and modified form.[10]

Herein lies the method to nature's madness. By fostering relentless struggle, nature ensures that what follows is in some way better – stronger, wiser, or hardier – than what preceded. This Darwin termed *natural selection*. Given two variants of a species, the odds favor the one best suited to its environment. When times are good, both variants may survive. When times are bad, life hinges on slight differences.

Of natural selection Darwin wrote to Harvard's Asa Gray, "What a trifling difference must often determine which shall survive, and which shall perish!"[11] In the Galapagos Islands, for example, "a fortis [finch] with a beak 11 millimeters long can crack a caltrop [seed]"; one with a beak only 0.5 millimeters shorter "will not even try."[12] Darwin understood all too well that "the smallest grain in the balance, in the long run, must tell on which death shall fall, and which shall survive."[13]

Darwin initially adopted Lamarck's term – *transmutation* – for evolution. The term elicits the mechanism of speciation: transformation via mutation. Thus, transmutation is the endgame of natural selection's trifold process: variation/mutation, anagenesis, and speciation. Over time, small changes accumulate to radically transform a species through anagenesis. When two variants of a species diverge to the point that individuals belonging to differing variations no longer choose to mate, or can produce fertile offspring when they do, then the variations have split into distinct species, and the result is *speciation*.

Descent through modification was Darwin's phrase for repeated speciation over time. His metaphor was the tree of life, pushing upward in time rather than downward, as "descent" might suggest. From the trunk, representing a common ancestor, a branch point signifies speciation, and the branches signify distinct species. New species go separate ways in time, each branching again when the pressures for adaptation accumulate to the point of further speciation. The topmost twigs of the tree correspond to species currently in existence. At times in the past, particularly at crisis points, conditions on Earth have fostered rapid speciation. At such junctures, old branches sprout new ones in profusion. Such events are termed *adaptive radiations* to indicate speciation in all directions, and at these times the tree of life more closely resembles an unruly bush than a tree.

There is a fourth tenet to evolution, implied by the upward direction of the tree. To make an omelet, it is not enough to chop onions, tomatoes, and green pepper into raw egg. One must then cook the mixture. Variation, struggle for survival, and natural selection form the raw ingredients of evolution. To cook the ingredients, one must add time.

Deep Time

At the outset of *Origin of Species*, Darwin cautioned, "He who . . . does not admit how vast have been the past periods of time, may at once close this volume."[14] Above all, Darwinism is a *"time* bomb."[15] It explodes the sweep of time as extravagantly as Copernicanism exploded the confines of space. Evolution's Earth is a Methuselah, not the whippersnapper of creationist rhetoric.

Deep time defies the comprehension of humans, whose years are numbered at threescore and ten. Geologists and evolutionary biologists invoke descriptive terminology to fathom the depths of time: *eons, eras, periods,* and *epochs.* These subdivide geologic time, as years, months, weeks, and days divide human time. Loosely, an eon is on the order of a billion years, eras encompass several hundred million years, periods span several tens of millions of years, and epochs tick off intervals ranging from ten to twenty million years. A thousand years is a drop in a bucket.

Evolution's recipe, with its long baking time, troubled Darwin, ever cognizant of any weaknesses of his theory. For evolution, whose processes are seemingly haphazard, to work its magic, Darwin intuited that hundreds of millions or billions of years are necessary. All nineteenth-century estimates of the earth's age fell resoundingly short of a billion years.

The French naturalist and mathematician Comte de Buffon (1707-1788) was first to estimate the age of the earth scientifically, by appealing to the new science of thermodynamics. Heating iron spheres to white heat, he recorded their cooling times. By comparing the cooling times of successively larger spheres, Buffon crudely extrapolated the earth's age, which he reckoned at 75,000 to 168,000 years. Although a mere eye blink relative to current values, Buffon's estimate raised eyebrows at the time by casting aspersions on the biblically based value of 6,000 years.[16]

In the nineteenth century, William Thomson, later knighted Lord Kelvin, estimated the lifespan of the sun, thereby providing an indirect bound for the age of the earth. A child prodigy and University of Glasgow professor by the age of twenty-two, Kelvin was a force to be reckoned with.[17] Aware that normal chemical processes could not fuel stars for long, he presumed that the sun derives its energy from gravitational collapse and arrived at a figure of 500 million years for its lifetime. Although orders of magnitude larger than Buffon's value, Kelvin's value still fell appreciably short of modern estimates.

Kelvin's estimate was a blow to Darwin, who wrote Lyell in 1868, "I take the sun much to heart." Three years later, in a letter to Wallace, Darwin remained troubled by the discrepancy between the thermodynamics of Kelvin and the deep time necessary for evolution. "I have not as yet been able to digest the fundamental notion of the shortened age of the sun and earth."[18] Kelvin and Darwin could not both be right. The safe bet was on Kelvin, but even towering geniuses can err.

Darwin died before the discovery of radioactivity by Bequerel and Röntgen, which led to accurate radiochronological determination of the earth's age. Both ahead of their time, Buffon and Kelvin remained ignorant of nuclear processes. Buffon's estimate missed the mark because he failed to account for a heat source within the earth herself, namely the spontaneous fission of radioactive elements. Similarly, Kelvin did not know of thermonuclear fusion, the energetic process that keeps the stars burning for eons. Detonation of the first H-bomb in 1952 laid bare the awesome mechanism of stellar fusion and revealed Kelvin's error. The price extracted for such knowledge, however, was high. Science had, as urged by Francis Bacon, tortured nature of her secrets, and in so doing had cast a mushroom cloud over the tree of life.

Despite unresolved uncertainty concerning the earth's age, Darwin presented voluminous evidence in support of evolution, which he organized into four principal categories: biogeography, paleontology, embryology, and morphology. Biogeography, in particular, regards the spatial distribution of species over geological time. For example, modern biogeographers, informed by the relatively new geological science of plate tectonics, use continental drift to explain the existence of closely allied species on different continents. However, it was a curious fact of local biogeography that intrigued both Darwin and Wallace, here related by Wallace: "Places not more than fifty or a hundred miles apart often have species of insects and birds at the one, which are not found at the other."[19] Similarly, in the Galapagos Islands, Darwin had noticed closely allied species confined to particular islands of the archipelago. Indeed, Darwin's "dangerous" idea was to recognize the possibility that the separate species radiated from a common ancestor, each species adapted advantageously to its specific locale.

Whereas biogeography deals with the living, paleontology deals with the dead. "The vertical column of geologic strata, laid down by sedimentary processes over the eons, lightly peppered with fossils, represents a tangible record showing which species lived when."[20] Paleontology is to time what biogeography is to space; it reveals that closely allied species are always clustered in close proximity in time. One never finds a wooly mammoth, for example, in strata strewn with the fossils of dinosaurs, for the simple reason that the two species occupied different geologic periods. Taken together, biogeography and paleontology reveal the spatiotemporal distribution of species from which the tree of life can be reconstructed.

Embryology offers independent corroboration of lineages established by other means, say, through paleontology. Embryology is that branch of biology that examines the stages of development that embryos pass through before birth or hatching, stages that tend to echo adult organisms at earlier stages of evolutionary history. Why, for example, are moths, flies, and beetles, which are so different as adults, so similar in their larval stages? More strikingly, who would suspect that barnacles are crustaceans, relatives of crabs and lobsters? And yet, the resemblance of their respective larvae is uncanny. More to the point, why does the human embryo pass through a stage like that of a reptile? Astutely, Darwin reasoned that "[t]he embryo is the animal in its less modified state" and that state "reveals the structure of its progenitor."[21]

Taxonomy is the science of the classification of biota. Early taxonomists relied almost exclusively on morphology, the study of anatomical shape and design. For example, all animals with backbones are classified as *vertebrates* and all flowering plants as *angiosperms*. The Swedish naturalist Carolus Linnaeus is generally regarded as the father of modern taxonomy, which dates to 1735. It was frequently said, "God created, but Linnaeus organized." Modern classifications based on the sequencing of DNA confirm just how good the early taxonomists were, correct in excess of 95 percent of the time. Thanks to the keen observations of taxonomists of his day, Darwin's original tree of life, constructed primarily on the basis of morphology, remains largely intact.

Two types of compelling morphological evidence for evolution are *vestigial structures* and *homologues*. Consider, for example, New Zealand, with its high concentration of flightless birds. Flight, an effective defense mechanism that birds have retained from their distant ancestor *Archaeopteryx*, a long-extinct feathered reptile, is energetically costly. Consequently, birds consume daily a sizable fraction of their body weight in food. Prior to the arrival of humans, New Zealand boasted a unique distinction: it had no mammalian inhabitants. Without predators, New Zealand's avian species, immigrants from nearby Australia, had no compulsion to fly and were rendered flightless by natural selection. In particular, the kiwi, New Zealand's national bird, has but vestiges of wings or tail.

Homologues, in contrast, are anatomical structures that manifest in different creatures in "every variety of form and function."[22] While the wings of bats and birds, the fins of whales, the paws of bears, and the hands of humans appear dissimilar, each is comprised of a five-digit skeletal structure that differs primarily in the length, weight, and strength assigned to each of the digits. One structure fits all, and presumably all derived from a common ancestor.

Circumstantial Evidence

Of *Origin of Species*, a Cambridge classmate wrote to Darwin, "The first perusal staggered me, the second convinced me, and the oftener I read it the more convinced I am of the general soundness of your theory."[23] But as critics are fond of noting, all the evidence was circumstantial. Mind you, circumstantial evidence can be damning. Many a murderer has swung from the gallows on circumstantial evidence. Yet creationists are quick to seize upon weaknesses, real or imagined. The contemporary defender of creationism Phillip Johnson, for example, asserts in *Darwin on Trial* (1991), "The prevailing assumption in evolutionary science seems to be that speculative possibilities, without experimental confirmation, are all that is really necessary."[24]

Johnson demands direct, real-time evidence of evolution, which Darwin did not provide. Why not? Until quite recently, real-time evidence was deemed impossible. Darwin reasonably assumed that the pace of evolution is too slow for observational evidence within the short span of a human lifetime. In *Origin* he wrote, "We see nothing of these slow changes in progress, until the hand of time has marked the lapse of ages, and then so imperfect is our view into long-past geological ages, that we see only that the forms of life are now different from what they formerly were."[25]

Consider the classic example of "fast" evolution: the horse. The fossil record suggests that a "mere" 1.5 million years elapsed between its archaic form and the modern horse.[26] If evolutionary processes require countless millennia to work their magic, then the demand for direct evidence of evolution during a human lifetime is a disingenuous ploy.

In 1949, evolutionist J. B. S. Haldane coined a new scientific unit to honor evolution's progenitor. A universal unit of biological change, the *darwin* represents a one percent change in some measurable characteristic – say, the length of a finch's beak or a horse's femur – per million years. Fossil evidence suggests that evolution proceeds at an *average* rate of about 0.1 darwins; that is, about 1/10 of one percent per million years. Darwin's primary source of data, the fossil record, suggests that evolution proceeds in tandem with geology, on a glacially slow time scale.

Because Darwin could not produce direct evidence of evolution, he found the best available analogy for its key mechanism, natural selection. *Origin* opens with a discussion of selective breeding, that is, artificial selection. Not simply an innocent bystander, Darwin built a coop at Down House and took up the breeding of fancy pigeons.

Everyone has some familiarity with selective breeding. To see the impressive results of artificial selection, one need travel no farther than the poultry or rabbit exhibits at the county fair. Or consider the domestic breeding of dogs. In a few short centuries, artificial selection has produced from one species variants as diverse as bulldogs, greyhounds, chihuahuas, and St. Bernards. Artificial selection is natural selection on steroids. And it can be used to modify not only the physical characteristics of the creature but its desires as well, as any aficionado of sheep dogs will appreciate.

Like time-lapse photography of growing flowers, selective breeding compresses the scale of evolutionary time. Artificial selection yields biological change rates measured in thousands of darwins. By contrast, until recently, a natural selection rate of one darwin would have been considered exceptional.[27] Even Huxley conceded, "Mr. Darwin does not so much prove that natural selection does occur, as that it must occur."[28] But in the same breath he added, "No other sort of demonstration is attainable."

And so the matter rested until the early 1970s: creationists demanding the impossible and evolutionists unable to deliver. And then the extraordinary happened: science caught evolution red-handed. The Grants changed the face of evolutionary biology.

The Smoking Gun

In 1973, Peter and Rosemary Grant, evolutionary biologists from Princeton University, went to live in one of the few places on Earth where one just might observe natural selection in real time: the Galapagos Islands. For a miserable summer, they settled on Daphne Major at the center of the Archipelago, a volcanic rock so inhospitable that few species survive there: barnacles, cactus, weeds, scorpions, and finches. Where life is harsh, biology is simple. And where biology is simple, one has a small chance of sorting out the interplay of forces.

For the next two decades, the Grants or their graduate students returned to the same hellish outpost summer after summer, meticulously documenting everything they could learn about Darwin's finches and their food sources. On the off-season back in Princeton, they would crunch the numbers to spot trends. Their extraordinary story is told in Jonathan Weiner's Pulitzer Prize-winning *The Beak of the Finch* (1995), a must-read for any evolution enthusiast or detractor.

The Grant's first four seasons were routine. Nothing terribly noteworthy transpired, giving them ample opportunity to whet their observational skills. The summer of 1977, however, was anything but typical. Daphne Major experienced near-total drought. When only 24 millimeters of rain fell in the entire year, the finches' seed supply collapsed, and 85 percent of all *fortis* finches perished.[29]

When the Grants analyzed the survivors, they were astounded. Surviving *fortis* were 5 to 6 percent larger on average than the birds that died. More to the point, the survivors sported beaks 0.39 millimeters bigger on average, a change just barely recognizable to the human eye but more than sufficient to give birds with bigger beaks a selective advantage in cracking the hardest seeds. "It was the most intense episode of natural selection ever documented in nature."[30] In terms of Haldane's unit of measure, the average rate of change in beak length for the drought year was 25,000 darwins. But that was only the start of the story. In 1983, record rainfall from an El Niño event transformed Daphne Major from desert into jungle. With food abundant, the birds bred like crazy. The Grants' data again revealed huge one-year rates of change, this time 6,000 darwins, but in the opposite direction of the changes wrought by the drought. Tracking the climate, selection pressure had reversed course, favoring smaller finches with smaller beaks.

Why had others never before observed such phenomenal change rates? "The Grants [were] the first scientists equipped with enough patience, stubbornness, ground support and sea support, enough computer power, airplane power, and staying power to watch the process actually happen."[31]

Biologists had missed the rapid action for two reasons. First, the earth's climate is relatively stable; yearly fluctuations tend to average out over time. Short-term rates of 25,000 darwins average over deep time to minuscule trends of 1 darwin or less. Second, the fossil record – where each sedimentary layer corresponds to a different era – had long been the primary source of data about evolution. Imagine filming the progress of evolution by time-lapse photography, with just one frame every million years. Peering back into the fossil record is like this. To catch evolution in the act, one needs to shoot a frame at least every generation, or better yet, every mating cycle. Until recently no one had looked closely enough.

Now that the Grants and others have reenergized evolutionary biology, the hand of evolution is visible everywhere: in the speciation of sticklebacks in the lakes of British Columbia, in the SOS response of bacteria, in human cancer cells, and in the unpredictable variations of the swine flu and HIV viruses.

For an organism to see, it needs more than eyes. It needs the neuronal connections that extract meaning from sensory data. By analogy, evolutionary biologists have just acquired those connections.

Not Without Peril

If you really want to witness evolution in action, find a species with a short reproductive cycle. Your favorite bacterium will do; garden-variety bacteria reproduce every twenty minutes. Grow a colony in a Petri dish. Then subject the colony to major stress: extreme heat or cold, or an antibiotic. A funny thing happens. The colony mutates wildly in all genetic directions. The randomness and number of mutations increase the odds that a few bacteria will resist the stressor. And from the hardy few springs a new colony, each member of which carries the adaptation. The entire process – called the SOS response – plays out before one's eyes in a matter of days.

The SOS response of bacteria would have delighted Darwin, but it keeps epidemiologists awake at night. In 1942, when mass-produced penicillin first started saving lives, it was a miracle drug, particularly effective against sometimes-fatal staphylococcus infections. But within four years of the commercial introduction of penicillin, hospitals reported penicillin-resistant strains of staphylococcus. Today, epidemiologists fret over the very real possibility of superbugs for which no drug is effective because "antibiotics exert a powerful evolutionary force driving infectious bacteria to evolve powerful defenses."[32] Among present-day superbugs are strains of *Staphylococcus aureus* (MRSA) resistant to an entire spectrum of powerful antibiotics. Drug companies are hard-pressed to stay a step ahead of bacteria, which have had at least three billion years of practice at evolution.

Humans ignore evolution only at their peril. Consider, for example, the year 1940, when farming in the United States changed dramatically with the introduction of synthetic pesticides. Principal among these new wonder chemicals was dichlorodiphenyltrichloroethane (DDT), first used widely against *Heliothis virescens*, a common moth whose larvae are fond of the cotton boll. DDT was amazingly effective. It killed everything in its path: insects good and bad, and many of the birds that ate the insects. It killed the pesky moths, too, but not quite all of them. The survivors were DDT resistant. So farmers upped the dosage, creating a vicious cycle that produced more pesticide-resistant moths and ever-higher concentrations of pesticide. The end result? Cotton crop losses escalated from 7 percent before the advent of pesticides to 13 percent after.[33] Meanwhile, cotton growers unwittingly poisoned the soil with a toxic brew of chemicals.

In 1962, Rachel Carson's *Silent Spring* sounded an environmental clarion call heard round the world. It documented the unintended ravages of pesticides and led ultimately to the ban of DDT. In particular, Carson exposed the noxious affinity of DDT for the reproductive systems of animals, particularly birds and mammals. When contaminated by DDT, birds characteristically lay thin-shelled, vulnerable eggs. By the time it was banned in 1972, DDT had brought America's symbol, the bald eagle, to the feathered edge of extinction. Concurrently, with the widespread introduction of pesticides, the incidence of cancers of the human reproductive system began a telltale increase.

What DDT was to the *Heliothis*, chemotherapy is to cancer. It kills most of the offending cells, but those that survive pass on a type of immunity. Too often, the cancer returns, resistant to further chemo. Evolution also complicates the fight against diseases such as AIDS and swine flu. The problem? Viruses mutate faster than the immune systems of their hosts can adapt. Faced with a rapidly moving target, epidemiologists are often stymied in developing effective vaccinations or antiviral medications.

There was a time when individual belief regarding evolution was just that: a private matter of faith. That time is no more. Denial is peril for all, not just for those in denial.

Of Mice and Men

Darwin got a few things wrong and was ignorant of others. A contemporary of the reclusive Gregor Mendel, Darwin had no knowledge of the mechanism of heredity that Mendel, an Augustinian priest, had worked out in a monastery garden.

Darwin was wrong too about the pace of evolution, but then every evolutionist has been wrong until quite recently. Evolution is not the plodding process it has appeared to be. As the Grants have documented, biological species are amazingly responsive to selection pressures. Species are biological PT boats, not battleships. They can turn on a dime, given the right conditions.

What is astounding, however, is how much Darwin got right. Today the theory of evolution is corroborated by evidence Darwin could not have dreamed of, from fields as diverse as population genetics,[34] genomics,[35] and plate tectonics, to name just three.

Consider genomics. Darwin's assertion that men and apes descended from a common ancestor did not sit well in his day. Yet even Darwin could not have anticipated how closely the species are related. Human beings and mice, our more distant mammalian ancestors, share 85 percent of the same DNA. It's not a guess. Both genomes have been completely sequenced. Men and apes share 98 percent. There is something wholesomely humbling about a shared legacy. In his notebooks Darwin mused, "Man in his arrogance thinks himself a great work, worthy the interposition of a deity." This can and has led to a dangerous type of hubris that the earth belongs to us alone. Darwin concluded his entry, "More humble and I believe truer to consider him created from animals."[36]

Constructing the chronology of life from a myriad of dissociated facts is akin to assembling a jigsaw puzzle with tens of thousands of pieces. The trouble is, most pieces have gone missing. Half the problem then is to locate the stray pieces, but that alone is insufficient. The pieces must then be inserted at the correct place, and each must fit cleanly against adjacent pieces.

By the onset of the twenty-first century, a seamless picture had begun to emerge. We can now tell the story of human origins through the language of science. Contributing to this great endeavor are numerous branches of science and their subspecialties: archaeology, paleontology, climatology, oceanography, cosmology, population biology, radiogeochronology, genetics, genomics, taxonomy, particle physics, embryology, biochemistry, and plate tectonics. The net effect of the first two Copernican revolutions has been the emergence of a scientific creation story that rivals in awe and enchantment the creation myth of any culture.

The Story of Planet Earth

A little less than 14 billion years ago, the universe burst into existence from a "primordial atom," unrolling time and space and spewing forth into spacetime the seeds of all that we know and all whom we love. The cataclysm of cataclysms, the Big Bang detonated so violently that radio telescopes still detect its faint afterglow. Expanding rapidly, the fireball cooled precipitously from its initial temperature in excess of 1.5 trillion degrees Kelvin, leaving an "undifferentiated soup" of particles.[37]

In the beginning, things happened quickly. Roughly ten seconds after the Bang, there was light. Photons dominated the universe. Fourteen seconds after the Bang, at three billion degrees Kelvin, the first stable atomic nuclei formed. At three minutes, as temperature plummeted through the one-billion-degree threshold – seventy times hotter than the center of our sun – helium 4 nuclei could and did form in great abundance. Still the universe remained too chaotic for the formation of heavier nuclei or stable atoms.

At 700,000 years, the universe had cooled sufficiently for free protons to grab electrons to form the lightest and most abundant element: hydrogen. At this point, helium contributed a quarter of the total mass of the entire universe; the rest, but for traces of other constituents, was hydrogen. At this juncture, the celestial pantry held all the raw ingredients for stars.

And there the universe would have rested, gaseous and undifferentiated, had it not been for its imperfections – minuscule irregularities in the distribution of matter. Gravitational attraction has a slight advantage where matter is most dense. Consequently, gravitational "hot spots" greedily attracted yet more matter in a runaway process that concentrated hydrogen and helium. At a certain critical mass, under the relentless heat and pressure imposed by gravitational compression, the thermonuclear furnace of hydrogen fusion ignited. Wherever that happened in the early universe, a star was born. Indeed billions of stars were born and bound by gravity into galaxies, Kant's island universes, separated by vast distances of interstellar space. One such galaxy was the Milky Way.

Early stars cycled through life and death. Some died with a whimper, others with a bang. Billions of years ago, near the edge of the Milky Way, a massive star succumbed, collapsing into itself and rebounding violently as a supernova. Supernovae are smelters for all heavy elements. In one particular smelter formed the ingredients for planet Earth.

Five billion years ago, a nebula of gaseous hydrogen, peppered with the debris of a supernova, contracted into a spinning disk. At its center, light burst forth from a newly ignited star, the sun. Dust and debris orbiting the sun accreted into protoplanets: *planetesimals*. In a cosmic demolition derby, countless planetesimals collided recklessly, building up planets and their moons. Some 4.6 billion years ago, a renegade planetesimal struck and cleaved the nascent Earth, creating her moon.[38]

The inner planets of the solar system, composed of heavier elements, formed first. As the earth coalesced, so did her sisters and brothers: Mercury, Venus, and Mars. Less dense, the gaseous giants – Jupiter, Saturn, Uranus, and Neptune – accreted more slowly. Detritus subjected the planets and their moons to relentless bombardment, slowly increasing their masses. Our moon's pockmarks attest to the pummeling she and the young earth endured. Meteorites, primarily of iron, nickel, and carbon, augmented the earth's mass, diversifying her chemical portfolio. With sufficient mass, the earth attracted gaseous elements: hydrogen and helium. Her early atmosphere, however, was not to last, largely stripped away by the solar wind.

Four billion years into the past, the withering bombardment of detritus subsided, but Earth, a bubbling stew, was far from calm. As crust formed on her molten slurry, heavier elements, chiefly iron, sank to the earth's center. To this day, her iron core remains molten and by flowing generates a magnetic field, shielding the earth's atmosphere and her inhabitants from the solar wind. Volcanoes erupted incessantly, releasing water vapor and a witch's brew of gaseous compounds from the earth's interior: ammonia, hydrogen sulfide, carbon dioxide, and methane. Largely devoid of gaseous oxygen, the earth's early atmosphere was noxious and seemingly inhospitable to life.

Around 3.8 billion years ago, the earth's crust solidified, bringing with it relative geologic stability. The oldest terrestrial rocks, zircons from Western Australia, date from this period.

Simultaneously with geologic stability, atmospheric water vapor condensed to fill the oceans. Somewhere in those oceans, from a primordial soup rich in organic compounds, an event so singular occurred that the solar system, and possibly the universe, took on a whole new dimension. Life emerged from the matrix of matter. The first life on Earth was rudimentary: one-celled bacteria. Four billion years later, bacteria remain the dominant life form, contributing more biomass to the earth than plants and animals combined. *Prokaryotes*, bacteria lack a nucleus. Simplicity has its perks. Having no preprogrammed life span, bacteria can live indefinitely in favorable conditions.

Anaerobic extremophiles, the first microbial bacteria, thrived near undersea volcanic vents. Masters of metabolism by fermentation, *thermophilic* bacteria multiplied rapidly, quickly diminishing their food supply. Bacterial species then diverged rapidly – perhaps by SOS response – filling metabolic niches, each associated with specialized chemistry. One daring innovation cleaved a branch in the tree of life. Blue-green cyanobacteria – the precursors of plants – developed an ability to extract energy from sunlight through photosynthesis. Feeding upon hydrogen sulfide and carbon dioxide, early cyanobacteria expelled sulfur. Necessity is the mother of invention. When hydrogen sulfides became scarce, the adaptive microbes evolved to extract hydrogen from water, ejecting molecular oxygen as a byproduct. Thus, we have the ingenuity of bacteria to thank for our oxygen-rich atmosphere.

There is strength in numbers. Some microbes collected in multicellular communities. Thriving in filtered sunlight near the sea's surface, vast mats of blue-green bacteria expelled prodigious quantities of oxygen. New types of microbes abounded, each metabolically specialized and each excreting a different kind of mineral that would later give geologic character to the earth's crust. Yet all of this happened only in the sea. The land remained too inhospitable even for the most highly adaptive organisms. Eminent biologist E. O. Wilson paints an inhospitable portrait of a geologically young earth:

> Short-length ultraviolet radiation . . . beat down on the dry basaltic rocks. It assaulted organisms venturing there out of the sea, shutting down their enzymatic synthesis, opening their membranes to ambient poisons, and rupturing their cells. But in the water, safe from the lethal rays, microscopic organisms swarmed, . . . simple organisms . . . devoid of nuclear membranes, mitochondria, chloroplasts, and the other organelles that give structural complexity to the cells of higher plants and animals.[39]

Between 3 billion and 1.5 billion years ago, microbes experimented with symbiosis. For example, a fast-swimming spirochete paired with a fermenting bacterium in a partnership advantageous to both. In particular, the bacterium gained fast locomotion to its food sources. Somewhere along the way, symbiosis went magically awry. As bacterial forms proliferated, some forms preyed upon others. Occasionally, the predator penetrated the host, but the host tolerated the invader and adapted to its presence, transforming the erstwhile invader into an *organelle* (a cellular subunit that performs a specialized task). Among the first organelles to appear were *mitochondria*, cellular power plants. Endosymbiosis – a symbiotic relationship in which one organism resides within another – may possibly have given rise to the first cells with distinct nuclei: *eukaryotes*.

A tad more than one billion years ago, oxygen concentrations reached current levels, precipitating the demise of 90 percent of anaerobic bacteria. The earth rusted as reduced iron oxidized. Traces of rust remain in the hues of red observed in exposed banded-iron formations. Among the most spectacular are the Mars-like red-rock hills and layered mesas of Sedona, Arizona, a New Age mecca.

In the upper reaches of the earth's oxygen-rich atmosphere, radiation converted some diatomic oxygen to ozone, oxygen's triatomic allotrope. Ozone's presence, as nature's sunscreen, was fortuitous for life. With protection from extreme ultraviolet radiation by an atmospheric ozone layer, bacteria moved inland to occupy ponds, rivers, and soil. Life had its first tentative foothold on land.

With the emergence of eukaryotes came another marvelous invention: sex. By mixing and matching genetic material, sexual reproduction afforded survival advantages during periods of intense stress by providing more variation for natural selection to act upon. Following the advent of sex, there was a massive radiation of unicellular species.

One billion years in the past, the earth was mature but restless. Global rifting raised great mountain ranges separated by vast valleys and laced with cascading rivers. Life experimented in a plethora of forms. Still, life remained largely unicellular. Biology and geology were closely coupled. The earth's atmosphere, oceans, and soil yielded the raw ingredients of life. In turn, life secreted inorganic compounds, minerals of sixty or more types, raw ingredients for the earth's ever-changing geological layers.

Eight hundred million years ago, a swing in the climate plunged the earth into a deep freeze. The resulting "snowball earth" reflected most of the sun's radiant energy. Today's climate simulations reveal two stable equilibria. One stable equilibrium is our familiar climate; the other is the snowball earth of antiquity.

At the end of the deep freeze, life took a great leap. Multicellular arrangements became permanent, and life exploded in a variety of soft-bodied biota known as *Ediacaria*. These needed no shells as they had no predators, except time itself, which imprinted *Ediacaria* in the first fossil records. Among these soft-bodied marine animals were tubeworms, jellies, and coral. Later came *foraminifera*, aquatic organisms protected by shells, forerunners of today's mollusks.

Roughly 545 million years ago, a new age dawned on the mature planet Earth. Humans later marked this the Paleozoic Era. After taking her time for eons, nature seemed in a hurry. Geological periods clicked by like the pages of a thriller: Cambrian, Ordovician, Silurian, Devonian, Carboniferous, Permian, Triassic, Jurassic, Cretaceous, and Tertiary, each bursting forth in a cornucopia of new flora, fauna, and fungi.

The Paleozoic Era opened with the Cambrian explosion, the most massive adaptive radiation of species ever to grace the earth. In a wave of infusion, algae ventured onto land for the first time, their cells coated in wax-like substances for protection from dehydration. Millipedes appeared upon land also, the first true terrestrial animals. In the sea, nature ran wild. The first primitive chordata emerged with cartilaginous notochords, precursors of the spinal chord. The large phylum would later include fish, the first vertebrates, and a bit later sharks.

On land, mosses and ferns abounded; they abound to this day. Mosses anchored themselves to the earth with threadlike *rhizoids*, the prototypes of roots, yet incapable of transporting water. The first to sink deep roots and to drink from the soil were ferns. Along with plants and animals, fungi, which digest first, then absorb nutrients, contributed to the global metabolism. Of the three kingdoms, the plant kingdom contained the most complex organisms.

And then the unthinkable happened: 440 million years ago a great cataclysm extinguished 50 percent of all species. The cause remains a mystery. Prior to the spasm, the number of biological families had progressed steadily upwards. Many were lost. Nature needed 25 million years to recover, but she did so with splendor. From the chaos after the spasm emerged new niches and new inventions. Fungi and algae formed the win-win lichen partnership, whereby the plant, through photosynthesis, provided sugars for the fungi, and the fungi mined mineral nutrients for the plant. Lichens, which can live for thousands of years, contributed to speeding up geological processes on Earth by more quickly weathering rock with their acidic wastes. By their action, for example, the 375-million-year-old Appalachian mountains, once majestic, have been wearied and worn by time.

Sharks filled the sea, and on land the first primitive insects evolved. Dragonflies took to the air. By the end of the Devonian period, lobe-finned fishes had transformed into amphibians, with new capabilities to breath air and walk on four legs. Amphibians, however, did not make a full transition to terra firma, "their tadpole progeny keeping one evolutionary foot in the water."[40] The Devonian period closed in a second mass extinction, in which 50 percent of all species vanished forever.

The Carboniferous Period was oppressive, the earth's surface swampy and its atmosphere sultry. In greenhouse-like conditions, plants proliferated, aided by wind-born spores, and attained gargantuan size. Dead and decaying organic matter accumulated in carbon sinks, trapping the fossil fuels that humans would burn in great quantities far in the future. Thin-skinned amphibians, dependent upon pond or marsh for moisture, morphed into hardier reptiles, their liberation from water completed by a stunning reproductive innovation: internal fertilization. Unlike their amphibian ancestors, reptiles lay encased, protected eggs. The first age of reptiles, the Permian, gave rise to the precursors of snakes, lizards, and turtles. Some reptiles developed membranes and learned to glide. An unusual convergence of tectonic plates produced one massive continent: Pangea. On that continent, beetles arose and flying insects proliferated. And then the Permian Period ended with an event so devastating that all life seemed in jeopardy: 95 percent of all species and 50 percent of all families vanished. Nature rebounded, as always, with a flourish. Life begat the first mammals, rodent-size insect eaters. Then close on the heels of the Permian/Triassic extinction followed another, marking the Triassic/Jurassic boundary. Rebuilding the range of diversity lost in back-to-back spasms cost nature 100 million years. Climate change, asteroid strike, or massive volcanic eruption – the cause of so much devastation has yet to reveal its hand.

The Jurassic period, which began 210 million years ago, was the age of large reptiles. Dinosaurs, "terrible lizards," roamed the earth and dominated its life forms. Some, like the gigantic and herbivorous brontosaurus, consumed tons of vegetation daily, their digestion aided by cellulose fermenting bacteria resident in their stomachs. Others, like the carnivorous *Tyrannosaurus rex*, able to run upright on two legs and sporting teeth ten inches long for tearing the flesh of hapless victims, preyed upon the herbivores. Birds appeared on land, flight being a clever adaptation for escape from so many gnashing teeth. Amphibians thrived too. Salamanders and frogs, too inconspicuous for dinosaurs to notice or too small to eat, found niches.

Nature flowered during the Cretaceous Period, some 140 million years ago. The air filled with new scents, and the landscape blazed with new colors. Insects found both color and scent attracting and formed symbiotic relationships with plants. In return for the reward of nectar, the insects served as pollinators. Flowering plants in turn bore fruit, and fruit supplied a new source of energy for birds and mammals. The earth seemed to awaken.

A horrific event 65 million years ago – the so-called "K-T boundary" – violently punctuated the Cretaceous and Tertiary Periods. An asteroid ten kilometers in diameter plunged to Earth at the hypervelocity of 72,000 kilometers per hour. The impact carved out the Yucatan Peninsula of Mexico. The earth rang like a bell, great tsunamis washed all shores, and dust enveloped the planet. Acid rains washed the earth for years.[41] The devastation was staggering: dinosaurs disappeared from the face of the earth, as did all animals over 55 pounds in weight. Nearly 90 percent of all marine species vanished as well.

Nature reawakened to a new dawn and a new age, the age of mammals. Mammals radiated in all directions, filling niches vacated by the departed dinosaurs. Some were predators; some were prey. Some were insect-eaters, and some were herbivorous. The flowering, fruit-bearing world was an Eden for the latter, fruit being a convenient source of energy that did not fight back.

High on the tree of life, new branches formed, among them the primates. Mammals returned to the sea in the form of archaic whales. Grasslands ran riot, aided by the fertilization of grazers, including deer and horses. As the earth cooled, her seasons grew more distinct and stabilized. Her crust gave way to tectonic pressures, thrusting upward the Himalayas and Andes. Old oceanic circulation patterns changed, stirring up nutrients. Seals and sea lions flourished as mammals mastered the sea to avail themselves of its new bounty.

Ten million years ago, orangutans branched from the great apes. From the great ape line, another branch, *Australopithecus afarensis*, sprouted in eastern Africa some four million years into the murky past. The first bipedal hominid, *Australopithecus afarensis*, is the likely ancestor of any-

one who will read this. One such hominid, whose remains surfaced in
Ethiopia in 1974, has come to be known as "Lucy."

Some 300,000 years ago, from Lucy's lineage, arose *Homo neanderthalensis*, the first species to bury its dead. Between 200,000 and 100,000 years
ago, Lucy's line gave rise also to *Homo sapiens*, a hominid with a brain three
times the size of Lucy's. Originating in Africa, *Homo sapiens* scattered far
and wide in waves of migrations, arriving in the Middle East 70,000 years
ago, in Australia 50,000 years ago, in Europe and Asia 40,000 years ago,
and in North America 20,000 years ago.[42]

Six thousand years ago, human beings developed the capacity to
record language in the written word. Six centuries ago, in 1439, German
goldsmith Johannes Gutenberg invented movable type, allowing the mass
production of books and wide dissemination of knowledge. Three centuries ago, in 1687, Isaac Newton published one of the most important
books in human history, *Principia Mathematica*, which ushered in the Age
of Enlightenment. Forty years ago, in 1969, a massive rocket designed on
the basis of Newton's laws thundered into space carrying three *Homo sapiens* from planet earth and bound for our nearest celestial neighbor. Eight
days, 3 hours, 18 minutes, and 35 seconds later, the three earthlings returned bearing rocks from another world dated at 4.2 billion years of age.

Not Quite a Random Walk

Evolution is untidy. To borrow E. O. Wilson's turn of phrase, it proceeds
in fits and starts by a "succession of dynasties."[43] New starts typically
arise following massive cataclysms, when vast numbers of species are extinguished and whole dynasties destroyed save for a few stragglers. Today's crocodiles and turtles, observes Wilson, are the scant few survivors
of yesterday's reptilian dynasty. Among flying insects, flies, wasps, moths,
and butterflies are all new arrivals. Dragonflies, on the other hand, have
remained aloft for nearly 200 million years. Traumatic as they seem, extinctions provide opportunities for biological creativity. In the wake of
extinction there is "rapid experimentation followed by slow standardization."[44] When all is well, what need is there for change?

Evolution is untidy because it is largely a statistical affair. Imagine
a tetrahedral die whose four faces are labeled with the principal directions of the compass: N, S, E, and W. Place a pencil dot near the center
of a sheet of graph paper at the intersection of vertical and horizontal grid
lines. With each roll of the die, move one unit on the graph paper in the direction indicated, connecting the dots as you go. Repeat this process over
and over. Where will the pencil rest after a very large number of rolls? If
the die is fair, the pencil will likely have wandered a considerable distance
from where it started, albeit along a haphazard path with no preferred direction. Were the experiment to be repeated, the pencil point would most
likely wander off haphazardly in another direction. Indeed, the Brownian

motion of dust particles suspended on water represents a commonplace example of a "fair" random walk.

On the other hand, if the die is loaded, with, say, a slightly greater probability of rolling W, then the pencil's path should wander, albeit erratically, in a westerly direction.

Evolution is a random walk over deep time, but with a bias in the form of feedback. Feedback enters through natural selection. In any given environment, the organism best adapted is most likely to survive. Natural selection imposes direction on evolution. In the main, the directional bias favors the evolution of higher intelligence because intelligence is the mother lode of adaptation, a fact appreciated by Lamarck, Wallace, and Teilhard de Chardin.

Evolution is also nonlinear in time. Imagine the pages of a thirty-volume encyclopedia to represent the roughly 4.5 billion years of geologic time on planet Earth. The volumes, each of which contains 500 pages, fifty lines to a page, are arranged ten to a shelf on three bookshelves. Suppose that we embark to read the entire encyclopedia from front to back without pause. As we do so, let the turning of each page mark one tick of the geologic clock: the passage of 300,000 years.

We turn page one of volume one with the birth of the solar system, at which point the sun and inner planets are in their infancy. We continue reading and flipping pages until the middle of the seventh volume on the first shelf, when rudimentary, unicellular life emerges from the primordial soup. Despite an amazing profusion of unicellular forms specifically adapted to light, heat, and oceanic and atmospheric chemistry, life remains predominantly unicellular until nearly the end of the third shelf, at volume 25 to be precise. The Cambrian explosion occurs early in volume 27. At the very end of volume 28, the first rodent-like mammals appear. At the end of volume 29 or the very beginning of volume 30, the earth springs forth for the first time in glorious flower and unfamiliar scents. In the second half of volume 30, a great asteroid extinguishes the dinosaurs and all large creatures. On page 467 of volume 30, primates diverge, with orangutans branching from the great apes. Cro-Magnon man, the "type specimen" of the species *Homo sapiens*, appears within the last third of the last page of volume 30. At the beginning of the last three lines of the last page of the last volume, an unknown artist sketches the first known primitive paintings on the walls of a cave near Lascaux, France. Buddha, Jesus, and Mohammed were all born during the last half of the very last line. Newton was born more than midway through the last word. Men step upon the moon in the middle of the final letter of the last page of the last volume.

From a human perspective, evolutionary change is not linear; it is exponential in time. The pace of evolution accelerates at a fantastic clip. Peering backwards at the evolutionary panorama from the vantage point of the steep rise at its current epoch, humans find it tempting to conclude that we are evolution's end point and its ultimate purpose. Wilson, like

Darwin, warns against such hubris: "A complete recovery from each of the five major extinctions required tens of millions of years. In particular, the Ordovician dip needed 25 million years, the Devonian 30 million years, the Permian and Triassic (combined because they were so close together in time) 100 million years, and the Cretaceous 20 million years."[45]

Life on Earth is now in the midst of a sixth great extinction. Unlike all those preceding, which resulted from natural processes or cataclysms, the current extinction stems from the disproportionate impact of a single species: *Homo sapiens*. Through overpopulation, overhunting, overgrazing, and overharvesting; through deforestation and habitat destruction; through the introduction of alien species and hybridization; through chemical pollution of air, soil, and water; through warfare and its effects upon all life in its path; and through modification of the climate by the burning of fossil fuels, humans are effecting unprecedented changes in life on Earth. Cautions Wilson: "[The historical recovery times from extinctions] should give pause to anyone who believes that what *Homo sapiens* destroys, nature will redeem. Maybe so, but not within any length of time that has meaning for contemporary humanity."[46]

And so, what might humanity glean from its first two "Copernican" revolutions? When all the chaff is blown away, the message is simple: nature is not static; she is restless and eternally creative.

> Creation is not an act but a process; it did not happen five or six thousand years ago but is going on before our eyes. Man is not compelled to be a mere spectator; he may become an assistant, a collaborator, a partner in the process of creation.[47]

Whether our youthful species is up to such awesome responsibility remains to be seen.

INTERLUDE

Lifting the Veil

Figure 20: Symbolic of the deteriorated state of Newtonian physics at the beginning of the twentieth century. (Author.)

Eleven

Through the Looking Glass

All the world's a stage, and all the men and women merely players.

– **Shakespeare**

In the drama of existence, we are ourselves both actors and spectators.

– **Niels Bohr**

The twentieth century erupted with twin revolutions in physics: quantum mechanics, and then close on its heels, relativity. Relativity, the work of one incomparable man – Einstein – established a new physics for the very large: stars, black holes, galaxies, and the universe as a whole. Quantum mechanics, a collaborative effort of the world's most astute theoretical physicists, nearly all Nobel laureates, established the physics of the minuscule: atoms, nuclei, electrons, protons, and quarks. Despite its mathematical and conceptual complexity, relativity can be fathomed intellectually. Despite its mathematical and conceptual simplicity, quantum mechanics cannot. Events at the atomic scale are incomprehensible in the light of everyday macroscopic experience. "The universe is not only queerer than we suppose," quipped British evolutionary biologist J. B. S. Haldane, "it is queerer than we *can* suppose."[1]

The year was 1900, and the world's preeminent theoretical physicist was Berlin's Max Planck. Conservative by nature, family, and nationality, Planck was averse to novelty and speculation, including his own. The last thing Planck wanted was to precipitate a revolution in his chosen field.

It started innocently, early in his career. Planck had ruminated for several years on the subject of blackbody radiation. His motivation was initially mundane. Under a grant from the electric companies, Planck sought to improve the light bulb by generating more light with less heat. He was "green" before green was cool. This seemingly straightforward task had foiled Planck, leaving him frustrated. His analyses, based in classical physics, all failed to match the experimental results. In desperation, Planck entertained a radical idea, assuaging his conscience by an old trick – that he was merely "saving appearances" and not describing physical reality. He assumed energy to be *quantized*. That is, energy cannot be emitted in continuous dosages but only in discrete units or lumps termed *quanta*.

The Latin noun *quantum*, for "bundle," connotes an indivisible unit. If energy is quantized, then there exists a smallest unit of energy that resists further subdivision. By crude analogy, in the modern supermarket, economy of scale restricts the shopper to quanta of most food products. For example, one can purchase a case, a six-pack, or a bottle of beer, but one cannot buy an ounce or a milliliter of beer. Packages of those sizes don't exist.

Planck surmised that light, as a form of energy, must be quantized. Its quantum is a *photon*. Planck's hypothesis related a photon's energy E to its frequency (color) f by the simple formula

$$H = hf$$

where h, now termed Planck's constant, is a fundamental constant of nature. Astoundingly, this seemingly unjustifiable adjustment to classical physics immediately resolved a conundrum known as the ultraviolet catastrophe. The breakthrough earned Planck a Nobel Prize for physics in 1918. He lived to regret his success, for he had tugged the thread that unraveled the comfortable physics of Newton.

Quantum Quandaries[2]

Newtonian physics describes a material universe comprised of myriads of distinct particles interacting against a backdrop of absolute space and absolute time. The interactions are assumed to be causal and deterministic; that is, predictable. They are assumed also to be independent of a detached observer; the scientist who observes the "majestic clockwork" is regarded as objective in the extreme. He or she plays no role whatsoever in the unfolding drama. Nature is permeated by a stark dualism: subject and object, mind and matter, are disjoint, and never the twain shall meet.

For 250 years, Newtonian principles, the bedrock of science, gave rise to one technological marvel after another: steam engines, factories, automobiles, airplanes, and space travel. Newton seemed invincible and the edifice of classical physics unassailable.

The first crack in the Newtonian edifice appeared in the early 1800s when Thomas Young, an English physician and polymath, demonstrated that light has wavelike attributes, in contradistinction to Newton's "corpuscular" theory that light consisted of a stream of particles. Many of Young's experiments with light manifested color fringes or interference patterns, the telltale fingerprints of wavelike behavior. Subsequent experiments revealed that light propagates as a traveling wave, as do all forms of electromagnetic radiation.

In 1873, James Clerk Maxwell, a Scottish theoretical physicist, gilded the crown of the wave theory of electromagnetism. Maxwell's equations (Fig. 10, page 101) quickly became foundational to physics, revered for

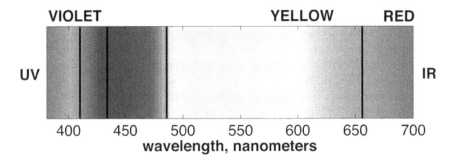

Figure 21: Visual solar spectrum with primary Fraunhofer absorption lines for hydrogen. UV and IR denote ultraviolet and infrared ends of the spectrum, respectively. (Author.)

their ability to describe, explain, and accurately predict all types of electromagnetic phenomena. Maxwell's success ensconced the wave theory of light deeply into the sacred canon of physics. To all appearances physics had reached maturity as a science, firmly established upon supposedly unshakeable foundations: Newtonian mechanics and Maxwellian electromagnetic theory.

Yet a troubling inconsistency had escaped the attention of most classical physicists: physics described nature as "schizophrenic." Newtonian mechanics dealt with particles, Maxwellian electromagnetics with waves. But particles and waves are mutually exclusive. Whereas particles are localized in space, waves are distributed throughout space. Particles cannot occupy the same space at the same time; waves can.

Planck's mentor had advised that he steer from physics, which presumably had lost its luster because everything of major import to the field had already been discovered. Fortunately Planck disregarded the advice and quickly discovered something new. At the Berlin Physical Society in December 1900, Planck concluded his talk with portentous words, "We therefore regard . . . energy to be composed of a very definite number of equal finite packages."[3] Apparently Newton had been right all along: light behaves as a stream of discrete particles.

Planck's radical theory of the quantum immediately resolved three "damned facts" that had niggled at physics: the ultraviolet catastrophe, the photoelectric effect, and the presence of Fraunhofer lines in the spectrum of an element. Each involved the nature of light, and each was inexplicable by classical physics.

Planck's attempts to improve the light bulb involved the study of so-called blackbody radiation. When one heats an object, it subsequently radiates energy back to its environment. The dominant frequency of the radiated electromagnetic energy depends upon the temperature to which the object is initially heated. Because humans perceive frequency in terms of color, we commonly speak of a "red-hot" coal or "white-hot" molten steel. The radiated energy is spread across contiguous bands of the electromagnetic spectrum, partly in the infrared spectrum, partly in the visible spectrum (hence the perceived color of hot objects), and partly as ultraviolet radiation beyond the visible spectrum.

Enter a quantum quandary. Classical physics predicted an absurdity: a heated object will radiate electromagnetic energy at all frequencies above the dominant frequency determined by the temperature of the object. If, however, the object radiates at *all* frequencies higher than its dominant frequency, the aggregate radiated energy is infinite. Thus, according to classical physics, one could pump a finite amount of energy into an object in the form of heat and get in return an infinite amount of energy in electromagnetic radiation, most of it in the ultraviolet range of the spectrum. Not a bad return on investment! But given that energy conservation – what goes in comes out – is foundational to physics, the ultraviolet catastrophe was a monumental thorn in the side of classical theories. Some shutoff mechanism must terminate radiation at higher frequencies, and Planck's quantization provided that elusive shut off. If radiant energy is emitted only in quanta, and if, according to his formula, a photon's energy is proportional to its frequency, then, at the cutoff frequency, there exists an energy hurdle too high for the photon to jump.

Einstein seized immediately upon Planck's innovation to explain the so-called photoelectric effect, which Philipp Lenard, a German experimentalist, had described in 1902. Lenard's experiment involved a clear glass tube, evacuated of air, with two conducting metal plates, one at each end of the tube. When the plates were charged using a battery, electrons tended to jump across the vacuum from the negatively charged cathode to the positive anode. But they wouldn't jump unless excited – by light – in which case they *might* jump. Curiously, whether or not the electrons jumped, and how many made the leap, depended upon the color and intensity of the light.

Let's render *color* and *intensity* in more technical terms. Wavelength is that property of light that humans perceive as color (Fig. 21). Red light lies at the high-wavelength (right) end of the visible spectrum, yellow to the right of middle, and violet near the low (left) end. Because the velocity of light, c, is constant, wavelength and frequency are inversely proportional: low-frequency light has a long wavelength; conversely, high-frequency light has a short wavelength. At wavelengths just above red, invisible to humans, lies infrared (IR) light. At wavelengths just below violet, also invisible to humans, lies ultraviolet (UV) light. According to Planck's law, the higher its frequency (the shorter the wavelength), the more energetic a photon of light. For this reason, sunbathers protect skin against energetic and damaging UV radiation but don't worry about relatively harmless IR radiation.

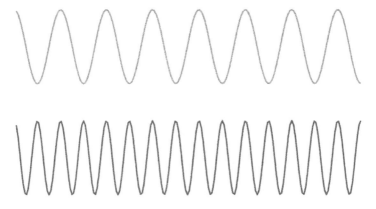

Figure 22: Two waves of the same amplitude but differing wavelengths. (Author.)

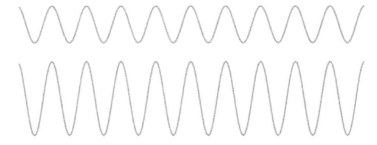

Figure 23: Two waves of the same wavelength but differing amplitudes. (Author.)

Amplitude is the technical term for what humans perceive as brightness or intensity. Figures 22 and 23 illustrate the properties of wavelength and intensity, respectively. The former compares waves of the same amplitude but differing wavelengths; the latter compares waves of the same wavelength but of differing amplitudes.

To classical physics, which treated light as wavelike, the photoelectric effect was puzzling. In a wave model, the color of light incident upon (i.e., hitting) the cathode should exert little effect on the number of jumping electrons. On the other hand, the light's intensity should have dramatic effect – the brighter the light, the more electrons it should knock loose. Furthermore, a wave model would predict that incident light waves should gradually dislodge electrons by a process of slow excitation.

What Lenard observed was altogether different than expected. For incident light at the red end of the spectrum, nothing whatsoever happened, no matter how bright the light. (For this reason, photographic darkroom lights are red.) But for incident light at the blue end of the spectrum, electrons jumped immediately, needing no time at all for excitation. And the brighter the blue light, the greater the number of leaping electrons. Einstein's 1905 explanation of Lenard's experimental results earned him the Nobel Prize for physics in 1921. According to Planck's formula, the "wallop" (energy) carried by an individual photon depends upon its frequency. Because of their higher frequency, blue photons pack far more "wallop" than red ones. Acting on the subatomic level like relatively massive billiard balls, blue quanta immediately eject electrons upon collision with the cathode of a photoelectric cell. Red quanta, being energetically wimpy, simply don't pack enough punch to eject electrons. Because collisions between pairs of photons and electrons occur singly – that is, one collision at a time – the likelihood of ejecting electrons does not increase by turning up the intensity of red light. However, for blue light, intensity matters.

While Germany's Nobel laureates came trickling out, in England Cambridge University's Cavendish Laboratory mass-produced them by beguiling nature to give up one secret after another. The illustrious James Clerk Maxwell, pioneer of electromagnetic theory, directed the Cavendish from its inception in 1871 until 1879. His successor, Lord Rayleigh, originated the esoteric branch of fluid mechanics known as hydrodynamic stability. Following Lord Rayleigh, J. J. Thomson – whose long and distinguished career produced a Nobel Prize in 1906 for the discovery of the electron – ably steered the Cavendish from 1884 until 1919.

Postdoctoral appointments at the Cavendish were coveted as practically tantamount to a Nobel Prize. So it was in 1895, when director Thomson took under his wing an aggressive young postdoctoral student from New Zealand. Radioactivity had just been discovered, adding exotic heavy elements to chemistry's pantheon. Ernest Rutherford, the postdoc, lost no time in putting the new elements through their paces. From experiments with thorium and uranium, he surmised that two fundamentally different types of radioactive emissions exist, which vary dramatically in their ability to penetrate material barriers. Rutherford dubbed these agents "alpha" and "beta" particles; alphas were the formidable ones. His star quickly rising, Rutherford left Cambridge to chair the department of physics at McGill University, where his experiments with radioactive half-life earned him a Nobel Prize in 1908.

For millennia, philosophers had debated the most basic question regarding matter: is it continuously subdivisible, or does there exist some small, indivisible unit? Before Socrates, Democritus had hypothesized the existence of "atoms" as fundamental building blocks of the material world. With the advent of chemistry came strong circumstantial evidence for the quantization of matter: chemical elements always react in ratios of whole numbers. But the age-old question wasn't fully settled until the twentieth century, and Rutherford played a big part.

From McGill, Rutherford returned to England as a professor of physics at the University of Manchester. Joining forces with Hans Geiger, inventor of the Geiger counter, Rutherford exploited alpha particles as projectiles to probe the atom and tease out its structure. What he found astonished him. Most projectiles penetrated unscathed through apparently solid gold foil. Occasionally, however, an alpha particle bounced back. Rutherford later reflected. "It was as incredible as if you fired a 15-inch shell at a piece of tissue paper and it came back and hit you!"[4] In 1919 Rutherford succeeded Thomson as the Cavendish's director, and under Rutherford, James Chadwick ferreted out the neutron, another constituent of the atom, garnering a Nobel Prize in 1935.

Electrons, protons, and neutrons: one by one the constituents of atoms revealed themselves. Everyone, however, was caught by surprise by the properties of atoms. Matter, the experiments revealed, is not very material. From the peculiar behavior of the gold-foil experiments, Rutherford deduced that atoms are mostly empty space, in fact 99.9+ percent emptiness. No other conclusion could account for the undisturbed passage through metal foils of so many of Rutherford's alpha particles. He surmised also that an extraordinarily dense nucleus must lie at the heart of an atom. Nothing else explained the startling return of an occasional alpha. Gradually, Rutherford developed a conceptual model of the atom as a tiny solar system, with Thomson's negatively charged electrons orbiting a positively charged nucleus of protons and neutrons as planets orbit the sun, vast emptiness between. The model, with inner space a microcosm of outer space, had a certain charm.

Although Rutherford's atomic model accounted for the atom's vast emptiness, it harbored a gaping theoretical flaw. Maxwell's equations predict that charged particles – like electrons or protons – radiate energy when accelerating. Orbiting electrons, by perpetually changing direction, accelerate. Even rough calculations revealed that whirling electrons should radiate energy at a fantastic rate and spiral into the atom's nucleus like an orbiting spacecraft caught in the drag of the earth's atmosphere. The atom would simply collapse onto its nucleus, instantly on a human time scale. In a word, Rutherford's atomic model was unstable. Real atoms, by contrast, are remarkably stable. In fact, they are virtually indestructible, unless subjected to otherworldly forces. Classical physics had no explanation for the durability of nature's minuscule building blocks.

Atoms were also associated with another inexplicable phenomenon. In 1814, by meticulous examination of light spectra, Joseph Fraunhofer, a German optician, identified 574 distinct dark lines distributed throughout the spectrum of solar light. Fraunhofer's compatriots, chemists Gustav Kirchoff and Robert Bunsen, partially revealed the mystery of the Fraunhofer lines. When heated in a gas flame, each physical element bears a unique color fingerprint hidden within the light emanated by the flame. A new and indispensable branch of science was born: spectroscopy.

Magically, Fraunhofer's spectral lines divulged the chemical composition of the sun – mostly hydrogen – from a vantage point 93 million miles distant. Later, in the 1920s, Hubble exploited the observed red shift in the hydrogen spectrum of galaxies to deduce the expansion rate of the universe. Yet despite the enormous utility of spectroscopy to physics, physicists at the turn of the twentieth century had no explanation for the phenomenon.

Enter a Dane to the rescue. During postdoctoral studies under Rutherford at the University of Manchester, Niels Bohr made an intuitive leap that still dazzles. While pondering the theoretical instability of Rutherford's model of the atom, Bohr divined a startling fix: he quantized the orbits. Electrons, unlike circling planets, were restricted to certain admissible orbits, each corresponding to a quantized energy level.

The Rutherford-Bohr model of the atom, proposed in 1913, immediately accounted for the Fraunhofer spectra of the elements. For the lightest element, hydrogen, the Fraunhofer lines were known as the Balmer series in honor of a Swiss schoolteacher who in 1885 had devised a simple algebraic formula that gave the precise location of each line in hydrogen's spectrum. Balmer's result had remained unexplained, however, for nearly three decades.[5] Bohr's fix of the atomic model stunned physicists. "In the simple case of hydrogen," for example, "one could calculate from Bohr's theory the frequencies of light emitted from the atom, and the agreement with the observation was perfect."[6] Not just good, but perfect (Fig. 21).

Bohr's atomic model accounted for spectral lines as follows. When a satellite loses energy, it gradually progresses through successively lower orbits. An electron behaves differently. When an electron loses energy, according to Bohr, that energy is shed as a photon. Borrowing from the successes of Planck and Einstein, Bohr understood that photon energies are quantized, which implied that the orbits of electrons are also quantized. Therefore, an electron cannot spiral into the nucleus like an aging satellite in earth orbit. It must jump discontinuously in a "quantum leap" from one orbit to another. When not jumping, it remains stably in an admissible orbit.

Each line of an element's spectrum corresponds to the precise frequency of the photon shed when an electron jumps from a higher admissible orbit to a lower one. Voilà! Atomic stability and spectral lines both explained. Not bad. But one might ask: where is the electron when it is between orbits? Eerily, at the instant of the jump, the electron is nowhere. It can never, for instance, reside halfway between two admissible orbits because that would require its energy to vary continuously rather than discretely, a violation of quantization. Rutherford found it all deeply disturbing because "Bohr's electrons seemed to choose not only the timing of their leap but the destination [orbit] too."[7] Erwin Schrödinger, a giant of quantum mechanics, was equally troubled. "If we are going to have to put up with these damn quantum jumps, I am sorry that I ever had anything to do with quantum theory."[8] It was enough to drive rational physicists mad.

The baffling quantum leaps of Bohr's electrons elicited a sense of déjà vu among physicists. Photon emissions from excited atoms are completely unpredictable, like decay events from radioactive elements. Probability, the bane of physicists, had first raised its ugly head with the discovery of radioactivity.

In 1895, Wilhelm Röntgen accidentally induced x-ray emissions while experimenting with fluorescence at the University of Wurzburg. The very next year, Henri Becquerel, working in Paris, discovered natural radioactivity, also by chance. He had inadvertently left a sulfate of uranium adjacent to photographic plates. Upon return he found the plates exposed despite their having been securely wrapped in opaque paper. For this chance discovery, in 1903 Becquerel shared the Nobel Prize with two compatriot physicists, the Curies, who isolated two previously unknown elements: radium and polonium, both of which were radioactive.

Radioactivity was fascinating and useful but also immensely baffling, a plethora of "damned facts." Classical physics presupposes strict causality: every consequence has an antecedent cause. But radioactive elements seem to exercise a kind of random spontaneity that violates causality. The laws for radioactive decay are probabilistic, expressed in half-lives. One can ascertain the probability of an aggregation of decay events from a lump of radium with great precision. But as to when an individual atom will shed its alpha or beta particle, one is powerless to say. Individual decay events occur randomly and apparently without cause.

Duality

Planck's quantum rescued physics from all the quandaries of 1900. It brilliantly "preserved appearances." But just beneath the surface of his deceptively simple formula, $E = hf$, lurked the shoal upon which classical physics was to founder: wave-particle duality.

On the left-hand side of Planck's equation, E expresses the energy contained in the photon, which acts like a *particle* in the photoelectric effect. The right-hand side of the equation relates E to the photon's frequency f, a property of waves, not of particles. Light therefore must possess a dual nature. Newton was correct. And Young was correct. But if waves and particles are irreconcilable, mutually exclusive notions, how can light possess properties of both? It was all quite perplexing. William Bragg, the 1915 Nobel laureate for x-ray crystallography, summed up the dilemma with tongue in cheek: "I teach the corpuscular (particle) theory of light on Mondays, Wednesdays, and Fridays, and the undulatory theory on Tuesdays, Thursdays, and Saturdays!"[9]

In 1924, things took an even more bizarre twist. If certain forms of energy, like light, manifest wave-particle duality, then, by Einstein's equivalence of mass and energy, matter might also be expected to manifest duality. Electrons, for example, were thought to be quintessential particles – that is, until the French aristocrat Louis de Broglie unveiled their wavelike tendencies. In his doctoral thesis, de Broglie proposed that electrons, like photons, possess a dual wave-particle nature. The revolutionary idea caught physicists completely off guard. In particular, de Broglie's thesis advisor, the eminent physicist Paul Langevin, was "stunned by the novelty" of the idea. De Broglie's boldness placed him in a bit of a quandary. "Like most people, physicists are generally fairly reluctant to accept any wild new idea, especially if there is not a shred of experimental evidence to support it. Predictably, therefore, de Broglie's examining committee in Paris was distinctly unsure as to what to do about the thesis."[10]

The committee hedged its bets. They forwarded a copy of de Broglie's thesis to Einstein for his comment. Fortunately for de Broglie, Einstein was sympathetic, and de Broglie received his Ph.D. in 1924. Three years later, experiments conducted independently in the United States and Scotland confirmed the wavelike nature of electrons, and in 1929, de Broglie won a Nobel Prize in physics for the discovery of "matter waves." Like Planck before him, de Broglie offered an innocuous little equation with monumental import:

$$\lambda = \frac{h}{mv}$$

Here, λ specifies the wavelength of an object of mass m and velocity v. The relationship suggests that all physical objects are "smeared" over space as waves, but it also explains why macroscopic objects are not perceived as waves. When m is large, the wavelength is exceedingly small, in which case the object appears localized. On the other hand, the wavelength of an electron is significant because its mass is so small (Fig. 24).[11]

And what of the little constant h that graces Planck's and de Broglie's formulas? First and foremost, it is an astronomically tiny number: $6.62607004 \times 10^{-34}$ in units of kilograms $\times$ meters squared per second. What does the dainty quantity signify? Fundamentally, Planck's constant expresses something about the granularity of energy and matter. The stuff of the universe is discrete, not continuous. One can divide and subdivide and further subdivide energy into smaller and smaller parcels – to very small parcels indeed – but the process cannot be repeated *ad infinitum*. There exists a smallest packet, the quantum, which cannot be further subdivided. But if energy is quantized, then, by Einstein's equivalence, so is matter. Thus Planck's constant provides indirect evidence for Democritus' hypothesized atoms, and its value quantifies the scale of these tiniest grains of matter, which no knife can further cleave. Democritus would have been astounded at their Lilliputian size. A grain of salt contains one billion billion atoms. It boggles the mind.

The quantum impresses not just for its diminutive size. Random jumps, vanishing electrons, acausality, probability, wave-particle duality, matter waves – with every advance in quantum mechanics, the quantum world seemed ever more topsy-turvy, causing physicists to lose sleep.

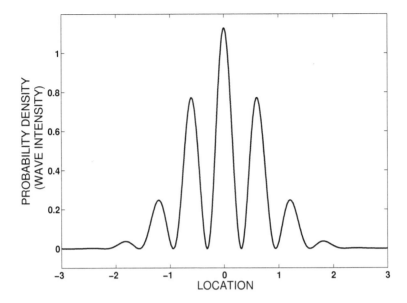

Figure 24: The figure illustrates two different concepts. First, it depicts the wave-function of a quantum object via its probability density distribution. In this case, the probability of finding the object – upon observation – between any two locations on the horizontal axis, say, between 1/2 and 1, is given by the "area under the curve." A long wavelength implies high uncertainty in the location of the object. Conversely, a short wavelength implies that the object is localized. Second, the distribution also depicts a typical interference pattern resulting from the interaction of two waves, say, water waves. In this case, the height of the curve corresponds to the resultant wave intensity as measured, for example, by a line of floating buoys. (Author.)

Indeterminacy

> I remember discussions with Bohr which went through many hours till very late at night and ended almost in despair; and when at the end of the discussion I went alone for a walk in the neighboring park I repeated to myself again and again the question: can nature possibly be as absurd as it seemed to us in these atomic experiments?[12]

The passage above, from Nobel laureate and quantum physicist Werner Heisenberg, reveals the genuine angst kindled within the scientific breast by some of the more bizarre implications of quantum theory. At the core of quantum mechanics, as foreshadowed by the probabilistic nature of radioactive decay, lies an immensely contentious issue: *indeterminacy*. Heisenberg was the first to couch the issue in precise mathematical terms.

By intellectual lineage, Heisenberg descended from Rutherford and Bohr, Heisenberg's scientific grandfather and father, respectively. A decade after laying his golden egg at Manchester under Rutherford's tutelage, Bohr had his own Nobel Prize in hand and a reputation to match as the "Great Dane" of physics. In 1922, Bohr presented his prize-winning atomic model to an eager audience at the University of Göttingen, where Heisenberg was an up-and-coming doctoral student. Intrigued by the subtle philosophical reflections that Bohr interspersed amongst his more sterile scientific facts, Heisenberg cornered Bohr after a seminar for a one-on-one discussion. It was a moment for the history books; Heisenberg's "real scientific career only began that day."[13]

In 1924, with a newly minted doctorate, Heisenberg left for Copenhagen to study under Bohr. Sixteen years his senior, Bohr cut an appealing father figure. Father-son relationships are often troubled, however, and the Bohr-Heisenberg relationship was no exception.[14] Bohr could be by turns stimulating or exasperating, profound or opaque, supportive or excessively demanding. Heisenberg learned to steer clear of Bohr's excessively sharp editorial knife. Upon completing his monumental "uncertainty principle," which eventually netted the 1932 Nobel Prize in physics, Heisenberg informed Bohr only afterwards that the paper had been mailed. Still, Heisenberg had bent considerably to his taskmaster's will. Grudgingly Heisenberg had adopted Bohr's favored term *uncertainty* in lieu of his original choice, a German word for "inexactness." Sometimes Heisenberg preferred yet a third option: indeterminacy.[15] Bohr's preference unfortunately stuck.

Regardless of its formal name, the uncertainty principle dominates the landscape of quantum mechanics, commanding attention from all directions. A direct consequence of the granularity of matter and energy quantified by Planck's constant, the principle is best grasped by a series of thought experiments.

Imagine having been pulled over for speeding, say, for driving at 65 miles per hour on a country road posted at 55. The cop who issued the ticket was hiding off the right side of the highway, obscured by bushes. You did not see the patrol car until well past it. Thus, the trooper's radar gun got you from behind. In traffic court, you attempt a creative quantum-mechanical defense:

> Your Honor, I was traveling at the posted speed limit of 55 mph. When the officer fired his radar gun, the colliding radio-frequency photons transferred their momenta to my car and bumped its speed to 65 mph. Had the officer not fired the radar gun, I would not have been speeding.

You, of course, must now pay the speeding fine and an additional fine for contempt of court because the argument is patently absurd. It contains, nonetheless, a germ of truth. Momentum *was* exchanged between the photons and your car. Moreover, the loss of momentum of the reflected photons caused a corresponding shift in their frequency – a Doppler shift – by which the policeman inferred your speed. But the effect on the car was immeasurably small because the car's momentum so vastly exceeded the combined momentum of a few photons.

Similarly, a disingenuous baseball pitcher claims to have been cheated out of a record for the world's fastest fastball. His argument asserts that, for all other contenders, the radar gun resided at the pitcher's mound, whereas for his trial, it was placed near the catcher. In the former cases, the reflected photons enhanced the ball's velocity slightly, whereas in his case, they impeded it from 105 to 104 mph, costing him the coveted title. Once again, the claim is true in principle and wrong in practical significance. The momentum of a moving fastball dwarfs that of colliding photons. The effect of the collisions on the ball's velocity, therefore, is undetectable.

Let's vary the experiment one final time. This time, we wish to detect the velocity of a moving *electron* by firing at it a photon from some source like a radar gun. The essential difference from previous cases is that electrons are exceedingly small, and thus each carries only a tiny amount of momentum. When the colliding photon strikes the electron, the collision occurs among virtual equals. As a consequence, the electron is knocked off its path, its trajectory so utterly altered that its original velocity or position cannot be inferred.

Heisenberg's genius was to express case three in the universal language of mathematics. His starting point was Newtonian mechanics. In classical mechanics, the prediction of a particle's future state – its velocity and position – hinges upon precise knowledge of its *initial state* at an instant in time called the *initial time*. Armed with this data and Newton's mechanics of motion, one could presumably predict, say, an electron's trajectory for all future time. But how does one obtain the electron's initial state?

For all their sophistication, physicists are not unlike a six-year-old boy who stumbles across a toad in his path and succumbs to the irresistible temptation to poke the toad, ostensibly to learn something about the creature in the process. Experimental physics is a formalized way of poking things and observing how said things respond. Each experiment requires three essential ingredients: an energy source, an object to be studied, and a detector. The experimental plan goes as follows: one fires a stream of energy at the object, some of that energy returns to the detector, and something is learned about the object from the pattern of returned energy. Rutherford's alpha-particle-and-gold-foil experiments were, in fact, vintage "poke-and-watch" experiments.

More familiar examples include x-ray photography, sonograms, and astronomy. In the former, the source of energy is high-frequency electromagnetic radiation in the form of x-rays, the object is the patient's broken arm, and the detector is a photographic film that guides the orthopedist in setting the fracture. In the second example, the source is high-frequency sound, inaudible to humans, the object is the mother's fetus, and the detector is a device that mathematically reconstructs the visual image of an unborn child from a tangle of reflected sound waves. In the final example, a distant star serves as both the object of interest and the source of energy. The detector is a photographic plate attached to a massive reflecting telescope isolated high on a mountain peak. In each case, a sentient being – a Nobel laureate, a doctor, a father, or an astronomer – monitors the detector.

Thus, Heisenberg reasoned, to establish the initial conditions of an electron, we could bombard it with photons and record the pattern of deflected or reflected photons. If Newtonian mechanics were fully in charge, we could infer from this pattern both the electron's position and its velocity. How does the situation change in the quantum world?

In the quantum world, we have some choice in the color (frequency) of the photon projectiles. Should we use "blue" photons, by which we mean highly energetic ones, or wimpy "red" ones? Suppose we first try plan B: use photons of low frequency or, accordingly, long wavelength, like the radio-frequency photons of a radar gun. An advantage of low-frequency photons is that they carry little momentum and, upon collision, scarcely disturb the electron's velocity, which can be inferred accurately. But what can be said of the electron's position? Unfortunately, very little. The problem is that the wavelength of the photon provides the natural "tick marks" of our distance-measuring "yardstick." For photons of long wavelength, the tick marks are exceedingly sparse. Therefore, we get only a crude measurement of position.

To circumvent this difficulty we revert to plan A, which involves high-frequency, short-wavelength photons, in which case we accurately measure the electron's position. What can be said of its velocity? Unfortunately, high-frequency photons carry large energies. By Einstein's equivalence of mass and energy, such photons act, relative to the electron, like massive bowling balls, which clobber it, knocking it far off its original trajectory and greatly disturbing its velocity. Therefore, we learn almost nothing about the electron's velocity.

There must be a way out of this corner. Let's try subdividing the high-frequency photons into smaller parcels that won't disturb the electron so violently. That is, let's use one-half, or one-fourth, or one-millionth of a photon. The central tenet of quantum mechanics, however, precludes this scheme: matter or energy come only in discrete packets that resist further subdivision. No matter how one approaches the problem, one is forced into the realization that the electron's initial position and velocity cannot simultaneously be known with high accuracy.

Heisenberg formalized the situation as follows:

$$(\Delta x)(\Delta v) > \frac{h}{m}$$

where Δx denotes the uncertainty in the position x of a body of mass m, and Δv denotes the uncertainty in its velocity v. Whenever the mass m is large relative to that of subatomic particles, as in the first two scenarios of our thought experiment, then both position and velocity can be measured with extraordinary precision. However, if the mass is very, very small, as in the third scenario, then position can be derived only by sacrificing knowledge of velocity, or vice versa. Both cannot be precisely determined simultaneously.

At first blush, the uncertainty principle appears to encode an epistemological difficulty. Presumably, there exists a well-defined physical reality independent of the observer. Surely, at any moment an electron *has* definite position and velocity. The difficulty lay, it was then thought, in the limitations of measurement – that is, in the observer's inability to accurately tease out these well-defined values.

Because it calls into question the essential nature of reality, quantum mechanics quickly intrudes into philosophy, and in 1958, Heisenberg confronted the issue head-on in his little gem *Physics and Philosophy*. As physicists agonized over the philosophical implications of quantum mechanics, Heisenberg and others came ineluctably to a startling revelation: the uncertainty principle should more correctly be branded the *indeterminacy* principle. The semantic shift from *uncertainty* to *indeterminacy* reflects a shift in the philosophical domain from epistemology to ontology. We cannot know the electron's position and velocity for the most disquieting of reasons: *quantum objects have neither definite position nor velocity*. They exist only in a probabilistic sense. Careful reenactment of Young's double-slit experiments left little else in the way of plausible interpretations of quantum events.

In a famous lecture series at Cal Tech in the 1960s, Richard Feynman, the late twentieth-century American wunderkind of quantum physics, summarized the results of modern double-slit experiments with an irresistible blend of clarity, humor, and wonder. What follows is a bare-bones summary. For greater detail and inspired illustrations, see *The Feynman Lectures on Physics*,[16] *The Quantum Universe*, and numerous other resources on the web.

Feynman's lectures on the quantum begin with descriptions of electron-gun investigations into wave-particle duality. Here, we take the liberty of substituting "photon" for "electron" in Feynman's descriptions, light being more intuitive than electrons to most people. The results of the two experiments are completely analogous, thanks to de Broglie.

Consider two thought experiments, Feynman asks: one with "bullets," which represent particles, the other with waves. These thought experiments can and have been conducted as real experiments in many guises.

For the particle experiment, a machine gun sprays BBs, somewhat randomly, in the general direction of a thick metal plate breached by two parallel vertical slits, each a bit wider than a BB. Behind the plate, a horizontal row of bins parallel to the plate accumulates the BBs that pass through the slits. After sufficient time, we examine a histogram of collected BBs. With slit 1 (on the left) open and slit 2 (on the right) closed, the BBs distribute themselves according to curve P1 of Fig. 25. The height of the distribution at each point represents the probability of a BB landing in a bin at that point. The peak of the distribution occurs just behind slit 1. Conversely, with slit 1 closed and slit 2 open, BBs collect according to probability distribution P2, the mirror image of P1. With both slits open, the BBs collect according to the double-peaked probability distribution P12, the average of distributions P1 and P2. There is no mystery here.

The second experiment involves water waves generated by forcing a buoy to move vertically up and down. The waves propagate toward a jetty that is breached by two openings, just as the slits breached the plate in the particle experiment. From the two breaches, circular wave fronts emerge on the far side of the jetty. In lieu of bins in the particle experiment, a row of buoys detects the amplitude of the waves that pass through the jetty. Each buoy records the maximum height obtained by the wave at that location. Unlike particles, two waves can coexist to *interfere constructively* or *destructively*. If their crests coincide, the peak amplitude is doubled. If the crest of one wave and the trough of another wave of coincide, the combined wave amplitude is zero. All variations between these two extremes are possible.

When breach 1 is open and breach 2 is blocked, the intensity distribution – wave energy versus location – looks identical to the pattern P1 in Fig. 25. Similarly, when breach 1 is blocked and breach 2 is open, the intensity distribution looks like P2 in Fig. 25. Still no surprises. In the third case, however, with both breaches open, wave interference leads to an intensity distribution similar to that in Fig. 24 and altogether different from P12 in Fig. 25.

Now that the differing behavior of particles and waves has been clarified, Feynman investigates the burning question: Is light wavelike or particle-like? For the final experiment, a light bulb emits monochromatic light of, say, blue color. The intensity can be varied from extremely bright to exceedingly dim. The detector is ordinary black-and-white photographic paper. A vertical opaque screen lies between the bulb and the photographic paper. The screen, parallel to the photographic paper, is pierced by two narrow, parallel slits separated by a distance of about one wavelength of blue light.

The experiment will make more sense to anyone who has ever worked in a darkroom. Photography is a quantum process by which the emulsion on photographic paper responds chemically to individual photons[17] of light. During development, grains on the paper reveal photon strikes by turning black. Grains not struck by photons remain white during development. Photography is also statistical. One cannot predict where the next photon will strike, only the probability of its striking at a given point

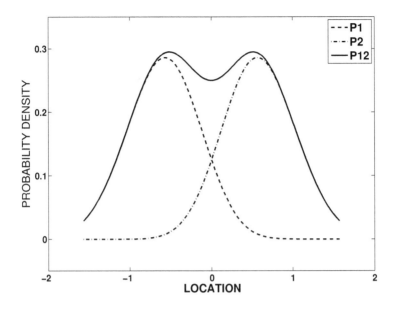

Figure 25: Probability density distributions of double-slit experiments with particles. Legend: dashed = P1, probability density for slit 1 open only; dashed-dotted = P2, probability density for slit 2 open only; solid = P12, probability density for both slits open. Probability densities P1 and P2 are halved for clarity. (Author.)

on the photographic paper. The probability of a photon strike varies inversely as the density of the negative. Where the negative is dense, few photons get through and the probability is low. Where the negative is nearly transparent, few photons are blocked and the probability is high.

In our experiment, the slotted screen serves as the darkroom "negative," impenetrable except at the slits. We begin by considering what happens when only slit 1 is open. Closest to the slit, the processed photographic paper is black. Very far away, it remains nearly white. Between these two regions are varying shades of grey that contain differing proportions of black and white grains. If P1 measures the blackness of the image, or equivalently, the probability of a photon strike, then P1 is distributed just as in Fig. 25. Similarly, with slit 1 closed and slit 2 open, probability distribution is given by P2 in Fig. 25. Still, no surprises. What happens when both slits are open?

Since the first experiments by Young in 1801, all such experiments have revealed an interference pattern as shown in Fig. 24. The probability distribution of intensity is similar to that for water waves, indicating conclusively that light propagates as a wave. Newton was wrong. Case closed!

Not so fast. In the modern form of Young's experiment, the photographic emulsion (or the digital charge-couple device) records only discrete quantum events, namely photons striking as particles. With both slits open, the probability distribution P12, however, is not the sum of P1 and P2, as it should be for particles. Feynman lays out this intellectual catch-22 in his lectures:

> Well that means presumably that it is not true that the [photons] go through [slit] 1 or [slit] 2 [like BBs] because if they did, the probabilities should add. Perhaps they go in a more complicated way. They split in half. . . . But no! They cannot; they always arrive in lumps [quanta]. . . . It is all quite mysterious. And the more you look at it the more mysterious it seems.[18]

Perhaps one could attempt to understand what is "really" happening by turning down the intensity of the light bulb, so low, in fact, that it emits a photon only once per hour on average. Surely individual photons must have to choose between slit 1 or slit 2 when both slits are open, and surely there will not be an interference pattern on the photographic paper.

Low-light photography requires long exposures. Underexposed images are grainy, the legacy of a statistical process with too few photons. Obtaining a sharp image requires an exposure of sufficient duration to provide statistical averaging. Long-time exposure has the added advantage that we can observe the emergence of the final image through time-lapse photography, stage by stage. Initially, the photons appear as a more-or-less random scattering of black dots, but over time a recognizable image builds up.

Figure 26 reveals the pattern on photographic material at successive stages. (Note that the negative image reverses black and white.) Astoundingly, the final stage reveals an interference pattern just like that obtained when the light intensity was high. But how can an individual photon interfere with itself?

The deeper one probes, the more perplexing it gets. For example, suppose we modify the apparatus with a camera or other sensor to determine, once and for all, which slit an individual photon passes through. Of course, to do so, we would need a source of energy to reveal the presence of the photon. If that source is just strong enough to detect the photon, the photon behaves as a BB, and the pattern on the photographic emulsion parallels a particle pattern. However, if the energy source is too weak to detect the photon, one observes an interference pattern. Thus, experiments designed to detect the particle-like attributes of photons reveal those attributes. Conversely, experiments designed to reveal wavelike attributes reveal photons behaving as waves. It appears that photons are very obliging; it is as if they "sense" whether one slit is open or two. However, no matter how clever the experiment, photons never behave *simultaneously* as wave and particle.

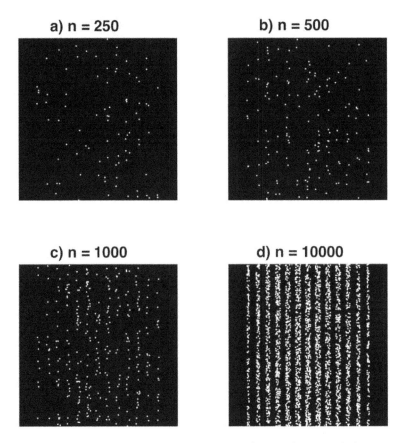

a) n = 250   b) n = 500

c) n = 1000   d) n = 10000

Figure 26: Results of the author's computer simulation of an actual electron-gun experiment performed by Hitachi in 1989. The interference pattern from n electrons ejected individually builds over time (frames a-d). Time-lapse photographic negatives reveal similar interference patterns for double-slit experiments with single photons. (Author.)

There is no concrete analogy in the macroscopic world for this sort of phenomenon. The closest common analogy is the widely documented Pygmalion effect observed by educators. Student performance rises or falls to match the expectations of the teacher. High expectations encourage high achievement, and conversely. Whether students, photons, or electrons, the objects of experiments seem to bend to our wishes. "What we observe," concluded Heisenberg, "is not nature in itself but nature exposed to our method of questioning."[19]

Energy and matter possess an irreconcilable duality, sometimes behaving as particles and sometimes as waves but never as both at the same instant. But what in the name of Zeus is a matter wave? In the words of Heisenberg: The electromagnetic waves were interpreted not as "real" waves but as probability waves. . . . [The probabilistic interpretation] was a quantitative version of the old concept of "potentia" in Aristotelian philosophy. It introduced something [somewhere] between the idea of an event and the actual event, a strange kind of physical reality just in the middle between possibility and reality.[20]

That is, matter waves are probability waves that quantify an object's *tendency to exist*! Until the moment of observation, quantum objects apparently have no real existence. Until observed, the electron is both nowhere and everywhere. Such interpretations, when first offered, shocked the world of physics.

Although no macroscopic physical analogy of quantum indeterminacy exists, here is a tantalizing mental analogy.[21] Imagine skiing down a steep slope. A tree grows in the middle of the run near the bottom, presenting a grave hazard. The tree lies immediately ahead if you maintain your current course. A number of possible scenarios exist, each associated with a probability. There is a nonzero probability of striking the tree and perishing or sustaining serious injury. You could veer to the right of the tree, an option with a probability of slightly less than 50 percent. With equal probability, you could veer left. The possibilities can be more finely graded, as in the very small probability of missing the tree by one mile to the right or the relatively high probability of missing it by six inches to the left. At the instant of your awareness of impending danger, all possibilities exist as potentia, each characterized by a probability. Time is of the essence. You decide to adjust course to the left. At the moment of conscious intent – that is, decision – all potentia but one dissolve and there remains a single reality: you miss the tree to the left. In quantum mechanics, this is known as "collapse of the wavefunction."

God Does Not Play Dice

Heisenberg pondered: "How [can] it be that the same [light] that produces interference patterns, and therefore must consist of waves, also produces the photoelectric effect, and therefore must consist of moving particles?"[22] The old controversy – particle or wave? – refuses to go away. Worse, it begins to look irresolvable. Physicists are accustomed to chewing on tough problems, with the promise of ultimate success. The toughest problems may take many attempts to crack, but eventually, like a dog's bone, they yield to persistence. But the quantum conundrum of duality will not crack; the more physicists chew, the more resilient the bone becomes.

"The strangest experience of those years was that the paradoxes of quantum theory did not disappear during this process of clarification," summarized Heisenberg.[23] Worse, each attempt to rectify quantum theory with conventional mechanics seemed to defile some sacred tenet of the latter. Each of the pioneers of quantum mechanics felt in some way that he had created a Frankenstein.

In 1926, Austrian physicist Erwin Schrödinger made perhaps the greatest contribution to the practical application of quantum mechanics: an evolution equation for the probability wave of an electron. For this, Schrödinger bagged the 1933 physics Nobel Prize. Schrödinger's contribution is all the more remarkable because he did not derive the equation; rather, he conjured it from a mere collection of clues, starting with de Broglie's relationship between an electron's wavelength and its momentum. "Where did we get that [equation] from?" Feynman marveled. "Nowhere. It is not possible to derive it from anything you know."[24] Nevertheless, upon Schrödinger's immensely practical equation rest the designs of many of the dazzling electronic devices of the modern world.

Another extraordinary feature of Schrödinger's wave equation is its range of applicability. Developed initially for the probability wavefunction of an orbiting electron, it works equally well when extended to the nucleus of the atom. The wave equation may indeed apply to *all* physical systems, and possibly to the entire universe. Talk about a useful equation!

Equations are predictive tools, so what does Schrödinger's equation predict? It models the propagation in space and time of the wavefunction associated with possible future states of a physical system. Given the state of the system now, what future states can it evolve to? Note that the very question itself represents a radical departure from the classical, deterministic view of nature. Schrödinger's wavefunction specifies not what *will* happen in the future but what is *possible* in the future, and it assigns to each possibility a probability of occurrence.

What is the sound of tree falling when there is no one to hear? Over this koan philosophers have beaten their breasts for millennia. But now physicists too ask similar questions. Where in its orbital is the electron when no one is looking for it? Everywhere and nowhere. The unobserved electron manifests only a tendency to exist. Somehow, the process of observation, presumably by a sentient being, forces the electron to materialize out of its shadow world of potentia. Observation collapses the wavefunction.

Such interpretations call into question the existence of a reality independent of observation. It is reported that Einstein seriously pondered how much sentience is needed to collapse the wavefunction. Could a mouse do it? Or a flea? "I don't like it, and I'm sorry I ever had anything to do with it,"[25] Schrödinger grumbled. Haunted by the ramifications of his own theory, Schrödinger devised a clever thought experiment to underscore the magnitude of the theory's absurdity. The *Gedanken* experiment has become legendary as the paradox of Schrödinger's cat.

Schrödinger's cat is a kind of magnifying class. It scales up quantum indeterminacy from the subatomic world and interjects it into the macroscopic world of practical experience. In the quantum world, between observations, probability states are *superimposed*, meaning that mutually exclusive events coexist as potentia. For example, in the double-slit experiment with both slits open, until the photon impacts the photographic plate, it exists as a probability wave with a 50-50 probability of passing through either slit. Schrödinger wondered: What if a quantum event triggered something macroscopic, like an accident that caused the death of his cat? Does this not lead to an absurdity? Until that quantum event is observed, the cat would coexist in a probability limbo, simultaneously alive and dead. To this day, Schrödinger's cat remains unresolved, a genuine paradox.

Complementarity

"I think I can safely say that nobody understands quantum mechanics," quipped Feynman.[26] No one struggled more untiringly to make sense of the nonsensical than Niels Bohr, who appreciated the inadequacy of language to describe quantum phenomena. From his valiant efforts to clarify the semantics of quantum processes emerged the "Copenhagen interpretation" of quantum mechanics. In August 1927, Bohr unveiled an early version of his synthesis at the famous Solvay conference in Brussels. Bohr's elucidations caused quite a stir, rankling Einstein and sparking opposition from many quarters. By the end of the conference, however, Bohr's view had more or less won the day.

At the heart of the Copenhagen interpretation is the notion of *complementarity*. Quantum objects possess an irreducible duality in that they manifest either as waves or particles. Wave and particle attributes are complementary in that they reveal different aspects of physical reality. However, they are mutually exclusive. A quantum object cannot simultaneously manifest as both wave and particle. How it manifests depends upon the experimental design. Experiments designed to measure the wavelike behavior of light or matter will measure that. Experiments designed to capture particle-like attributes will measure those. For a complete description of a phenomenon, one must include the experimental setup that measures it. It makes no sense to talk of a phenomenon independent of the measurement device.

Heisenberg expanded upon the Copenhagen interpretation. He and most theoretical physicists, Einstein excluded, concurred on the probabilistic interpretation of Schrödinger's wavefunction. Beyond that, opinions diverged, physicists being a perversely independent lot. Heisenberg contended that observation collapses the wavefunction. Not so Bohr, who saw no essential paradox with Schrödinger's cat. Schrödinger and Heisenberg, on the other hand, did. On this, however, virtually all physicists agreed: the Cartesian partition, which cleaved the world into the *res extensa* (matter) and the *res cogitans* (mind), has collapsed. "Nature cannot be described 'from the outside,' " as if by a spectator. "Description is dialogue, communication, and this communication is subject to constraints that demonstrate that we are macroscopic beings embedded in the physical world."[27]

The laboratory of science no longer cleanly separates subject and object. Articulated Heisenberg, "The very act of observing alters the object being observed when the quantum numbers are small."[28] The subject interacts with the object. Mind and matter are not distinct.

Since 1950, enticing alternatives to the orthodox Copenhagen interpretation have arisen. These include Feynman's multiple histories interpretation, David Bohm's theory of implicate and explicate orders, and the many worlds or *multiverse* interpretation, among others. The latter in particular seems the stuff of science fiction: infinitely many universes coexisting in parallel. For any quantum experiment, each possible outcome is fulfilled in one of these universes. We branch from universe to universe according to our conscious choices, each of which collapses a wavefunction.

Given the mental contortions to which quantum physicists have had to resort, it has been tempting to abandon quantum mechanics altogether for its stubborn refusal to submit to rational explanations. The problem, however, is that quantum mechanics remains the most successful physical theory ever devised; it has never failed. Quantum mechanics allows the derivation of fundamental constants of nature to extraordinary precision. It is the principle behind conventional and digital photography and the theory that has made the modern revolutions in electronics and information possible. And it has survived repeated attempts by the world's greatest physicists to poke holes in its foundations.

Entanglement

Einstein, whose audacity deposed Newton, grew intellectually stodgy in old age. With respect to quantum mechanics, he was positively reactionary. A die-hard determinist, Einstein rejected the statistical implications of quantum theory. In a letter to his friend and fellow physicist Max Born, Einstein confided his unease. "Quantum mechanics is very impressive. But an inner voice tells me that it is not yet the real McCoy. The theory produces a good deal but hardly brings us closer to the secret of the Old One. I am at all events convinced that *He* does not play dice."[29]

By 1935, Einstein believed he had found the Achilles heel of quantum mechanics. With two coworkers, he made a last heroic attempt to expose its fatal flaw by posing what became known as the EPR paradox.

Strictly speaking, paradoxes are unresolved, and the EPR paradox, unlike the paradox of Schrödinger's cat, has been resolved. Einstein lost, and quantum mechanics won. However, in 1935, when Einstein and coworkers posed their devilishly clever *Gedanken* experiment, a betting man might have put odds on Einstein's team. Quantum mechanics must be incomplete, Einstein felt instinctively. To Einstein, the probabilistic appearances of quantum phenomena were mirages, the result not of any propensity of the natural world toward statistics but rather of our incomplete knowledge of nature. He and others hypothesized that there must exist hidden variables of which the current theory remains ignorant. If the hidden variables were revealed, he argued, the apparent statistical predilection of nature would evaporate.

As the originator of the theory of relativity, Einstein held the *principle of local causes* to be inviolate. According to special relativity, the velocity of light, c, plays the role as a sort of universal speed limit, faster than which travel is forbidden by physical impossibility. Nothing – neither matter nor information – should be able to rove from point A to point B faster than a beam of light can make the transit. To illustrate, suppose we humans unwisely break the atmospheric test-ban treaty to detonate an H-bomb. Aliens presumably will be unaware of the ominous event until photons from the flash on Earth reach their distant planet. Einstein and most other physicists regarded as absurd the thought that an event at one point in spacetime could *instantaneously* affect an outcome in a distant region.

The EPR paper, its acronym from the initials of the three contributors – Einstein, Boris Podolsky, and Nathan Rosen – appeared under the title "Can Quantum-Mechanical Descriptions of Physical Reality Be Considered Complete?"[30] As originally conceived, their thought experiment involved simple, two-particle quantum systems. Quantum particles, whether electrons, positrons, or photons, involve a property called "spin." It is a helpful analogy (although not physically accurate) to think of such particles as spinning about an axis like a top. Spin, a form of angular momentum, is defined by an axis of rotation and an angular velocity. For subatomic particles, spin is quantized: spin magnitudes must be in whole or half multiples of Planck's constant h. For photons, spin is synonymous with polarization. Photons carry spin values of ± 1. Electrons, in contrast, carry spins of $\pm 1/2$. For these particles, no other spin values are permitted by nature.

Consider the simplest aggregate of quantum particles: a two-particle system consisting of one spin-up ($+1$) photon and one spin-down (-1) photon. The net spin momentum is thereby zero, the sum of the individual spins. By conservation of angular momentum, the net spin of the system must remain zero for all time.

And now things get interesting. As a quantum of angular momentum, spin is subject to Heisenberg's uncertainty principle, which imposes two enormous implications. First, until measurement, a particle's spin exists only as a probability, not as a reality. Second, the act of measuring the particle's spin *determines* its actual value. Now to Einstein's punch line. Consider the creation of a two-particle system of net spin zero and comprised of two photons, each traveling horizontally at velocity c in opposite directions. Suppose further that the two photons are now separated by an enormous expanse of space. Suppose finally that we measure the spin of one photon, which fixes its spin. According to quantum theory, preservation of angular momentum requires the other photon to adjust its spin *immediately* to the exact opposite value of its twin. But how does the twin know what spin value to assume? Presumably by some sort of transfer of information that must occur instantly, requiring superluminal communication and thereby violating the principle of local causes. Einstein believed that he held quantum theory by the soft parts. He concluded the argument:

> One can escape from this conclusion [that quantum theory is incomplete] only by either assuming the measurement of [particle 1 telepathically] changes the real situation at [particle 2] or by denying independent real situations as such to things which are spatially separated from each other. Both alternatives appear to me entirely unacceptable.[31]

Einstein euphemistically termed such mysterious "telepathic" or "superluminal" communication by the colorful phrase "spooky action at a distance." Recall that "spooky action at a distance" had troubled Newton about his own theory of gravitation. No physicist worth his salt was prepared to accept that events in one corner of the universe could instantaneously affect those in a remote corner. Einstein concluded not that quantum theory was wrong but that it was incomplete. He wanted the hypothesized hidden variables exposed to the light of day.

And so the EPR paradox rested for thirty years, a full-fledged, unresolved, perplexing conundrum about the nature of reality. Then, in 1964, John Stewart Bell, an Irish physicist associated with CERN (the European Center for Nuclear Research), proved a theorem directly related to EPR.

Bell's theorem is extraordinary for several reasons. Foremost, it is a *theorem*, not a *theory*. Unlike theories, which come and go based upon the most recent experimental evidence, theorems, which are mathematical, rest on the solid foundation of formal logic and carry 100 percent certitude. In particular, Bell's theorem established a mathematical inequality by which to experimentally test whether the principle of local causes is valid or the statistical predictions of quantum mechanics are valid. According to Bell's theorem, however, there had to be a winner because the two views of reality are mutually exclusive.

In the decade following the publication of Bell's theorem, numerous physicists reformulated and sharpened the argument. By the early 1970s, experimentalists at Lawrence Berkeley Laboratory had posed a version involving photon polarization that could be tested in the laboratory. Polarization has to do with the alignment of the plane in which light waves propagate when light behaves as a wave. Even non-physicists can appreciate polarization, provided they have worn sunglasses. White light is comprised of photons with a random assortment of polarizations. Polarized lenses transmit only those photons whose waves are oriented in a particular direction, say, vertically, reducing intensity and glare.

The flesh and blood incarnation of Einstein's thought experiment was elegant. When excited by an electric current, certain gases (neon, for example) emit photons in identical-twin pairs, each with exactly the same spin, another name for polarization. If quantum mechanics is valid, whatever one photon does when passed through a polarizer, its twin should do the same.

In 1972, Stuart Freedman and John Clauser of Lawrence Laboratory performed the long-awaited experiment to resolve the EPR paradox. By generalizing Bell's inequality, they directly tested Einstein's assertion that hidden variables could preserve local causes. The results were unequivocal: "Our data, in agreement with quantum mechanics, . . . provid[e] strong evidence against local hidden-variable theories."[32] Einstein lost, and spooky action at a distance prevailed.

The modern term for spooky action at a distance is *quantum entanglement*. The term suggests that quantum particles entangled "at birth" remained entangled forever despite the intervening distance. One is reminded of the psychic connections widely reported by human twins. Once connected, always connected, it seems.

Since the classic experiment by Freedman and Clauser, numerous first-rate experimentalists and laboratories have tested various aspects of quantum entanglement. In 1986, for example, Allain Aspect and coworkers in Paris performed a tantalizing variant of the EPR experiment in which "the 'decisions' as to which way to orient the polarizers were made only after the photons were in full flight."[33] Specifically, photomultiplier tubes captured the photons, indicating each capture with a click, analogously to how a Geiger counter clicks when capturing a particle from radioactive decay. Just ahead of each photomultiplier tube was a variable polarizer. Depending upon its orientation, each polarizer could either block the photon or transmit it.

When the two polarizers were oriented identically, Aspect's team observed that a click at one photomultiplier tube, indicating the arrival of a photon, was *always* accompanied by a similar click at the twin's detector. That is, the spin of the separated particles correlated at exactly 1.0; that is, a perfect 100 percent. On the other hand, when the polarizers were oriented at right angles to one another, a click at one detector was always accompanied by *no* click at the other. Either way, the spins of the two photons remained perfectly correlated, provided the pair had been created by the same quantum event.

Even more convincingly, Aspect and his coworkers created a photon "control group" of non-entangled photons with randomized polarizations and obtained altogether different results. Bottom line: for entangled photons, measuring the spin of one *always* and *immediately* precipitates the corresponding spin in the other, no matter how great their spatial separation. In 1975, shortly after the first experimental test of Bell's theorem, a prescient physicist, Henry Stapp, went out on a limb to write in a governmental report: "Bell's theorem is the most profound discovery of science."[34] At the time, it is doubtful that many concurred. Today, Stapp's words ring true. Quantum entanglement, now well established, is quickly making its way into a variety of strange applications, from cryptography to quantum computing.

Bell's theorem may very well be the Cullinan diamond of physics' many gems. By revealing quantum entanglement, Bell's theorem exposes subtle and mysterious interconnections that lie outside the universe's spacetime fabric. Had Einstein been buried rather than cremated, he would be turning in his grave.

Epilogue

Quantum mechanics not only collapsed the Cartesian partition, it clanged the death knell for Newtonian determinism.

From the 1960s through the 1980s, a new interdisciplinary field took the scientific world by storm: nonlinear dynamics, better known as "chaos theory" to the general public. Mathematical jargon couched in alluring terms – chaos, fractals, strange attractors, and bifurcations – wormed its way from the halls of academia into the public consciousness. Artistic books brimming with fractal images began to grace coffee tables. Trade books on chaos turned into best sellers. Nothing, however, caught the public imagination like the "butterfly principle," as science's "sensitive dependence to initial conditions" became known to the lay world.

Chaos theory revealed something heretofore unappreciated by science: for highly nonlinear systems – those with strong feedback mechanisms – small differences in initial data lead to drastically different evolved states over long periods of time. Nature has an affinity for the nonlinear; many of her mechanisms are nonlinear: gravitation and the strong and weak nuclear forces, to mention three. Nonlinear processes govern the weather, too, and for that reason, the flap of a butterfly's wings in Brazil on Sunday literally affects Peoria's weather early the following week.

Laplace's daemon, the hypothetical epitome of determinism (See Chapter 1), required substantial foreknowledge in order to predict the future, specifically, the initial conditions of all particles that comprise the universe. Ignoring the practicalities of such an assignment, is it possible in principle? At the quantum level, the uncertainty principle precludes precise knowledge of initial conditions because any attempt to measure position and velocity simultaneously contaminates at least one of them. And thus, on these two swords – quantum uncertainty and sensitive dependence – Laplace's daemon falls. Future states of a nonlinear universe are inherently unpredictable because unavoidable small errors in initial data wield enormous future consequences. Not only is the universe nonlocal, it is also nondeterministic.

In 2007, a group of researchers at premier institutions in Austria and Poland collaborated to perform yet another stunning EPR-like experiment. The experiment, by Simon Groeblacher and a host of coworkers, tested a new theorem (2003) by A. Leggett that further refines Bell's theorem.

Leggett's theorem yields a mathematical inequality that, when combined with Bell's inequality, permits the independent testing of Einstein's two assumptions: locality and reality. Strictly speaking, "Bell's theorem proves that all hidden-variable theories based on the *joint* assumption of *locality* and *realism* [emphases added] are at variance with the predictions of quantum mechanics."[35] Locality we have discussed previously. But what is meant by realism? Realism asserts that "All measurement outcomes depend on pre-existing properties of objects that are independent of the measurement."[36] In layman's terms, there's a real world out there independent of us, though we may see it differently because of variations in our measurement devices (i.e., our eyes and glasses). Realism is the assumption of most sane humans, scientists included, a few philosophers excepted. It was certainly Einstein's assumption when he referred to "real situations" in the classic statement of the EPR paradox.

The results of Groeblacher's team, published in *Nature*, once again validated quantum mechanical predictions. The authors concluded: "Our result suggests that giving up the concept of locality is not sufficient to be consistent with quantum experiments, unless certain intuitive features of realism are [also] abandoned." Independent reality, it now appears, has become the latest casualty of quantum mechanics. The universe is nonlocal, nondeterministic, and apparently "unreal" as well. Haldane was right: the universe is "queerer" than we can imagine.

Until quite recently, entanglement had been observed only at the quantum level and never for *mesoscopic* systems: systems comprised of large collections of atoms. However, in June 2009, researchers at the National Institutes of Standards and Technology in Boulder, Colorado, announced that they had succeeded in entangling the internal states of two distinct and separated mechanical oscillators, each comprised of multiple atomic ions, "extending the regime where entanglement has been observed in nature." Once again, the groundbreaking results were announced in the world's leading scientific journal: *Nature*. The researchers concluded with allusions to "possibilities for testing non-locality with mesoscopic systems."[37]

Who knows, perhaps entangled cats are next.

Twelve

Parallel Universes

The world is made of stories, not of atoms.

– Muriel Ruykeyser

To enter the world of the quantum is to "fall through the looking glass" and into alternate realities. But of all the mind-bending notions entertained by quantum physicists – complementarity, indeterminacy, entanglement, and more – parallel universes are most definitely otherworldly. What is new to science, however, is old hat in other domains of human experience such as mythology and religion. Brimming with stories of parallel realities, myth and religion tell of invisible mystical or spiritual worlds closely allied with the physical world of common experience, of alternate universes with subtle interconnections. Among the greatest modern mythmakers, a master of parallel universes, is J. K. Rowling, the creator of Harry Potter.

Rowling's exquisite storytelling beguiles readers by bringing to life an alternate reality: the magical world of Hogwarts School of Witchcraft and Wizardry. What child or adult does not revel in the exploits of Harry Potter, Hermione Granger, and Ronald Weasley? What reader does not shrink at the mention of He-Who-Must-Not-Be-Named or hang upon every word of Hogwarts' aged and profoundly wise headmaster Albus Dumbledore?

Potter lives in a world parallel to our familiar one, the mundane world of ordinary animals and humans without magical powers, fittingly called "muggles." Hidden from but intertwined with the muggle world exists an alternate reality of fantastical creatures – unicorns, hippogriffs, giants, goblins, animagi, and werewolves – as well as seemingly ordinary boys and girls with extraordinary predilections for magical arts: potion-making, conjuring, transformation, and apparating.

Rowling's gift, which has effected her own transformation from poverty to more wealth than the Queen of England, is the ability to transport (*apparate* in her lingo) the reader to a parallel universe, so strange and yet so utterly credible. Although Rowling's setting – Hogwarts School – is new to readers, her theme is as old as humankind, for the notion of parallel worlds is the basic stuff of all mythology.

In the highly acclaimed PBS documentary of the 1980s, *The Power of Myth*, mythologist Joseph Campbell offered interviewer Bill Moyers a luminous description of myth as "the secret opening through which the inexhaustible energies of the cosmos pour into human cultural manifestation."[1] Campbell's image of myth as a secret portal is magical, for in it he invokes mythology itself.

For example, in my youth, I once sat spellbound around a campfire, in the thrall of a Hopi creation myth. Retold by a sympathetic national park ranger, it made an impression that lingers intact forty years later. In the story, Shrike, a dumb but determined bird, leads a beleaguered people from one world to a higher one through a hole in the sky, accomplishing by determination alone a feat that neither Eagle, with his power, nor Crow, with his cunning, could accomplish. Since then, I have come to appreciate that such portals – connections from one world to another – figure prominently in the creation myths of diverse peoples. In North America particularly, magical portals permeate the mythologies of descendants of the Anasazi. Common to ancient pueblo sites throughout the American Southwest, *kivas*, sunken circular rooms dug into mother earth, offered sacred space for spiritual ritual. And in the floor of each *kiva*, well away from the center, rests a *si pa pu*, the small opening where spiritual energies from another world percolate up to our own.

The universal myth, therefore, tells of an alternate world, full of potential and power, that lies just behind our troubled human world. While these worlds seem disjoint, that is only illusion, for umbilical cords – portals – connect the two, and through these passages energy flows. Our job on Earth is to find the portals, which are often hidden. The chief function of religion is to provide hints as to the locations of the portals and to entice adherents to an alternate reality – the spiritual world – where harmony reigns and time ceases to matter if it exists at all.

The notion of parallel worlds is ubiquitous. Plato distinguishes the world of "forms" from the world "that flows." Jesus contrasts "the world" and the "kingdom of heaven." Among the oldest references to a parallel and transcendent reality, the *Tao Te Ching* contrasts the *Tao* and the *Te*: "*Tao* is the Supreme Reality, the all-pervasive substratum; it is the universe and the way the universe operates. *Te* is the shape and power of the *Tao*; it is the way the *Tao* manifests, it is the *Tao* particularized to a form or a virtue. *Tao* is the transcendent reality; *Te* is the immanent reality."[2]

Lao Tzu's "Tao" or Buddhism's "unmanifested" is the ground of being, the timeless, spaceless emptiness from which all manifested things emerge. It is the world of forms in Plato's thinking. To the Jew or Christian, the unmanifested is God.[3] To the Muslim, it is Allah. All is created in the image of God, and every manifest object or creature is a facet of God.

In the Harry Potter tales, Rowling paints muggles not so much as ungifted as unaware. Scattered all about, hidden in plain sight, are portals admitting entrance to the magical world, but they are overlooked by muggles preoccupied with the hustle and bustle of wearisome existence. The portal most familiar to Harry Potter fans is Platform $9\frac{3}{4}$ at King's Cross Station, London. As the name suggests, Platform $9\frac{3}{4}$ resides between rail platforms 9 and 10, for rail service to and from Cambridge. Thousands of harried travelers pass daily by the portal to the Hogwarts Express without ever noticing its existence.

A second, more obscure portal leads to Diagon Alley, the snaking medieval passageway that splinters off a bustling London Street via a magical door in a seedy pub, the Leaky Cauldron. Along Diagon Alley, strange shopkeepers hawk all the necessities of the wizarding world: wands, robes, spell books, potions, and post owls for sending messages.

Similarly, in C. S. Lewis' second Narnia tale, *The Lion, the Witch, and the Wardrobe*, the portal to Narnia lies in a wardrobe in an out-of-the-way country manor, where four siblings have been sent to live with an eccentric uncle to escape the London blitz. Only children, who have the time and innate wonder to explore the wardrobe, gain admittance to Narnia. Perhaps this is the meaning of Jesus' injunction to seek the kingdom of heaven as a child. Most adults are just too damn busy to imagine or explore.

Most of us adults muddle through life, prisoners of ordinary reality and blinkered consciousness, "whilst all about [us], parted from [us] by the flimsiest of screens, there lie potential forms of consciousness entirely different." Occasionally however, the veil lifts. "We may go through life without suspecting their existence," continued William James, the great turn-of-the-century psychologist, "but apply the right stimulus, and at a touch they are there in all their completeness."[4]

Myth upon myth addresses the relationship between parallel planes of existence: the physical plane and the spiritual plane, the latter of which props up and energizes the former.[5] Each plane demands a different mode of consciousness. To appropriate more neutral language, let us refer to the two planes as physical reality and mystical reality. Like an amphibian that needs both land and water to fulfill its "amphibianhood," the enlightened human, observed the Roman mystic Plotinus, must be at home in both realities.[6] Man, it is said, is like a frog. In its youth, the tadpole lives entirely in water. When the heavy tail of unawareness drops away, the mature frog inhabits two worlds, the world of water and the world of air, the latter signifying mystical reality.[7] The great crisis of our age is that so many have attained adulthood still bearing the heavy tail of unawareness. And those with the heaviest tails croak the loudest.

"There are more things in heaven and earth, Horatio, than are dreamt of in your philosophy," counters Hamlet to his loyal but excessively rational friend. Ordinary experience, recognizes Hamlet, harbors far more than meets the eye. Indeed, to the enlightened, there is nothing "ordinary" about ordinary experience. Science, like Horatio, prides itself in rationalism, assiduously avoiding the mystical. Quantum mechanics, however, has foiled rationalism. At the level of the quantum, the veil between science and mysticism seems precariously thin. Quantum theory sounds a scientific wake-up call: be alert to the portals, to mysterious interconnections between particles and between worlds.

Consider, for example, wave-particle duality, which hints at an irreconcilable dichotomy of perception. New to science, duality is ancient to philosophy. To the Vedantic philosophers of Indian antiquity, two views of reality compete within the human psyche: dualism and nondualism. Dualism implies more than one, and characterizes a reality of multiplicity and materiality. Multiplicity implies uniqueness, and uniqueness implies difference and separation. Hinduism's dualistic world is the manifest world of Buddhism and the world "that flows" to Plato. The dualistic world, therefore, is a particle world, a world inhabited by individuals, separate but occasionally interacting: quarks, photons, named and unnamed planets with named and unnamed moons, asteroids, chemical elements, and individuals of distinct biological species. At the human level, dualism connotes a world comprised of individuals separated by skins, tribes separated by ethnicity, and countries separated by boundaries and often by fear. Despite our masses, in the Western world, where materiality dominates and materialism is a religion, dualism imposes the background static of isolation and alienation.

Hinduism, however, also teaches that dualism is *maya*, a kind of "illusion" or "delusion." Collective delusion makes us see the world as many. That delusion is *lila*, the play of God. Behind and supporting the manifest world lies the Unmanifest, the ground of being, the One. We are not separate after all. There is underlying unity. Words fail those who have experienced such Oneness, but poetry comes close. "There is but one reality, like a brimming ocean . . . boundless as the sky, indivisible, absolute. It is like a vast sheet of water, shoreless and calm."[8]

Similarly, Bohr's notion of *complementarity* has a familiar ring. By the term complementarity, the oracle of quantum theory reminds us of dual modes of apprehending "reality," neither sufficient in itself for a complete picture. "Reality" has facets. Particle descriptions capture some of its aspects; wave descriptions capture others. To comprehend more fully, we need both pairs of glasses. In human rather than scientific terms, we need reason and intuition, glasses that see things as individuated and separate and glasses that see things as interrelated.

The greatest challenge of quantum mechanics to Western thinking is its calling into question the very existence of an objective material reality. Quantum particles exist only as probability waves until observed. Observation, an intentional act by a sentient being, collapses the wavefunction, transforming potentiality into actuality. Is there an analog in human experience for such mysterious workings? Yes: the power of intention. Intention transforms dreams and potential into action. In words attributed to Goethe:

The moment one definitely commits oneself, then providence moves too. All sorts of things occur to help one that would never otherwise have occurred. A whole stream of events issues from the decisions, raising in one's favor all manner of unforeseen incidents and meetings and material assistance which no man could have dreamed would come his way. Whatever you can do or dream you can, begin it. Boldness has genius, power and magic in it. Begin it now.[9]

Few fully appreciate the power that setting intention may have in mobilizing the universe to a purpose. Perhaps it is intention rather than measurement per se that collapses the wavefunction. To some, the rending of the veil spells trouble. To Planck and Einstein, quantum mechanics was a Trojan horse whose acceptance into the canon of physics nearly effected the demise of its host. Even those, like Heisenberg, who did not view its consequences as quite so dire nevertheless understood full well that quantum mechanics had precipitated "the completely unexpected realization that a consistent pursuit of classical physics forces a transformation in the very basis of this physics."[10] Quantum mechanics not only undermines Newtonian determinism, it dissolves the Cartesian partition, casts doubt on the primacy of materialism, and abandons the principle of local causes in favor of quantum entanglement.

To others, however, the rending of the veil is welcome and overdue. Quantum mechanics goads the most jaded to marvel at a universe as mysterious as it is vast. Quantum mechanics hints that materialism is not all there is, and physics, steeped in materialism, is slowly awakening. "The universe is a communion of subjects, not a collection of objects."[11] Quantum physicists, albeit reluctantly, are beginning to agree.

In esoteric spiritual traditions, enlightenment is a kind of expectant waiting, the recognition of the universe's ability to surprise us, to bring us serendipitously into a state of awe. Enlightenment involves living in the moment so as not to miss the surprise when it comes, in full awareness that one could stumble at any instant upon a portal and be *apparated*, even if momentarily, to an alternate reality. In that moment, the illusion of multiplicity and materiality drops away. In that moment of the mystical experience, the tadpole loses its tail.

Gifts of grace, mystical encounters can be neither purchased nor earned. One can prepare, however. Such gifts are more likely to be given when one is ready to receive. This, of course, was one message of Jesus' parable of the ten virgins, who expectantly await the arrival of the bridegroom. In all spiritual traditions, meditation – stilling the mind – is a common way to ready the portal. And so is the wise setting of one's intentions. As in the moment of observation, when the wavefunction collapses and brings into existence one reality from the myriad of possibilities, so our intentions shape our potential. In the infinitely wise words of Hogwarts' Albus Dumbledore, "It is our choices, Harry, that show what we truly are, far more than our abilities."[12]

PART III

The Third "Copernican" Revolution: Psychology & Spirituality

Figure 27: Hubble Space Telescope image commonly dubbed "the Eye of God."
(NASA, NOAO, ESA, Hubble Helix Nebula Team, M. Meixner [STSCI], and T. A.
Rector [NRAO].)

Thirteen

The Arrows of Time

Time is invention, or it is nothing at all.

– **Henri Bergson**

"In any attempt to bridge the domains of experience belonging to the spiritual and physical sides of our nature," noted Arthur Eddington, "time occupies the key position."[1]

For millennia, the nature of time has preoccupied our deepest thinkers, scientists and philosophers alike: Plato, Einstein, Freud, Whitehead, and Bergson, to mention but a few.[2] Time beats at the heart of philosophy, whose central problem is "the relationship between Being and Becoming,"[3] observed Whitehead, and whose central task "is to reconcile permanence and change."[4]

Integral also to science, time provided the key that unlocked relativity. In the euphoria of the moment, it might have seemed that Einstein's union of space and time resolved the issue of time for science and for philosophy. Of the marriage, Einstein's former math professor Hermann Minkowski waxed poetic: "Henceforth space by itself and time by itself are doomed to fade away into mere shadows, and only a kind of union of the two will preserve an independent reality."[5]

But relativity did not reveal the nature of time; it merely deepened the mystery. Although conjoined, space and time are not equal partners, at least in the mind's eye, where time remains resolutely other than space. One can tango in space, but not in time. By this it is meant that in space a dancer may move left and then right, or forward and then back. She cannot, however, move forward in time and then back. Time marches unwaveringly onward, obstinately refusing to retreat. Time, unlike space, is unidirectional: an arrow. This little fact, as we shall see, is exceedingly odd.

As Einstein suspected, time may be a trick played by the mind. In human consciousness at least, time flows in an orderly progression: past to present to future. Stated another way, time, in human perception, flows in a direction *opposite* to the experiential sequence of potentiality, intent, and actuality. By this we mean that the future is characterized by potential, pregnant with variegated possibility, not yet manifested. Our decisions, the manifestations of our intentions, always take place in the present moment: the "now." The act of deciding selects one path into the future from among the many possibilities. Once a path is selected, it becomes actuality, and once consigned to the past, our actuality cannot be undone. As the poets know, in the irrevocability of the past lies much unhappiness. "For of all sad words of tongue or pen, the saddest are these: 'It might have been!' "[6]

The philosopher who turns to science for insight into time's intrinsic arrow is primed for bitter disappointment. Time's directionality cannot be found in physics, for the simple reason that all the evolution equations of physics are time reversible. In the words of eminent mathematical physicist Roger Penrose:

> All the successful equations of physics are symmetrical in time. They can be used equally well in one direction as the other. The future and the past seem physically to be on completely equal footing. Newton's laws, Hamilton's equations, Maxwell's equations, Einstein's general relativity, Dirac's equation, the Schrödinger equation – all remain effectively unaltered if we reverse the direction of time (by replacing the coordinate t, which represents time, by $-t$).[7]

Consider, for example, the differential equations that model the orbital trajectory of a planet about the sun. Solving these equations, which are rooted in Newton's laws of motion and his theory of gravitation, yields elliptical orbits, in accordance with Kepler's observations of planetary motion. If one reverses the sign on time t in the governing equations, one still obtains an elliptical orbit, but the planet traverses the ellipse in the opposite direction. Either solution is equally plausible. Similarly for electrons that propagate through magnetic fields according to Maxwell's equations – reverse the direction of time, and the electron happily reverses its direction of flight. And similarly too for the probability waves of Schrödinger's equation: past and future are interchangeable. No physical law cares one whit about the direction of time.

That the laws of physics fail to reveal time's unidirectional nature is puzzling and hints of wondrous inner workings in the cosmos that remain hidden to the probing eyes of science.

A Quiver of Arrows

In truth, there are multiple temporal arrows. In his best-seller, *A Brief History of Time*, Big Bang cosmologist Stephen Hawking lists three distinct arrows: the psychological arrow, the thermodynamic arrow, and the cosmological arrow.[8] The fun begins when one asks: are the arrows related, and if so, how?

The psychological arrow refers to the directional flow of time in human perception: from past to present to future, as discussed above.

The thermodynamic arrow, about which much more will be said shortly, refers to the flow of heat with respect to time. Heat, like water, flows "downhill," from a hotter object to a colder one. Let's relate the psychological and thermodynamic arrows, for which we invoke a simple thought experiment. Imagine a thermodynamic system comprised of a hot object, say, a heated stone, adjacent to a cooler one, say, a carton of milk, inside an insulated cooler. Infrared photographs of the system are taken at two instants in time a few hours apart. False-color infrared photography reveals temperature as color, say, hot as red, cold as blue, warm as yellow. Imagine that you are now given two photographs of the system, one of its initial state, the other of a much later state. Having not been told which is which, you are nevertheless asked to order the photographs with respect to time. Photograph A shows two yellow objects in the cooler; photograph B shows a red object adjacent to a blue one. With absolute certainty, you know that photograph B preceded photograph A because, as psychological time progresses, heat flows from hotter to colder. To Hawking, the thermodynamic and psychological arrows are identical, for reasons to come later.

As former Lucasian chair of mathematics at Cambridge University – a seat once held by Newton – Stephen Hawking occupied until recently the most revered academic post in the world. And as a mathematical cosmologist, Hawking is most interested in time's cosmological arrow. Since the Big Bang, the universe has undergone expansion. More precisely, in the psychological past – to relate the psychological and cosmological arrows of time – the universe was smaller than it is now. In the psychological future, it will be larger. Whether the universe will continue such expansion indefinitely or ultimately reverse course and contract in a "Big Crunch" is a matter of current scientific interest. The fate of the cosmos depends sensitively upon the prevalence of so-called dark matter, for which there is no direct means of observation at present. Should the gravitational tug of dark matter ultimately slow and reverse the cosmos' expanding trend, the cosmological arrow would flip, turning opposite to that of the psychological arrow. But in the present epoch of the universe, the cosmological and psychological arrows, according to Hawking, are aligned.

To Hawking's quiver of three arrows, we add a fourth: the evolutionary arrow. To a greater or lesser degree, each of the major champions of evolutionary theory – Jean-Baptiste Lamarck, Darwin, Alfred Russel Wallace, and Teilhard de Chardin – perceived evolution as directional. Although subject to random events such as mutations, evolution is not merely a random walk in time devoid of direction. The feedback mechanism of natural selection biases the outcome strongly in a particular direction (as discussed in Chapter 10). Teilhard, the most modern of evolution's proponents above, articulated the evolutionary trend by the notion of *complexity-consciousness*. As psychological time marches forward, evolution proceeds from simpler organisms toward organisms of greater biological complexity with concomitantly higher intelligence and self-awareness. Complexity and intelligence provide strong advantages – principally adaptability – to an organism struggling to survive.

To summarize: Time's quiver contains four arrows, all currently aligned. Before delving more deeply into the relationships among the arrows, let's briefly consider another aspect of time: its central role in human conflict.

Throughout the course of Western history, in philosophy and in science, two opposing worldviews have continually clashed, often fiercely. Let's term these worldviews *static* and *dynamic*. In the former view, some process, once completed or perfected, is no longer subject to change. In the latter view, the process, never completed nor perfected, is subject to perpetual change. Aristotle, a proponent of the former view, espoused the immutability of the heavens; Galileo's telescope revealed otherwise. Of remarkable ferocity, the "Galileo affair" pitched Galileists against Aristotelians. The Church was strongly in the latter camp. Similarly, in the early days of the science of geology, catastrophists argued bitterly with uniformitarians over the nature of geological change. In the early twentieth century, the Scopes Monkey Trial pitted creationists, guardians of the static worldview, against evolutionists, the new dynamicists. And in the second half of the twentieth century, steady-state cosmology (initially supported by Einstein) vociferously opposed Big Bang cosmology.

In an international bestseller, *Order Out of Chaos*, devoted wholly to elucidating the nature of time, Nobel Prize-winning chemist Ilya Prigogine casts the opposing worldviews in softer terms in the hope of reconciliation: "being" versus "becoming." Strangely, the first faint glimmers of reconciliation are to be found in the unlikeliest of places: thermodynamics – the science of heat.

The Science of Heat

"It was the best of times, it was the worst of times . . ."[9] The scientific advances of the seventeenth and eighteenth centuries gave birth in the nineteenth to the Industrial Revolution. In Dickens' immortal words, the bustling and grimy age of industry represented simultaneously the flowering of the Enlightenment and human exploitation on an unprecedented scale. Factory workers and child laborers suffered Dickensian conditions that snuffed out many a life prematurely. And yet the genie of science had unleashed the enormous potential of mechanization. At the heart of the new mechanical age was a marvelous invention: the steam engine.

What the computer was to the late twentieth century, the steam engine was to the nineteenth. It pervaded and transformed every aspect of the Victorian economy from which it emerged.[10] Steam engines powered looms, propelled ships, and revolutionized transportation by the advent of the steam locomotive.

Thermodynamics – the study of the transformation and transfer of heat energy – arose, haltingly at first, from attempts to better harness the energetic processes of steam engines. An international cadre of luminaries sprang up to create and formalize thermodynamic theory: Sadi Carnot, a Frenchman; James Prescott Joule, an Englishman; Rudolph Clausius, a German; William Thomson (Lord Kelvin), a Scot; J. Willard Gibbs, an American; and Ludwig Boltzmann, an Austrian.

To scientists steeped in classical physics and engineering, thermodynamics seemed at first strange and counterintuitive. The key to the theory's development lay not in an aggregation of details but rather in successive steps of abstraction, by which the lumbering steam engine was gradually stripped of all its mechanical properties to reveal its quintessence, its scientific "soul."

Carnot, whose idealized steam engine consisted only of a reservoir of heat, a cold "sink," and a flow of "caloric" (Carnot's term for heat), took the first crucial steps of abstraction. Surprisingly, from such a simple conceptual model, Carnot was able to derive an upper limit for the efficiency of any heat engine, not just the steam engine. Today the class of devices known as "heat engines" includes the internal combustion engine, the diesel engine, and the jet engine, in addition to the steam engine. More broadly, heat engines also encompass all metabolizing organisms, including human beings; hence the connection between thermodynamics and psychological time.

Struck down by cholera in his prime, Carnot died at the age of thirty-six. Fortunately for humanity, his abstractions fell into the hands of one who could fully appreciate them: Lord Kelvin, who had applied thermodynamic theory to estimate the age of the earth. Kelvin first noticed the equivalence of heat and *work*, which have the same physical units. From this equivalence came the modern definition of *energy* as the ability to do work. Energy takes many guises: kinetic (the energy of motion), potential (the energy associated with gravitation), chemical, nuclear, thermal, and so on. Energy's fundamental trait: it is everlasting. Although it may be transmuted from one form into another, energy can neither be created nor destroyed. Thus, through the efforts of the early thermodynamicists, energy joined mass and momentum in the pantheon of conserved quantities. Conservation of energy is thermodynamics' first law:

Energy can be neither created nor destroyed. It can only change forms.

Like energy itself, the first law can be stated in many forms, as in the following alternative: the change of energy stored in a system is the heat energy added to the system minus the work accomplished by the system. What therefore is the essence of a steam engine? It is a mechanical device that transforms the energy in heat into useful work.

To Carnot's abstract heat engine, Kelvin added another component: output in the form of work. The quintessential heat engine, according to

Kelvin, consisted of four basic components: the heat reservoir, the cold sink, the flow of heat energy between the two, and the conversion of some of the stream of flowing energy into work. Kelvin's genius was recognizing that, of the four components of the idealized engine, the cold sink was key. Heat must have somewhere to flow. Without the cold sink, it remains trapped. This Kelvin expressed in a universal law:[11]

All viable heat engines must have a cold sink.

In 1850, roughly concurrent with Kelvin, Clausius proffered a similar statement in a paper by the German title *Ueber die bewegende Kraft der Waerme* ("On the Motive Force of Heat"):[12]

Heat does not flow from a cooler body to a hotter body.

Kelvin's and Clausius' statements were equivalent forms of an entirely new scientific law: the second law of thermodynamics. The reader may be forgiven for the mistaken impression that the second law is ho-hum, matter-of-fact, and dull. On the contrary, of the new law Eddington wrote: "From the point of view of the philosophy of science, the conception associated with the [second law] must, I think, be ranked as the great contribution of the 19th century to scientific thought."[13]

Buried within the second law is a glimpse of the Holy Grail of physics and philosophy: the nature of time. Indeed, Eddington concluded that the second law of thermodynamics holds *the supreme position* [emphasis added] among the laws of nature."

> If someone points out to you that your pet theory of the universe is in disagreement with Maxwell's equations – then so much the worse for Maxwell's equations. If it is found to be contradicted by observation – well, these experimentalists do bungle things sometimes. But if your theory is found to be against the second law of thermodynamics, I can give you no hope; there is nothing for it to do but to collapse in deepest humiliation.[14]

Despite the somewhat tongue-in-cheek tone of Eddington's observation, it recognizes that the second law expresses something fundamental about the inner workings of the universe. Within the second law lies the difference between being and becoming.

To Hell in a Handbasket

Central to the second law is a new concept: *entropy*. It's a slippery notion, so we'll sneak up on it.

When one first encounters the notion of entropy, it seems utterly foreign. Previous physical notions – mass, momentum, and energy – fit a common pattern. Each is conserved. Entropy, on the other hand, is not; it naturally tends to increase.

To get a handle on entropy, let's more carefully revisit the psychological arrow of time. Imagine having the break shot in a game of pool. The balls have been nicely racked into a triangular arrangement and await the opening break. The cue ball strikes the rack; a cascade of rapid collisions follows, with balls scattering helter-skelter. Some ricochet off cushions; with luck, one or two drop into a pocket.

Now imagine that on the second shot, the sequence of events from the first shot is exactly reversed in time, so that the balls reassemble themselves into a perfect triangular rack! Were this actually to happen, both players might need a good stiff drink because such an event would be as shocking as, and rarer than, seeing a ghost. Remarkably, as extraordinary as this second-shot scenario might seem, nothing whatsoever in the laws of physics prevents its occurrence.

The poolroom vignette reveals something significant about the progression of time in human perception: time flows in a direction in which events are ordered from less probable to more probable. Temporal awareness in human perception is associated with probabilities. Entropy therefore quantifies the probability state of a system. Physical systems of low probability have low entropy; those with high probability have high entropy. Consequently, entropy tends to increase over (psychological) time.

Entropy is commonly associated with disorder, as a measure of disorder. However, that association can be problematic. For example, disorder seems subjective; it lies in the eye of the human beholder. Entropy, on the contrary, like mass or velocity or temperature, is an objectively quantifiable property of thermodynamic systems. Before being more specific about what entropy measures, let's run another illustrative vignette, adapted from popular works by Stephen Hawking and his contemporary and friend, Roger Penrose.

Imagine a wine glass, filled with red wine and sitting precariously at the edge of a table. A truck rumbles by on a nearby street. Vibrations tip the delicate balance of the glass and its contents. The glass falls to the floor and shatters into dozens of fragments, some small, some large. The wine runs along the wooden floor. Some seeps between the cracks of the floorboards. Some eventually soaks into the untreated wood.

Now suppose the mishap has been recorded with a high-speed camera, which can play back the scene again and again at full speed or in slow motion, and in forward or reverse motion. We air two screenings of "The Accident" and ask an audience of humans which version is most plausible. By a flip of the coin, the first version plays "The Accident" backwards, with the frames reversed in time. The audience sees wine gushing upwards from the floorboards and gathering itself into a blob in the air to settle into a glass that has just assembled itself at the floor from scattered fragments. The glass and its contents then appear to leap up to the table, where the glass assumes an unstable position just at the table's edge.

In contrast, the second screening plays "The Accident" in the way that it appeared originally to the film crew. The audience is then asked which version represents "reality." Presumably, the audience will unanimously choose the latter version as more consistent with the individual and collective views of reality of the participants. Chances are, members of the audience may even laugh at the former version of "The Accident" as absurd. Such things do not happen in reality as we perceive it. Glasses do not reassemble themselves from fragments and leap upward. Order does not naturally increase. Entropy does not naturally decrease.

Curiously, no physical laws prevent the version of "The Accident" in which the glass leaps back upon the table. It is not prohibited by quantum mechanics. It is not prohibited by relativity or by the laws of electromagnetism. And it is not prohibited by the conservation of mass, momentum, or energy. In general, classical physics does not prohibit the glass from reassembling itself. What then does prevent the former version of "The Accident" from materializing in "reality?" The second law of thermodynamics, the law of entropy. The former version is too highly improbable.

It is tempting to think that the glass has lost energy as it falls from the table, and therefore has insufficient energy to leap back. But this is an illusion. Energy, we remind the reader, is conserved. The potential energy possessed by the glass and its contents when it was sitting upon the table is converted to kinetic energy as it falls, and to heat energy when it is fractured. Each of the fragments, and the floor itself, will be ever so slightly warmer after "The Accident" than before. The potential energy before "The Accident," however, was concentrated. The heat energy after "The Accident" is diffuse, the result of random thermodynamic fluctuations of the atoms and molecules of the glass, whose fragments now lie scattered all about. In order for the glass to leap back onto the table, all the trillions of atoms and molecules of the glass and its contents must "conspire" to fluctuate in a coordinated action at exactly the same instant, much like all the passengers and the driver of a car might shove in the same direction at the same time to push-start its stalled engine. It is one thing for five passengers of an automobile to give coordinated action. It is quite another for trillions of atoms and molecules, each listening to the beat of its own drum, to spontaneously act in concert. Thus, the glass leaping back to the table is not impossible; it is, however, highly improbable – so improbable that the event is unlikely to occur in a human lifetime; so improbable, in fact, that it is unlikely to occur in the lifetime of the universe. The "forward" version of "The Accident" was the "correct" one because it corresponds to the most probable ordering of events.

In the previous scenario, the probability state of the system before and after the accident is associated with the form in which the predominant energy exists. All the energy is accounted for, but its form has changed dramatically, from potential energy before the accident to heat energy afterwards.

Loosely speaking, therefore, entropy measures the quality (availability) of the aggregate energy within a thermodynamic system. Specifically, the more concentrated (available or accessible) the form of energy, the lower its entropy. Conversely, the more diffuse (unavailable) the form of energy, the higher its entropy. Symbolic of the modern age, the automobile is a ubiquitous transformer of energy. By way of illustration, what happens in terms of energy and entropy when one makes a trip by automobile?

First, one fills the tank with gasoline, a highly concentrated form of chemical energy. The gasoline is injected into the cylinders, where it mixes with air and is ignited by the spark plugs. Combustion produces high heat. The cylinders, therefore, are the high-temperature reservoirs of Kelvin's abstract heat engine. With heat comes pressure, which drives down the cylinders one by one in succession, thereby turning the crankshaft and converting some heat energy to work. The work is transmitted through the drive train to the wheels to propel the car at highway speeds. Kelvin correctly observed that there must be a cold sink in addition to a heat source. Where does the heat energy flow? Ultimately, for an automobile the cold sink is the atmosphere, at a nominal temperature of 70 degrees Fahrenheit, give or take several tens of degrees. Heat is dumped from the cylinders to the atmosphere via the radiator. On hot days, one's automobile is less efficient than on cool days because the differential between the cylinder-head temperature and the atmospheric temperature is diminished.

When all is said and done, and the car is parked back in the garage with a nearly empty tank, where has the energy gone that was initially stored in the gasoline? In one sense, nothing has happened to it. Energy is conserved, and a careful audit will reveal that all of it can be accounted for. Some energy exited as waste heat through the exhaust pipe. Some was transformed into work to propel the car. Some was transferred to the cold sink and warmed the atmosphere via the radiator. Some was bound in the chemical products of combustion, chiefly carbon dioxide and water vapor. Ultimately though, the lion's share of the energy once stored in the gasoline winds up as low-grade, distributed heat: a slight warming of the nearby atmosphere. Even the work that propelled the car ultimately ends up as low-grade heat, through the rolling resistance that heats the tires and the wind resistance that imposes drag upon the car and ever so slightly warms the nearby air. At the end of the trip, from the point of view of energy, all is present and accounted for, but its form has changed dramatically. What was once highly concentrated in gasoline is now diffuse, scattered all about in the atmosphere, unrecoverable for all practical purposes.

From the point of view of entropy, things look very different. The initial state of the system – here meant to include the car, the highway, and the air near the highway – is one of low entropy because much of the energy of the system is concentrated and stored in the car as gasoline. At each step of the work cycle of the engine, some energy is scattered abroad as low-grade heat. For all practical purposes, low-grade heat cannot readily be recovered. Because it is unavailable for further use, low-grade heat energy is associated with high entropy. As the trip proceeds, and more and more chemical energy ends up as diffuse atmospheric heat, the entropy of the system increases over time.

The automotive example illustrates both of thermodynamics' fundamental laws:

> *First Law*: Energy is conserved.

> *Second Law*: The entropy of an isolated system is nondecreasing.

At first blush, thermodynamics, the essence of engineering, does not seem terribly sexy. So let's back up to 1865, the year in which Clausius first expressed the two laws in succinct form. Here they are in German:

1. *Die Energie die Welt ist konstant.*
2. *Die Entropie die Welt streibt einen Maximum zu.*

Clausius' original statements are stunning in their sweep. The German *der Welt* translates as "the world," or equally, "the universe." Thus, both laws are cosmological in their implications. Each says something about the functioning of *the universe*! First, energy is conserved throughout the universe. Second, on the whole, the *entropy* of the universe increases toward a maximum. Without entropy, as we shall see, I would not be here writing, nor would you be there reading.

Because opportunities for confusion abound, let's carefully define the terms *system*, *open system*, *closed system*, and *isolated system*.[15] By *system*, we mean a real or imaginary "box" of sufficient size to contain the heat engine or thermodynamic process of interest and some or all of its environment. The region outside the box is the *surroundings*. The modifiers *open*, *closed*, and *isolated* refer to the permeability of the boundaries of the system. By *open system*, we mean that the thermodynamic system may exchange mass *and* energy with its surroundings. That is, mass and energy may pass through the boundaries of the box. By the term *closed system* it is implied that energy may be exchanged with the surroundings, but that the box is impermeable to mass. Finally, by *isolated system*, it is meant that the box is impermeable to the transfer of both mass and energy across its boundaries.

Each of these concepts, of course, is an idealization. Most real thermodynamic systems fit more or less into one of these categories, but only if appropriate simplifying assumptions are made. Furthermore, where one establishes the boundaries of the box is somewhat arbitrary, which implies that the classification of the system as to open, closed, or isolated can be manipulated somewhat according to where one draws its boundaries.

That said, let's classify as best we can three familiar thermodynamic systems: a freezer, the planet Earth, and the universe as a whole. In the first example, we'll let the box be the freezer itself. Because food may be placed into the freezer for storage and retrieved at a later date for dinner, we conclude that mass may be exchanged between the system and its surrounding environment. Moreover, because the freezer will not operate unless it is plugged in, electrical energy must pass into the box from outside its boundary. Moreover, via its cooling fins, heat energy is also exchanged with the surroundings. Therefore, the freezer, as idealized, represents an open system. Consequently, the second law of thermodynamics, which applies only to isolated systems, does not apply to a freezer.

In the second example, we imagine a box in space, big enough to house planet Earth and its atmosphere, but not containing the sun, the moon, or other planets. In its early days, Earth, bombarded constantly by asteroids and meteorites, increased in mass over time. Today, such events are rare, and mass increase from a slow influx of small meteorites is negligible. On the other hand, Earth receives a prodigious and continuous flux of radiant energy from the sun. Thus, practically speaking, system Earth could be classified as closed. Its mass is essentially fixed, but the imaginary system boundary is transparent to heat and light.

As the second law does not apply to closed systems, it does not apply to planet Earth as a whole. This is extremely fortunate for all inhabitants of the planet because the sun's energy is the ultimate source of all food in the food chain. Photosynthesis, the process by which plants concentrate diffuse energy into the form of sugars, is the root process of the food chain. From the energy stored in plant sugars the herbivores survive. And by the energy stored in the bodies of herbivores, the carnivores also thrive, including most humans. Were Earth an isolated system, life would be impossible.

Finally, it can be argued that the universe as a whole is self-contained and hence isolated. Here the second law applies, and entropy increases on average. Like a watch wound up at the moment of the Big Bang, thermodynamically speaking, the universe is running down. While net energy remains constant, its availability diminishes with time. The collapse of a massive star into a black hole, for example, represents an extreme increase in entropy. Black holes, the gravesites of large stars, are associated with high entropy because the energy contained within their event horizons is essentially unavailable.

Some rephrase the second law pessimistically when it is applied to cosmology: *The universe is going to hell in a handbasket.* What started with a bang may end with a whimper, the probable fate of the universe being "heat death." Thermodynamic destiny ordains a future that is cold and formidable, with the universe's energy imprisoned inside black holes or useless as waste heat, too weak and diffuse for photosynthesis to support life.

The less-dire technical term for heat death is *equilibrium*. At equilibrium, the heat reservoir and the cold sink of a thermodynamic system have reached the same temperature. Heat no longer flows between them. Indeed, nothing much happens at equilibrium, and time ceases to be relevant. Within galaxies, on the other hand, current conditions are far from equilibrium (FFE). Galaxies are populated by hundreds of billions of stars, each a colossal source of heat and light with a lifespan of billions of years. And at the center of each galaxy, it is believed, lies a massive black hole, the galaxy's heat sink. A galaxy therefore, is a FFE heat engine on a sub-cosmic scale, capable of gargantuan "work." What, therefore, is the work of a galaxy?

In FFE conditions, miraculous things can and do happen in confined localities. Order can emerge spontaneously from chaos.[16] In turbulent fluid flows, for example, large-scale structures arise – stable macroscopic patterns that persist despite the turbulent (random) microscopic nature of the flow. The red spot of Jupiter is one such example of a stable large structure in an otherwise chaotic flow. FFE chemical reactions sometimes produce a "chemical clock," the color-coordinated synchronization of flashes among trillions of reacting molecules, a symphony of individual molecules responding in unison to an unseen conductor.[17] Studies of chemical clocks, flocks of birds, and hordes of fireflies lead Cornell University applied mathematician Steven Strogatz to conclude, "Synchronization is the most pervasive drive in nature. . . . I'm not saying the law of entropy is wrong. It's not. But there is a countervailing force in the universe: a tendency toward spontaneous order."[18] To date, science has no fully satisfactory explanation for the choreographed flights of large flocks of birds, for the prevalence of large-scale structures in turbulent flows, for the synchronization of millions of fireflies, or for the coordination of billions of reacting molecules in chemical clocks. But nevertheless, these things happen. Despite the universal trend toward disorder, order can arise in systems that are not isolated and are FFE.

As interesting as Jupiter's spot and chemical clocks might be, these phenomena pale relative to the cosmological significance of FFE conditions. "A universe far from equilibrium is [necessary] for the macroscopic world to be a world inhabited by 'observers' – that is, to be a living world."[19] In other words, FFE conditions are essential for the emergence of life in the cosmos. Life, at least as we know it in our corner of the universe, is associated with a type of molecular order that seems to defy the second law. As the physical universe marches inexorably toward heat death, life swims upstream against entropy, extracting order from a universe in decline. Life persists at the nexus of countervailing trends.

Reversibility

The association of entropy with disorder is largely due to the powerful insights of Austrian physicist Ludwig Boltzmann. Sadly, Boltzmann's life was itself a delicate balance between order and chaos. Susceptible to severe mood swings and depression, Boltzmann took his life in 1906, a great loss to his family and the scientific world and an unfortunate victory over order by chaos.

Still, Boltzmann's legacy is substantial and enduring. After earning a doctorate in 1866 in the kinetic theory of gases, Boltzmann assumed a succession of prominent academic positions in Graz, Heidelberg, and Berlin, where he studied under luminaries such as Kirchoff and Helmholtz. His unique combination of training, experience, and personality prodded Boltzmann into an extraordinary scientific contribution, the establishment of an entirely new discipline: *statistical mechanics*. As the name implies, statistical mechanics seeks to determine the properties of thermodynamic systems based upon the average behavior of large numbers of individual atoms and molecules. Boltzmann shares credit for the idea with J. Willard Gibbs, a gentlemanly American, who, after also studying in Heidelberg under Kirchoff and Helmholtz, independently formulated the theory.

As Boltzmann and Gibbs understood, the atoms and molecules of solids, liquids, and gases remain perpetually in chaotic motion. Temperature quantifies the frenzy of that motion. Molecular frenzy can be glimpsed by the naked eye in the phenomenon of Brownian motion, by which dust particles suspended on water jiggle about randomly and incessantly. Recall that, in his *anni mirabilis* of 1905, Einstein surmised that Brownian motion resulted from the underlying restlessness of water molecules, and he estimated the size of those molecules on the basis of careful observations and analysis of Brownian motion.

The statistics of Boltzmann and Gibbs makes clear why heat naturally flows from hot to cold. When more excited (hotter) molecules lie next to less excited (cooler) ones, it is more probable that the less excited ones become more active from the constant jostling of their companions, who in the process lose some of their energy.

Similarly, when a perfume bottle is uncorked and left upon a table, the sweet scent wafts to every corner of the room. Through thermodynamic fluctuations – incessant collisions on a molecular scale – the densely packed perfume molecules ultimately wind up distributed randomly among the molecules of the air in the room. The laws of probability make it highly unlikely that those same perfume molecules will behave like sheep and later aggregate back into the bottle. Neither do the molecules of coffee cream spontaneously regroup into a swirl at the center of the cup and leap back into the carton. Thus, from a statistical point of view, the second law of thermodynamics can be recast:

**An isolated system tends to become less ordered
with the passage of time.**

The association of entropy with disorder affords some comic reformulations of the second law:[20]

> An untended dorm room will rapidly degenerate into a pigsty.
>
> It is easier to scramble eggs than to unscramble them.
>
> Refrigerators don't work unless plugged in.

Beneath the humor lie hints about the mysterious inner workings of the universe and the unidirectional flow of time. The first and second statements above reflect the natural tendency of isolated systems, including the universe as a whole, toward increasing disorder, or more correctly, toward states of higher probability. The third statement dices things more finely. It addresses what the second law does *not* say. It does not say that heat cannot flow from cold bodies to hot bodies or that order cannot be created out of disorder. But doing so requires an input of concentrated energy from outside, in which case the system is no longer isolated. A refrigerator transfers heat from colder bodies (the carton of milk) to hotter ones (the room), but it doesn't do so spontaneously. It must first tap into an outside source of energy via the electrical cord and the outlet.

Although entropy is commonly associated with manifest disorder, as hinted previously, this interpretation, in some instances, can be misleading. As a counterexample to the norm, consider the cosmological process of stellar accretion by which stars and planets are born. During accretion, interstellar clouds of dust and gas slowly collapse due to weak gravitational interaction to form celestial objects. To the human observer (if he or she could live so long and see so far), it would appear that random distributions of matter give way to organization, that particles gather themselves into stars and planets, almost as if the shattered glass in "The Accident" had leapt back onto the table! Appearances are misleading. In most common examples, which usually involve thermodynamic systems without significant gravitational effects, entropy and disorder are synonymous. But in gravitating systems, the potential energy of the scattered particles and atoms gives them, in aggregate, lower entropy when dispersed than when collapsed into a star or planet, whereby some of the potential energy has been lost as heat and radiated away.

The safe bet then is to associate entropy with probability rather than with disorder. For either thermodynamic or gravitating systems, the following statement of the second law always holds:

**An isolated system tends to evolve in time from states of lower
probability toward states of higher probability.**

The entropy of an isolated system is *nondecreasing*, which implies that it must either increase or remain constant. Thus far, we have ignored the possibility that entropy can remain constant, in which case the internal processes of the system are said to be *reversible*. Generally, reversible processes idealize those that do not occur in nature because natural processes normally involve frictional losses. Sometimes, however, nature fools us. The flow of electrical current in a superconductor is lossless. The exchange between kinetic and potential energy of a comet as it traverses a huge elliptical orbit around the sun, speeding up as it gets closer, slowing down as it recedes, is very nearly reversible because there is so little friction in the void of space. For this reason, astronomers can predict a comet's return with great precision.

Often, idealized reversibility is useful for practical purposes, including engineering. Reversibility establishes the limits of what is nearly possible. One useful thermodynamic idealization has culminated in the hybrid automobile, whose efficiency doubles that of conventional autos. Reversible processes represent unattainable goals toward which engineers nevertheless strive, satisfied when they get just a bit closer to the goal than previously possible. And in the case of hybrid automobiles, that striving has led many innovations in efficiency, including regenerative braking, which converts kinetic energy to chemical energy during deceleration.[21]

Reversible processes are ideal; they are also time symmetric, by which we mean that they appear "natural" to humans in either temporal direction. For example, the orbiting of a planet about a star is, for all practical purposes, a reversible process. If the direction of time were reversed, the planet would orbit in the reverse direction, but nothing would appear amiss to an observer. Unlike the wine-glass incident, both scenarios – forward or reverse time – are equally plausible to the observing human. Either way, the cyclical conversions between kinetic and potential energies are complete, without loss in so-called "waste" heat. Reversible processes are in essence "timeless" because the state of the system does not fundamentally change over time. Hence, reversibility is the thermodynamic analog of "being."

We come then to a strange fact: the flow of time in human perception – "Becoming" – is related to thermodynamic irreversibility. More precisely, the psychological arrow of time corresponds to time flowing, in our less than thermodynamically perfect world, in the direction of increasing entropy. In *A Brief History of Time*, Hawking argues that psychological time results from the thermodynamic irreversibility of memory storage and retrieval within the human brain. Human memory, Hawking reasons, functions similarly to computer memory and hence is irreversible.

To be specific, in computer memory, information is encoded into strings of bits, binary integers that take on only one of two possible values: 0 or 1. For example, the one-byte binary equivalent of the decimal integer 11 is 00001011. Switching a bit's value in memory from 1 to 0 or vice versa is like throwing a tiny electronic switch, where the bit "0" signifies that the switch is "off" and the bit "1" signifies that it is "on." With a sufficiently long sequence of bits, one can store all the photographs in a camera, all the words in this book, all the books in a library, or all the information recorded in the entirety of human history. When information is stored in computer memory, overwriting what was there previously, some of the bits are reset from their previous values. Some "off" switches are flipped to "on," and vice versa. The process requires electrical energy, which winds up as waste heat that is exhausted to the room by the computer's cooling fans. Waste heat is essentially unrecoverable, and hence the entropy of the computational system (say, a laptop running on battery power and the room in which it resides) increases. Memory storage and retrieval is irreversible.

Similarly, thinking requires considerable energy, as anyone who has ever studied for an important exam – interspersed by frequent trips to the refrigerator – knows. Thus, thinking, like computing, is an irreversible process that takes concentrated energy (say, snack food) and converts it to low-grade energy in the form of waste heat. Hawking concludes: "Just like a computer, we remember things in the order in which entropy increases. [Thermodynamic] disorder increases with time because we measure time in the direction in which disorder increases. You can't have a safer bet than that."[22]

More fundamentally, Hawking is claiming here that the psychological and thermodynamic arrows of time are one and the same. *Human perception of time is a direct consequence of the second law of thermodynamics and the fact that human beings are biological heat engines.*

Thermodynamics has given us a new analogy for the cosmos. "The great clockwork" of the Newtonian worldview yields to "the cosmic heat engine" of the modern analogy. In the old worldview, ruled by Newtonian mechanics and Maxwellian electrodynamics, all processes are time reversible. They proceed equally plausibly in forward or backward time. Isolated systems comprised of particles that interact reversibly ultimately repeat their patterns, or nearly so, as proven by the twentieth-century French mathematician Henri Poincaré. Hence, Aristotle could look at the night sky and reason that the heavens are immutable. The heavens change, but the changes are cyclical. The patterns, reasoned Aristotle, are fixed for eternity. Reversible processes are indeed timeless. Thermodynamic reversibility is the philosopher's state of "being."

Irreversible processes, on the other hand, "flow." Their patterns never repeat. When far from equilibrium, thermodynamic systems are incubators of creativity; order arises spontaneously from chaos, disperses after a time, and arises again in an altogether different manifestation. Irreversibility reigns in a cosmos subject to the second law. The universe is a cosmic heat engine, whose work is continual creation. The universe is an engine of "Becoming." "Physics and metaphysics . . . [come] together today in a conception of the world in . . . process."[23]

Love

Processes that flow against the tide of entropy are termed *counterentropic*. Counterentropic activities do not violate the second law because they never occur in isolated systems. Of all the counterentropic processes of the universe, the evolution of life on Earth is by far the most glorious and mysterious of which we are aware. In chapters to come, we explore a question whose answer may very well be the Holy Grail of both science and philosophy: What is the relationship between the psychological/thermodynamic, cosmological, and evolutionary arrows of time?[24]

To prepare for what lies ahead, let us now leave the broad universe to refocus our attention on a small corner: our home, our yard, and our family. What counterentropic processes do we encounter in our little neck of the woods?

Many familiar experiences are counterentropic, deriving order from apparent chaos. The restoration of an antique car is but one. More likely than not, the Model T Ford found in the old barn will no longer run. Its tires have dry-rotted. The engine oil has gelled to gunk. The body, having lost its sheen and then its paint, is rusted through in spots. The chrome has long since peeled from the bumpers. If the Model T is fortunate enough to fall into the hands of an antique auto enthusiast, its fortunes may reverse. With sufficient tender, loving care, the owner, by expenditures of time and energy, can reverse the years of deterioration to bring the old Ford back to life, the pride of the county fair and its owner.

Home ownership too is largely low-intensity warfare against the entropic ravages of the universe. Left alone, paint peels. Exposed to the elements, weather-beaten siding rots and decays. Aged shingles, brittle from years of harsh sunlight, break and expose the underlying roof, which springs leaks. Foundations crack under the relentless pressure of tree roots and open themselves to termite tunnels. Wet walls mildew. Left forlorn, within a hundred years or so, a house will degenerate so completely as to be largely recycled by mother earth. Painting, reroofing, cleaning the gutters, and treating the foundation are all energetic, counterentropic processes designed to temporarily halt and reverse the natural tendency to decay.

In the living world, gardening is counterentropic. A garden is an island universe of order where disorder would prefer to reign. Left to its own, the untended plot of earth degenerates into a chaotic tangle of vines and weeds. Pulling weeds, mulching, planting flowers, collecting June bugs one by one, and sculpting plots: these are counterentropic inputs of energy from the tender. The soul of gardening is wonderfully captured in David

Mallett's "The Garden Song," whose lyrics can be found on the internet along with performances by various artists.

Perhaps the most counterentropic process in ordinary human experience is that of parenting. The young of humans are helpless, incapable of fending for themselves for many years. Why they are helpless is a trick of evolution. Among all the inhabitants of the earth, humans are best adapted to change. But being resilient and responsive to change requires a large brain, and a large brain implies a large skull, which makes childbirth difficult.

Thus, much of the growth of the human brain occurs outside the womb. During the first two to three years of life, the human child is extraordinarily vulnerable, utterly dependent upon the parent for food, clothing, and hygiene. Without the intervention of the parent, the child would succumb quickly to hunger and filth. Parenting continues to a lesser degree at least through adolescence, the parent gently but constantly "nipping the heels" of the child, steering him or her away from danger and toward what is wholesome and good.

Whether one restores an old car, preserves one's home against the ravages of nature, tends a garden, or parents a child, one is engaged in a labor of love. To anthropomorphize thermodynamics, love is the extreme counterentropic sport: an expenditure of energy for the sake of someone or something other than oneself. The salmon swims upstream for the sake of future generations, spawning its final burst of energy.

Evolution too is quintessentially counterentropic, on a cosmic scale. By evolutionary processes, complex living organisms have been gathered together from a diffuse matrix of dust, debris, and gas scattered chaotically across the universe by the Big Bang and subsequent supernovae. Paradoxically, the increase in complexity-consciousness of the biological universe runs counter to the inexorable slide of the physical universe toward equilibrium. And as we shall see, the former is possible only because of the latter. Evolution is all about "Becoming." Whatever its mysterious mechanisms, evolution seems the labor of love of an invisible gardener whose garden is the whole universe.

Fourteen

What Is Life?

There is nothing over which a free man ponders less than death; his wisdom is, to meditate not on death but on life.

– Spinoza

In February 1943, during a bleak season enshrouded within a bleaker moment of human history, in a country known for its melancholy, the early rays of a new dawn peeked over the dark horizon. Amid the chaos of war, the portentous dawn went largely unheralded. An entire decade ticked by until the scientific import of Schrödinger's glimpse into the future was fully appreciated.

Anticipating the Nazi *Anschluss*,[1] in September 1939 quantum physicist Erwin Schrödinger, an Austrian, fled to Dublin, where he was received with open arms by the faculty of Trinity College, long "a refuge for scholars and a nucleus of civilization beyond the reach of invading barbarians."[2] There Schrödinger remained happily for seventeen years, riding out World War II and its aftermath in a tranquility that, from the perspective of older age, seemed "shameful" to him, given the horrors that war inflicted upon Europe, his native Austria, and the world.[3]

At Trinity, in partial recompense to his generous hosts, Schrödinger delivered a series of lectures "to an audience of about four hundred which did not appreciably dwindle."[4] And from these lectures, Cambridge University Press published, in 1944, a little book of ninety pages. *What Is Life?*, reprinted more than a dozen times, is now an immortal scientific classic whose positive influence upon the human community is incalculable.

Schrödinger began the lectures with an apology for venturing, as a physicist, into biology, for which he had no formal training.

We [have] inherited from our forefathers the keen longing for unified, all-embracing knowledge. . . . But the spread, both in width and depth, of the multifarious branches of knowledge during the last hundred odd years has confronted us with a queer dilemma. We feel clearly that we are only now beginning to acquire reliable material for welding together the sum total of all that is known into a whole; but on the other hand, it has become next to impossible for a single mind fully to command more than a small specialized portion of it.[5]

Prior to the nineteenth century, rare individuals – Renaissance men and women – could still command the essence of all collected Western knowledge. Not so by the twentieth century – the Age of Information, when the acquisition of knowledge exploded with exponential growth. Schrödinger found himself squarely on the horns of the twentieth century's catch-22: we lack in wisdom because we drown in knowledge. Undeterred, he plowed on: "I can see no other escape from this dilemma (lest our true aim be lost forever) than that some of us should venture to embark on a synthesis of facts and theories, albeit with second-hand and incomplete knowledge of some of them – and at the risk of making fools of ourselves."[6]

Schrödinger's risk paid off. *What Is Life?* ranks "among the most influential scientific writings of the [twentieth] century." Among the legions inspired "by this highly original and profoundly thoughtful physicist"[7] are contemporary mathematical physicists Freeman Dyson and Roger Penrose and "many scientists who have made fundamental contributions in biology, such as J. B. S. Haldane and Francis Crick."[8] At Cambridge University's Trinity College in 1985, Dyson delivered a sequel to Schrödinger's lectures under the title *Origins of Life*. Recognizing the vast ripple effect of his predecessor's thought, Dyson began: "Schrödinger's book was seminal because he knew how to ask the right questions."

> What is the physical structure of the molecules which are duplicated when chromosomes divide? How is the process of duplication to be understood? How do these molecules retain their individuality from generation to generation? How do they succeed in controlling the metabolism of cells? How do they create the organization that is visible in the structure and function of higher organisms?[9]

Throughout the illustrious history of science, asking the right questions has distinguished genius from mere insight and has pushed open the frontiers. Once the questions are posed, answers ultimately follow. Dyson continued: "[Schrödinger] did not answer these questions, but by asking them he set biology moving along the path which led to the epoch-making discoveries of the subsequent forty years: to the discovery of the double helix and the triplet code, to the precise analysis and wholesale synthesis of genes, and to the quantitative measurement of the evolutionary divergence of species." In short, Schrödinger set the stage for cracking the genetic code of life.

Immensely Simple Questions

"Why are atoms so small?" ponders Schrödinger at the outset of *What Is Life?* It was, of course, a back gate to his real question: Why are living organisms so large relative to atoms?

The answer lay in a simple thought experiment: imagine existence as a tiny, rudimentary organism comprised of a few atoms. One would experience life under unrelenting and endless bombardment, constantly jostled about by the thermodynamic fluctuations of the atoms and molecules of one's environment, whether air or water. Such a life would be unstable at best, with little opportunity for adjustment or adaptation. Such a life could not support reliable sensation or thought. Indeed, such is the life of many present-day bacteria, so minuscule that Brownian motions knock them about ferociously and incessantly.

Relative stability, it would seem, is necessary for evolution, and stability comes about by size. Larger organisms – those comprised of millions, billions, or trillions of molecules – are relatively insulated from random molecular fluctuations by the law of large numbers. Consider the frog. Although the individual water molecules in a pond dance about frenetically, the average velocity of pond water is nil, and a creature the size of a frog swims about quite placidly, unperturbed by the molecular frenzy of its pond.

Schrödinger next set out to speculate about the structure of the genetic material. Unlike Darwin, Schrödinger was privy to Mendelian genetics. He knew, for example, that the genetic material was consigned to long molecules resident in the nuclei of cells. He knew that humans possess 23 pairs of *chromosomes* and that distinct *genes* on the chromosomes encode individual traits. He knew too that the particular mix of genes inherited from the parents determines the physical characteristics of an individual according to which traits are dominant and which recessive.

Schrödinger knew that ordinary cells are *diploid*; that is, they carry two copies of the genetic material, one from the father and the other from the mother. In marked distinction, *gametes* – the eggs or sperm involved in reproduction – are *haploid* and contain but a single copy of the genetic code. By Schrödinger's day, the processes of cell division, visible under powerful microscopes, had been observed with fascination and characterized in great detail. During *mitosis* – the division of an ordinary cell – the genetic fibers (to use Schrödinger's archaic term for DNA) uncoil, stretch, and replicate. Following replication, the cell divides, each daughter cell containing a complete, dual set of chromosomes. On the other hand during *meiosis*, the cell division of gametes, *two* successive cell divisions follow the genetic replication phase, thereby creating four haploid cells rather than two diploid ones. All this Schrödinger knew and understood.

Schrödinger did not know the molecular basis of the genetic fibers. But he knew enough to make an educated guess that was prophetic: that genetic fibers are "aperiodic crystals," by which he meant long molecules whose constituent atoms never exactly repeat a discernible pattern.

Ordinary crystals were familiar to physicists. Crystals are the stuff of metals, minerals, and even ice. In crystalline structures, atoms arrange themselves in a regular lattice, in patterns that repeat periodically.

By way of analogy, consider the difference between rational and irrational numbers in mathematics. A rational number, for example, 2/7, is any number that can be expressed as the ratio of two wholes. When expressed as a decimal fraction, rational numbers repeat a block of digits. For example, $2/7 = 0.285714285714$ ... ; the block 285714 repeats *ad infinitum*. Consequently, it is not necessary to remember all infinity of digits. It suffices to remember the first six and that those digits repeat thereafter. That is, because the decimal expansion of 2/7 is *periodic*, all relevant information is contained in the first period. The second block carries no new information beyond that contained in the first.

Irrational numbers are different. The decimal fraction for the square root of 2, for example, is 1.414213562373 . . . , without repetition. Similarly, the decimal expansion for π, now known to millions of digits, begins with 3.141592653589793 but never repeats a block of digits. As decimal expansions, irrational numbers are *aperiodic*, devoid of replicated patterns. Thus, each additional digit carries new information. By analogy, rational numbers are to periodic crystals what irrational numbers are to Schrödinger's aperiodic crystals.

Because its genetic code conveys an extraordinary amount of information about the organism, Schrödinger surmised that genetic molecules must be aperiodic. Patterns should not repeat because repetition wastes information. No new information is contained in repeated structure. Similarly, Schrödinger's prowess as a physicist shed laser light upon another dark corner of biology: the mechanism of mutation. Mutation was key to the theory of evolution. Darwin, however, knew nothing about its mechanisms. He knew only that mutations were spontaneous, random, heritable changes in the attributes of organisms. By Schrödinger's era, the locus of mutation was clear: mutations originate in the genetic fibers of reproductive cells. The mechanism, however, remained mysterious.

As a quantum theorist, Schrödinger correctly inferred that mutations are quantum events. A body of circumstantial evidence supported the interpretive leap. First, mutations, like quantum events, are random; they cannot be predicted with certainty. Second, mutations, like quantum events, are small. These two facts combined would explain why mutations are normally confined to one locus on a gene. Moreover, nature provided abundant natural stimuli for mutation. Thermodynamic fluctuations could occasionally provoke mutation by spontaneously altering molecular structure. More frequent mutations would result from background radiation, by the bombardment of cosmic rays originating outside the solar system or from the radioactive decay of elements within the earth's crust and atmosphere. Finally, Schrödinger's smoking gun – exposure to artificially produced x-rays dramatically increases the odds of mutation – strongly corroborated the quantum hypothesis.

For the Darwinian process of natural selection to work efficiently, mutations must be relatively rare. If mutations are too frequent, then natural selection cannot effectively sort out advantages from disadvantages. Ideally, the genetic code should most frequently be mutation-free and only rarely contain a single mutation. Hence, the genetic molecule must be relatively stable. Stability comes by size. That is, the genetic fiber must contain a huge number of atoms as a safeguard, through the law of large numbers, against too frequent mutation.

Within a decade of Schrödinger's lectures, thanks in part to the biological insights of the wayward physicist, science was poised to decode the gene. The race was on for the structure of DNA: deoxyribonucleic acid.

The Double Helix

It was a glorious year for Britons – 1953, that is. Television carried live to the world the pageantry of Queen Elizabeth II's coronation. Britain's ninth expedition to Mt. Everest succeeded spectacularly when New Zealand's Sir Edmund Hillary and Sherpa Tenzing Norgay set foot atop the world on May 29. And just a month earlier, on April 25, two researchers at Cambridge University's illustrious Cavendish Laboratory published in *Nature* one of the most important scientific discoveries of all time: the structure of Schrödinger's genetic fiber. Of these momentous events, one was tainted, and remains so to this day. At issue are ends versus means in the pursuit of science.

Science can be collaborative, or it can be a blood sport. The contests for the theory of evolution and the structure of DNA are studies in contrast within the scientific paradigm. The latter was as ruthless as the former was gentlemanly. Alfred Russel Wallace, Darwin's competitor, happily conceded primary credit for natural selection to Darwin. Darwin's first inclination, on the other hand, when confronted with Wallace's letter bearing ideas uncannily similar to his own, was to give Wallace credit. Although competitors, the men remained friends to the end, with Wallace serving as a pallbearer at Darwin's funeral. In contrast, who deserves credit for the 1962 Nobel Prize in physiology remains, to a degree, contested because of how the battle was fought to reveal "the secret of life."

Biological cells contain three different types of long organic molecules: proteins, ribonucleic acid (RNA), and DNA. All three have been known for a long time, and all three were early candidates for encoding the secret of life. Proteins were first described at the University of Utrecht by the Dutch chemist Gerhardus Johannes Mulder, who adopted the term *protein* coined by the Swedish chemist Jöns Jakob Berzelius in 1838. Proteins, also known as polypeptides, are among the most complicated molecules known to man. Immensely long linear chains of amino acids folded into globules, proteins are the handymen of the living cell. They perform a vast array of functions. Some, like collagens, contribute structure to hair and skin. Others, like hormones, transport messages throughout the bloodstream. The vast majority of proteins act as enzymes, the catalysts of biological reactions and vital to the metabolic functions of living cells.

In 1869, Johannes Friedrich Meischer, a Swiss physician and biologist, first isolated DNA from white blood cells. The monumental discovery, underappreciated at the time, took place at the University of Tübingen just a year after Meischer had similarly isolated RNA.

DNA is a nucleic acid, now known to contain the genetic instructions for the development and function of a living organism. Eukaryotic organisms store most of their DNA within their nuclei. Some is found also in independent organelles such as mitochondria and chloroplasts, which perform specialized cellular functions, such as energy regulation. Nuclear DNA encodes the blueprint for the overall organism and much more. Among its many functions, DNA encodes instructions for the synthesis of proteins. The genetic blueprint is organized into genes, which typically encode a single trait or a single protein. Most importantly, DNA replicates during mitosis, ensuring that each new cell obtains an exact copy of the original blueprint. It does so by the most ingenious structure – that of two intertwined, mirror-image strands arranged as a double helix – but that is getting ahead of the story.

Proteins and DNA have a symbiotic relationship. DNA tells the cell how to manufacture each type of protein. Some proteins in turn, through catalytic action, unzip DNA during mitosis. Neither could function without the other.

Replication and catalysis: the two interrelated functions of life's long molecules. Proteins catalyze but do not replicate. DNA replicates but does catalyze. RNA, on the other hand, is a shape-shifter that both replicates and catalyzes. Like DNA, RNA consists of a long chain of nucleotides, but the *ribose* building blocks each contain one more oxygen atom than do the nucleotide molecules of DNA. Unlike DNA, RNA molecules are typically relatively short and single-stranded, although double-stranded variants of RNA exist that can unzip like DNA. Most interestingly, RNA appears to be older than DNA. It is found in the most rudimentary biological forms: prokaryotic cells, viruses, and *bacteriophages*, viruses that infect bacteria. Evidence is accumulating that RNA was the first biological macromolecule and the original genetic material. However, it appears that evolution has favored the specialization of RNA's two principal functions: catalysis and replication. Proteins took the former function, and DNA the latter.

Although proteins, RNA, and DNA were chemically isolated more than a century ago, their precise structures and functions were unknown until relatively recently. One held the key to life, but which was a tossup.

In the mid-to-late 1940s, most geneticists were barking up the wrong tree. It was generally believed that proteins, the most complicated of the three candidates, must harbor genetic information.[10] In hot pursuit of the structure of proteins was Cambridge's Cavendish Laboratory, directed by William Lawrence Bragg, the youngest Nobel laureate in history. At the age of just twenty-five, Bragg had received the coveted award for pioneering work, conducted with his father, in the development of x-ray crystallography.

Bragg quickly exploited the powerful new tool to probe the crystalline structure of proteins. In 1936, a diffident twenty-two-year-old chemist joined the Cavendish's up-and-coming crystallography group. Jewish by ancestry, Catholic by choice, Max Perutz had fled his native Vienna ahead of the Nazi *Anschluss*. Like his compatriot Schrödinger, Perutz found both sanctuary and intellectual ferment in a sheltering university. Eternally grateful, he never left Cambridge. Too inexperienced to understand the near impossibility of his Ph.D. assignment, Perutz set his sights on resolving the structure of hemoglobin, an enormously complex protein. By 1948, Perutz ran the crystallography laboratory, having against all odds earned Bragg's respect by coaxing hemoglobin – biology's "molecular lung," the oxygenator of blood – to reveal many of its secrets.[11]

As the newly appointed head of the Medical Research Council's (MRC) Molecular Biology Unit, Perutz quickly assembled a world-class research team. The MRC would soon churn out Nobel laureates.

In 1949, on a gut reaction, Perutz hired Francis Crick. Crick was then a thirty-five-year old physicist whose Ph.D. studies had been interrupted by the Second World War, his formidable theoretical skills diverted to the development of magnetic land mines. Crick's progress toward the Ph.D. further slowed when he turned from physics to biology, having fallen under the spell of Schrödinger's *What Is Life?* In Perutz' garden, late bloomer Crick was about to blossom into full glory.

Crick's personality was magnetic. His booming laugh resonated down the halls of the laboratory. He had a penchant for novel interpretations of experimental results. He also sported a protean ego in whose shadow lesser scientists cowered. Bragg considered him a thorn in the side. Still, Crick's powerful insights and breadth of knowledge made him popular as a sounding board for ideas in ferment. However, despite enormous potential, six years after the war, Crick still had produced nothing particularly noteworthy.[12] That was about to change.

A bit more than an hour away from Cambridge by train, at King's College, London, Maurice Wilkins conducted similar work, also using x-ray crystallography, to unravel the structure of DNA. Recent experiments by the bacteriologist O. T. Avery at New York's Rockefeller Institute[13] had suggested rather strongly that DNA, not protein, carried the genetic code. This frustrated Crick, whose broad-ranging scientific interests encompassed all things new and exciting. Crick wanted to be where the action was, but by stodgy protocol and the British sense of fair play, DNA was the domain of King's College, and protein was the domain of the Cavendish. DNA was off limits to Crick.

Making matters worse, Crick and Wilkins were old friends who had often discussed science over dinner in their younger days. For Crick to infringe upon Wilkins' turf would be exceedingly unseemly. Still Wilkins' apparent lack of enthusiasm for DNA rankled Crick. If Avery were right, DNA was genetics' Rosetta Stone. Whoever cracked the code was virtually guaranteed a Nobel Prize.

After three years of training in x-ray diffraction crystallography in Paris, Rosalind Franklin returned to her native England in 1951 as a research associate in John Randall's laboratory at King's College. It happened to be the same laboratory in which Wilkins worked on DNA. Franklin's status at King's was ambiguous, at least to Wilkins. Wilkins presumed Franklin to be his assistant; after all, science at the time was largely a gentleman's game, and Franklin, as a woman, was an interloper. Franklin, however, would have none of it. She considered herself to be Wilkins' peer. The relationship was strained from the outset. But one thing was clear: of the two, Franklin was by far the better crystallographer. J. D. Bernal, who had trained Perutz at the Cavendish, called her stunning x-ray photographs of DNA "among the most beautiful x-ray photographs of any substance ever taken."[14]

Crick's British sensibilities were undermined by the arrival at the Cavendish of a newly minted American Ph.D. named James Watson. Watson brought with him the competitive, no-holds-barred mentality of American science. A wunderkind, Watson had earned a B.Sc. in zoology from Chicago and a Ph.D. in the same field from Indiana University, both by the tender age of twenty-two.

Watson arrived at the Cavendish following a year of postdoctoral work in Copenhagen, an obligatory stop for promising young scientists. In Copenhagen Watson had received basic training in the microbiology and biochemistry of DNA, initially with little enthusiasm. However, at a zoological symposium in Naples in the spring of 1951, Watson chanced upon Wilkins and witnessed for the first time x-ray photographs of DNA. The serendipitous meeting dramatically reset Watson's research plans. Convinced of the crucial role crystallography would likely play in revealing the structure of DNA, Watson wooed Wilkins, hoping for an invitation to collaborate. Wilkins, however, appeared uninterested. Never mind. If Wilkins would not extend an invitation, Watson would create one.

Informing the National Research Council (NRC), which supported his research fellowship in Copenhagen, that he wished to transfer to Max Perutz' x-ray group at the Cavendish, Watson cavalierly packed off to Cambridge before receiving an answer. When the NRC refused his request, Watson decided nonetheless to remain at Cambridge, living on the savings accrued from his relatively lucrative fellowship in Copenhagen. Upon arrival at the Cavendish, Watson began hot pursuit of biology's Holy Grail: the structure of Schrödinger's genetic molecule.[15]

At the Cavendish, Watson and Crick spontaneously struck up both a friendship and a scientific collaboration. Before long, Watson enjoyed dinners at the "Green Door," Crick's charming apartment. The tiny but endearing Green Door had a curious connection to intellectual life in Cambridge. There Max Perutz had lived happily as a newlywed. When the Perutz family grew too large, Max gladly relinquished the Green Door to Crick and his wife Odile, both fond of entertaining. The Green Door frequently lavished its bohemian hospitality at dinnertime; Watson was a regular at the table. Conversations ranged widely but always included work. Watson chipped away continually at Crick's sensitivities regarding DNA, and before long he had effected the desired outcome: a permanent shift in Crick's research focus from proteins to DNA.

Given Wilkins' relative lack of fire in the belly, the Cavendish's chief competition came from the New World, from America's Linus Pauling. Pauling, the world's premier organic chemist at the time, was a force to be reckoned with. Pauling had already solved a significant protein problem. While Watson was returning from the earlier conference in Naples, he had received a firsthand report of Pauling's most recent coup. In a masterful lecture, Pauling had revealed his discovery that several prototypical protein chains were configured as single "alpha" helices. With dramatic flair, Pauling had built his presentation to a crescendo. The talk climaxed in the unveiling of a scale model of a protein molecule.

Pauling's discovery was an embarrassment to the Cavendish, particularly to Perutz, whose laboratory was far better equipped than Pauling's. Lamented Perutz, "It is a terrible shame that we missed this model through a combination of unfortunate circumstances and blunders for which we shall never forgive ourselves."[16] Perutz immediately set to the work of redemption. If Pauling were correct, the structure would almost certainly reveal a telltale fingerprint in an x-ray photograph. It did, and Perutz fired off the corroborating result to *Nature* in time to grab a small token of Pauling's glory. To Perutz, the photograph that confirmed the conjectured alpha-helical structure of proteins was "the most thrilling discovery of my life."[17] Bragg, however, was chagrined by Pauling's coup. Pauling had scooped the Cavendish too many times. It would not happen again.

Word of Pauling's showmanship made a deep impression on Watson. Within a few days of their initial meeting, Watson and Crick had devised a game plan. They would "imitate Pauling and beat him at his own game."[18] Watson soon secured a copy of Pauling's text *Nature of the Chemical Bond*. As 1951 ended, Crick gave Watson a second copy as a Christmas present.

Wherein lay Pauling's genius? What logic had Pauling used? What atoms could sit next to one another? What models could be built with consistent chemistry? The questions hinted that DNA was far more daunting than had been presumed. It stood to reason that DNA should have a helical structure like protein. But protein consisted of a single helix. Wilkins suspected that DNA consisted of multiple intertwined helices. DNA had its own peculiar complications. Some proteins contained but a single polypeptide, but DNA always contained four nucleotides: adenine, guanine, cytosine, and thymine, known symbolically as A, G, C, and T, respectively. Almost certainly, as Schrödinger had inferred, the ordering of the nucleotide bases was irregular. The problem was further complicated because the Cavendish's implicit charter did not permit crystallographic examination of DNA. Crick and Watson must rely on the x-ray photographs of Franklin and Wilkins, and those two, unfortunately, were hardly on speaking terms.

In November 1951, in the hopes of learning something relevant to the DNA problem from recent x-ray photographs, Watson attended a lecture in London by Franklin. Following the poorly attended lecture, Watson and Wilkins lunched in Soho. Wilkins was surprisingly upbeat despite Franklin's pessimistic assessment that offered little hope of a near-term solution to the DNA problem. She had mentioned nothing about Pauling's use of structural models as a complement to crystallographic examinations. Nor did she believe that DNA was helical.

The next day, Crick met Watson at Paddington Station, and the duo set off for Oxford to pick the brain of Dorothy Hodgkin, arguably the best crystallographer England had produced to date. Having arrived late in the day, they didn't learn much of use from Hodgkin, who was devoting her considerable energies to the study of insulin, the first protein to be sequenced. However, the time in transit was seminal.

No sooner had the train doors closed than Crick began to quiz Watson about Franklin's talk. To Crick's amazement and annoyance, Watson had taken no notes. In self-defense from Crick's withering barrage, Watson seized upon the one thing he thought he could remember from the talk: the water content of DNA.[19] From this one detail and a few minutes of reverie and scribbling, Crick had an epiphany. Only a very few structural arrangements of DNA were compatible both with Franklin's x-ray data and with a theory that Crick himself and his laboratory colleague William Cochran had devised. It appeared that DNA could consist only of two, three, or four helical strands.

Finding the structure of DNA was looking for a needle in a haystack, but the size of the haystack had just shrunk to manageable size. By the end of the train ride, Crick and Watson were euphoric, having whiffed the scent of victory.

The duo devoted the following week to assembling models. Although DNA chains were impossibly long, only a snippet of the chain was necessary to establish its basic structure. By afternoon tea on the Monday following the Oxford excursion, a promising arrangement, presumably consistent with Franklin's x-ray data, began to emerge: three intertwined helical chains. By Tuesday morning, confident of success, Crick and Watson placed a call to Wilkins and asked him to come up to Cambridge immediately for his assessment. Wilkins seemed unusually indifferent and noncommittal. Later in the day, however, Wilkins rang up with the word he would arrive at 10:10 the very next morning.

Wilkins arrived on Wednesday as scheduled, with an entourage that included, to Crick and Watson's chagrin, Rosalind Franklin. After a summary by Crick, in which Wilkins' group remained impatiently polite, Franklin cut to the chase. The model was incompatible with the water content of DNA. The fractional water content that Watson had conjured from memory following Franklin's lecture was off by a factor of ten! Wilkins' group was not happy with having been called out on a wild goose chase.

Neither was Bragg happy with the debacle. Crick, the *enfant terrible* of the Cavendish, yet to complete his Ph.D., had been a thorn in Bragg's flesh on more than one occasion. Recently the two had clashed when Crick accused the director of having stolen an idea. The Cavendish could afford no further embarrassment and would tolerate no unauthorized research. Crick and Watson were forthwith forbidden to pursue DNA further.

Crick returned to his Ph.D. studies. Watson, as a cover, took up the study of the tobacco mosaic virus (TMV), a prototype of RNA. The study of RNA had obvious advantages. RNA could reveal much about DNA, but RNA was out of Wilkins' jurisdiction.

A buzz ran through the laboratory at the word that Pauling was to visit England to present his protein work before the Royal Society. This was followed by disbelief when the U.S. Department of State unexpectedly pulled Pauling's visa. Pauling had openly advocated peaceful coexistence with the Soviets, too radical an idea during the Red Scare days of the McCarthy era. The Department of State had retaliated by hobbling its premier chemist. Another scientific casualty of McCarthyism, Salvador Luria, Watson's Ph.D. advisor, was forbidden to attend a microbiology meeting in Oxford where he was to have talked on "The Nature of Viral Multiplication." Luria tapped Watson as a stand-in, whose job it was to update the assembly on American progress with bacteriophages. A week prior to the meeting, Watson received a letter from the research group at Cold Spring Harbor summarizing their most recent findings, among them a gem. Viral DNA, with almost no protein content, could infect bacteria. Here was new evidence that DNA, not protein, carried the genetic code.

Soon thereafter, chemistry provided another tantalizing clue. At Columbia University, Austrian-born biochemist Erwin Chargaff had painstakingly cataloged the chemical ingredients of DNA and had found the proportions of the bases adenine (A) and thymine (T) to be virtually identical, almost to the molecule. The same was true for cytosine (C) and guanine (G). Such regularities were too unusual to be the result of mere chance. Moreover the ratios of A to C (or A to G) varied by organism, an indication that the nucleotides might encode genetic traits specific to the organism.

Watson reported the new result to Crick. From the deep recesses of his brain, Crick summoned an almost forgotten but salient fact. For three decades an unsubstantiated theory had existed about how genes duplicate during cell division: the theory of chemical complementarity. By analogy, one compound was like a lock and the other like a key. Only the two, and no others, could be paired. Equivalently, by photographic analogy, one compound was the negative and the other the positive. From the negative, the positive could be constructed or vice versa. Chargaff's results suggested something he himself had not realized: A and T were complementary "lock and key" nucleotides, as were C and G.

Despite the fact that all fingers pointed to DNA as the genetic molecule, decoding its structure stalled for an entire year. Crick, whose star had gradually risen, worked to complete his dissertation and received a job offer from Brooklyn. Wilkins took up the recreational pastime of fencing and devoted his professional attention to interference microscopy. He reported that Franklin remained unconvinced that DNA was helical and that she refused to share her data with him. Nevertheless, she let slip her belief that the sugar-phosphate backbone of DNA must reside on the outside of the molecule rather than on the inside. With the aid of a new x-ray device that could take twenty photos for every one made by the previous instruments, Watson obtained passable x-rays of the tobacco mosaic virus. These convinced him and Crick that the structure of the TMV was indeed helical. Nothing more could be learned about DNA from the TMV. To make further progress, the clandestine DNA investigators needed Franklin's photos, which were not forthcoming. So Watson turned his attention to the sex lives of bacteria.

In the fall of 1952, Pauling's son Peter took a position at the Cavendish, where he was assigned to share the office occupied by Watson and Crick. The threesome got along well enough. When science temporarily ran aground, they found the topic of women an entertaining diversion. At the end of the year, however, Peter walked in one morning and dropped a bombshell. His father had discovered the structure of DNA. A manuscript would be forthcoming. Peter's announcement realized the greatest fears of his office mates. There was slim chance the titan Pauling had got it wrong.

By Christmas faint hope remained alive at the Cavendish that Pauling had blundered, for the obvious reason that Pauling had yet to publish. Moreover, Wilkins didn't seem particularly worried. His spirits were in fact buoyed by Franklin's announcement that she would be leaving King's College to work in J. D. Bernal's laboratory at Birkbeck College. In the bargain she would relinquish all DNA work.

Then two copies of Pauling's manuscript arrived at the Cavendish: one addressed to Bragg and the other to Pauling's son. Adamant that Crick's ban from DNA work should remain in force, Bragg withheld the paper from Francis. It was a futile gesture. Stepping momentarily from his office, Peter Pauling returned to find Crick and Watson engrossed in his personal copy of his father's manuscript.[20] Astoundingly, the structure Pauling proposed was a triple helix with the backbone at the center, virtually the same structure that Crick and Watson had abandoned fifteen months earlier! Moreover, it soon became clear that Pauling's nucleic acid was not acidic. The master had stubbed his toe. Crick and Watson were still in the game, but time was now of the essence. When Pauling realized his error, he would stop at nothing to redeem himself.

Watson and Crick estimated a grace period of at most six weeks before Pauling was back in full swing. After toasting Pauling's failure at Cambridge's most popular and celebrated pub, the Eagle, the duo returned to DNA with a sense of utmost urgency.

Wilkins had to be alerted of the situation, and a personal visit was preferred to a phone call. Watson, who needed to be in London for other reasons, was sent as emissary to King's. There he learned from Wilkins that, in anticipation of Franklin's departure, Wilkins and a colleague had been quietly replicating Franklin's work. The admission was a backhand compliment to Franklin, whose photos were the gold standard to which Wilkins must aspire if DNA work was to thrive in her absence. Watson's pulse began to race when Wilkins disappeared into another room and reappeared with one of Franklin's new photos, apparently with neither her consent nor knowledge. In scientific folklore, the infamous photograph is known as photo 51. It was of DNA in the so-called B configuration, surrounded by water. The evidence for DNA's helicity was overwhelming. Moreover, it strongly reinforced Franklin's contention of the molecule's exterior backbone.

On the train back to Cambridge, Watson's mind raced through the pros and cons of two versus three intertwined helices. By the time the train pulled into Cambridge station, he was convinced that DNA might very well be a double helix.

Crick was not yet in when Watson arrived at the Cavendish early the next morning. Needing to ventilate scientifically, Watson went straightway to Bragg's office. To his relief Bragg's position had softened. Pauling had preempted the Cavendish too many times; here was opportunity to turn the tables. Bragg lifted the moratorium on DNA research and instructed the machine shop to assist with model building. Then Crick sauntered in. He remained skeptical, unconvinced of the plausibility of a double-helical structure. Crick argued that the group should explore physical models with both two and three helices, with rigorous attention to chemistry in the process.

Within a few days, the machine shop had begun turning out components for the models. Watson devoted a day and a half in the futile attempt to build a two-chain model with an interior backbone, against Franklin's assertion that the backbone was exterior. Watson's reluctance to tackle the exterior backbone was not simply intransigence or pique at Franklin. An exterior backbone introduced the greater complexity of packing irregular sequences of bases into the interior. However, when Watson's initial modeling attempt failed, he was left no other choice than to follow Franklin's lead.

Meanwhile Wilkins, never one to yield to excitement, had decided to put off building models for six weeks, until after Franklin's transfer to Birkbeck. Delicately, Watson and Crick asked Wilkins if they might proceed with model building with the latter's blessing. The blessing was forthcoming, but it was also irrelevant. With time of the essence, model building would go on at the Cavendish regardless of whether or not Wilkins offered his blessing.

Crick continued to side-rail some energy to his dissertation. Watson, on the other hand, was wholly consumed by attempts to arrange the base pairs of DNA. His one diversion was the cinema, where, for a few hours at a stretch, his mind was mercifully devoid of molecules. With the genetic mechanism always lurking as a backdrop to his efforts, Watson constructed a double helix with like-to-like bonding of the base pairs on the interior of the helix; that is A to A, T to T, and so on. Each identical helix was a template for the other, a necessity for genetic duplication. Crick, the spoiler, deftly shot down the like-to-like model. It would not satisfy Chargaff's law by which, in terms of the number of molecules, A = T and C = G.

A shortage of model pieces imposed an agonizing lull in the model-building frenzy. Unable to bear the wait, Watson cut out cardboard surrogates for the base pairs. The next day, while shuffling the cardboard pieces about on a flat desk, the breakthrough came. All the chemical bonds formed naturally if the base pairs were complementary: A to T, and C to G, rather than like-to-like. Moreover, this arrangement naturally satisfied Chargaff's law.

Over the next few days, a model was carefully constructed from parts fabricated by the machine shop. Crick plunged headlong into the operation along with Watson. A plumb line was brought in to align the model's axis. Measurements verified that molecular alignments were consistent with the bonding angles allowed by chemical theory. Visitors from other parts of the laboratory filed in and out, each checking the model for inconsistencies of one sort or another, according to their particular areas of expertise. One by one they nodded approval and filed out.

The helical model that emerged was a thing of beauty, like a well-made spiral staircase (Fig. 28). More aptly, it was a twisted ladder. Long sugar-phosphate molecules framed the two exterior backbones of DNA, the uprights of the ladder. When twisted, the ladder's uprights intertwined. Complementary base pairs formed the rungs of the ladder: A paired with T; C paired with G. The model was indifferent to the ordering of the pairs on the rungs. Any order was permissible, including Schrödinger's aperiodic ordering.

Word began to leak that the Cavendish had a DNA model that passed scrutiny. After cautioning against "crying wolf," the cause of their former embarrassment, Crick couldn't help spilling the beans to the clientele of the Eagle. Lawrence Bragg slipped by to view the model as soon as he

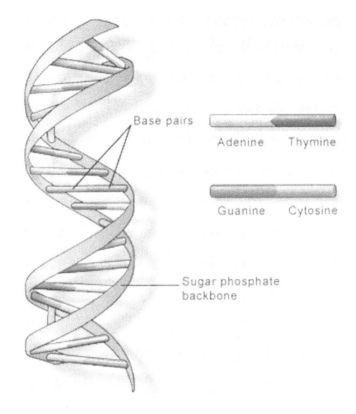

Figure 28: Double helical structure of DNA with base-pair coupling. (National Institutes of Health, U.S. National Library of Medicine.)

was sufficiently recovered from a bout of the flu. Bragg was immediately impressed by the complementarity of the two chains.

It was time to tell Wilkins, which Crick and Watson both dreaded. Their fears were unfounded. A minute was all Wilkins needed to assess the model. It looked good. The remaining step of validation was a careful comparison of the x-ray diffraction patterns in the King's College data with the predictions of the Crick-Cochran theory. Wilkins returned to London to measure reflection angles. To the great relief of Watson and Crick, "there was not a hint of bitterness in his voice."[21] The two were even more shocked when Wilkins called back two days later to say that both he and Franklin concurred: their data corroborated the double-helix structure.

The animosity between Franklin and the other three evaporated overnight. Her dogged belief in the exterior backbone had not been willful stubbornness but the result of first-rate crystallographic technique. She had been vindicated, and in the moment of triumph accepted as an equal, which is what she had wanted and deserved all along.

Wilkins advocated quick dissemination of the discovery by joint publications in the journal *Nature*, with two supporting papers by the King's College group to follow the groundbreaking paper by Watson and Crick. By then word had filtered out to Pauling. As great men can afford to be, he was magnanimous in defeat, essentially conceding the trophy of a Nobel Prize to the Cavendish team.

Less than a month after the submission of the manuscript, one of the most celebrated scientific papers of all time appeared in the April 25, 1953, edition of *Nature*. By J. D. Watson and F. H. Crick, it carried the modest title, "A Structure for Deoxyribose Nucleic Acid." It ran to only 900 words and in the fine print of the journal filled but a single page. It acknowledged the contributions of King's College in general and those of M. H. F. Wilkins and R. E. Franklin in particular.

The paper is a model of clarity and conciseness and is surprisingly accessible to a general audience. It begins with immortal words: "We wish to suggest a structure for the salt of deoxyribose nucleic acid (D.N.A.). This structure has novel features which are of considerable biological interest." The third paragraph from the last contains perhaps the greatest understatement ever to grace the annals of science: "It has not escaped our notice that the specific pairing we have postulated immediately suggests a possible copying mechanism for the genetic material."

Indeed, nature had revealed one of her deepest secrets. The double helix has been likened to a spiral staircase and to a ladder. It has also been likened to a zipper. During mitosis, the two helices unzip, separating along the base pairs, one base from each pair remaining attached to its backbone. Each helix is then a perfect template for rebuilding its complement. When the rebuilding process is complete, the cell, with two complete sets of DNA, divides, each half retaining a complete double helix. It is the biological equivalent of transferring an exact copy of a file on a computer.

The computer analogy is apt in more ways than one. In Freeman Dyson's Tarner lectures of 1985, he suggested that, if cells are "hardware," then the DNA contained in their nuclei is "software." The latter instructs the former how to replicate and what functions to perform. The analogy is far from a forced one. A computer program consists of a set of instructions, compiled into a binary code of 0's and 1's. Similarly, life is regulated by a chemical code: a four-character sequence of the base pairs A, C, G, and T.

Crick and Watson, Franklin and Wilkins, had found the chemical structure of the genetic code. They had not, however, decoded it.

Epilogue

Serendipitously, the Second World War produced a new tool for all manner of decoding: the digital computer. In England, at Bletchley Park, the site of the top-secret Code and Cypher School, mathematician and logician Alan Turing adapted computing to a purpose of grave import, breaking the Nazis' cryptographic codes. Upon success, to which the Nazis were never the wiser, the tide of the war turned. The same technology, once mature, would ultimately decode the biological Rosetta Stone. But five decades would have to pass before computing was up to such a gargantuan task.

In 1958, at the age of thirty-seven, Rosalind Elsie Franklin died from ovarian cancer, a likely consequence of years of x-ray exposure.

In 1962, the Molecular Biology Unit of Cambridge's old Medical Research Center moved into a spacious new five-story building. With the new digs came a new name: the Laboratory for Molecular Biology (LMB). Max Perutz continued at its helm. Fittingly, that same year the Nobel Committee honored four of the LMB's associates. "For their discoveries concerning the molecular structure of nucleic acids," the Committee bestowed its physiology prize, proportioned equally, to the Cavendish's Crick and Watson and to their King's College comrade Maurice Wilkins. The prize in chemistry went to Max Perutz and colleague John Kendrew for their analogous work on the structure of proteins. Collectively the awards ushered in the new age of molecular biology, now at its zenith.

Nobel Prizes are not awarded posthumously. Neither are they granted to more than three recipients at a time. Hence, the 1962 Nobel for physiology named only Crick, Watson, and Wilkins, with no mention of the invaluable contributions of Rosalind Elsie Franklin.

In 1968, James Watson published *The Double Helix*, an explosively controversial account of the race for the structure of DNA. Many were aghast at its tone and revelations. Watson's tale of cutthroat science forced a debate about the ethics of scientific endeavor that continues to the present. Among other things, *The Double Helix* aired publicly for the first time that Watson had relied upon unpublished work obtained from the rival crystallography laboratory at King's College without the permission of its director, John Randall. Many of the main characters in the story of DNA were tarred with Watson's brush. Wrote Wilkins to Perutz, "I think that you will agree that during this period described by [Watson], *none* of us behave quite impeccably."[22] Perutz' integrity was also called into question. He had apparently concealed from Randall the role Wilkins played in divulging Franklin's photo 51. In soul-searching his role as director of the crystallography lab at Cambridge at the time, Perutz conceded, "[A]s a matter of courtesy, I should have asked Randall for permission to show [the report] to Watson and Crick, but in 1953 I was inexperienced and casual in administrative matters and, since the report was not confidential, I saw no reason for withholding it."[23]

Fifty years to the month from the appearance in *Nature* of Watson and Crick's historic paper, the international Human Genome Project (HGP) was completed in 2003, culminating nearly a century of genetics research that began modestly in 1911 when an undergraduate researcher, Alfred Sturtevant, mapped the genes of the fruit fly (*Drosophila melanogaster*) so as to keep track of mutations in successive generations. "Sturtevant's very first gene map can be likened to the Wright brothers' first flight at Kitty Hawk. In turn, the HGP can be compared to the Apollo program bringing humanity to the moon."[24]

Like the Apollo Program, the colossal HGP involved thousands of scientists, billions of dollars, and thirteen years of intense effort. As in the days of Apollo, marshalling such massive resources required the intervention of big governmental agencies: the United States Department of Energy and the National Institutes of Health (NIH). And like the Apollo Program, whose benefits were anticipated to serve "all mankind," the HGP was motivated by a mixed bag of parochial and humanitarian objectives. Among the latter was the hope that knowledge of the genome would "lead to revolutionary new ways to diagnose, treat, and someday prevent the thousands of disorders that affect us."[25] As the project unfolded, its humanitarian possibilities swept up the world. The United Kingdom (especially the LMB), France, Germany, China, and Japan each contributed to the success of the HGP, as did a host of other nations.

The primary objective of the HGP was to completely decipher the sequence of base pairs of a prototypical human genome. Doing so required the world's fastest supercomputers to crunch around the clock for years. Under the able direction of Francis Collins, the current director of the NIH, milestone after milestone fell ahead of schedule. Fortuitously, full sequencing of the human genome was completed just in time to add fanfare to the commemoration of the fiftieth anniversary of the publication of "A Structure for Deoxyribose Nucleic Acid." The human genetic code – the double helix – was found to contain some three billion base pairs, organized into 20,000 to 25,000 genes. As Schrödinger had astutely predicted, the base-pair sequence is very long – presumably for stability – and aperiodic because it carries a whopping amount of information. However, the number of human genes – far fewer than had been expected – came as a big surprise. "Many things on your dinner plate have more genes than do you," quips Collins.[26]

Maurice Wilkins and Francis Crick both died in 2004, having lived to see completion of the HGP. Odile Crick, Francis' artistic wife, who drew the famous sketch of the double helix for its unveiling in *Nature*, died in July 2008. James Watson survives, having culminated a long, distinguished, and occasionally controversial career, first at Harvard University, then at the National Institutes of Health (NIH), and later as president of Cold Spring Harbor Laboratory. While at NIH, Watson was instrumental in initiating the HGP that was brought to completion by Collins.

Despite the male chauvinism that permeated Watson's account of the DNA story, the epilogue of *The Double Helix* offered a touching tribute to Rosalind Franklin.

By then all traces of our early bickering were forgotten, and we both [Watson and Crick] came to appreciate her personal honesty and generosity, realizing years too late the struggles that the intelligent woman faces to be accepted by a scientific world which often regards women as mere diversions from serious thinking. Rosalind's exemplary courage and integrity were apparent to all when, knowing she was mortally ill, she did not complain but continued working on a high level until a few weeks before her death.[27]

Rosalind Elsie Franklin died never knowing the pivotal role that photo 51 played in revealing the secret of life.[28] But history has not forgotten her. Francis Collins plays a handcrafted guitar, the fretboard of which is decorated by a mother-of-pearl inlay of a double helix. When Collins solicited names for his cherished guitar from colleagues at NIH, one name surfaced above all others: Rosalind.

Fifteen

A World Aflame

Out of this world we cannot fall.

– **Giordano Bruno**

In recompense for Thomas Huxley's dogged defense of the theory of evolution, natural selection clearly favored Huxley's progeny. Or so it would seem, for the Huxley dynasty boasts an astounding array of scientists and thinkers: Leonard Huxley, a noted biographer; Andrew Huxley, a Nobel laureate in biophysics; Aldous Huxley, playwright, poet, and author of the dystopian *Brave New World* and the religious classic *The Perennial Philosophy*; Francis Huxley, a renowned anthropologist; and Julian Huxley, knighted for extraordinary contributions to the evolutionary synthesis.[1] It was Sir Julian who made the observation proffered in the Introduction concerning *Homo sapiens'* unique place in the natural order, which bears repeating:

**Man discovers that he is nothing else
than evolution become conscious of itself.**

The Problem of Consciousness

At the scientific heart of the third Copernican revolution lies the problem of consciousness. The self-awareness apparent in the higher canopy of the tree of life attains a kind of critical mass in the human being: humans reflect upon their own perceptions, thoughts, and emotions. Humans study their own consciousness through the great microscope of science. And what is seen through that lens? Nothing at all!

Actually, there are two problems of consciousness: the "easy" problem and the "hard" problem. The easy problem concerns how sensory perception correlates with neural activity and brain function. Twenty-first-century imaging techniques allow modern Magellans – cartographers of the neural realm – to map brain activity in exquisite detail. The National Institutes of Health (NIH), for example, now funds BrainMaps, an ambitious project akin to the Human Genome Project, whose goal is to create a digital atlas of brain function at a submicron level of resolution. Progress

277

is rapid, and it is virtually certain that the "easy" problem will be fully solved.

The "hard" problem of consciousness is altogether something else. In a nutshell: "Sensation is an abstraction, not a replication, of the real world."[2] How do physical stimuli generate *subjective* experience? For example, when converted to electrical impulses by the retina and processed by the visual cortex in the occipital lobe of the brain, electromagnetic radiation at a wavelength of 700 nanometers is experienced as red. We haven't a clue why RED.

The mind is not a *tabula rasa*, Kant concluded. Uninterpreted sensory input, he argued, would be useless, "nothing to us" and "less than a dream." In today's lingo, uninterpreted sensory data would be noise devoid of music, pixels devoid of image, or caresses devoid of care. The mind does not simply record reality; it shapes reality.

In its October 2007 issue, *Scientific American* summarized our progress on the "hard" problem:

> How brain processes translate to consciousness is one of the greatest unsolved questions in science. Although the scientific method can delineate events immediately after the Big Bang and uncover the biochemical nuts and bolts of the brain, it has utterly failed to satisfactorily explain how subjective experience is created.[3]

No matter how hard one looks, one does not find experiential consciousness springing forth from matter. Under no rock that science overturns does mind lurk like a salamander. In no scientific theory is consciousness either explicit or implicit. In a spontaneous flash of classroom insight, Erin, an honors student, captured the greatest quandary of contemporary science:

What is consciousness if you cannot poke it with your finger?

The previous chapter dealt in part with the extraordinary contributions to biological science of the physicist Erwin Schrödinger, through his prescient classic *What Is Life?* In 1956, Schrödinger again ventured intrepidly outside his field and into a scientific no-man's-land: consciousness. His Tarner lectures at Cambridge were published as *Mind and Matter*, a sequel befitting *What Is Life?* On page one, Schrödinger asks, "What kind of material process is directly associated with consciousness?" Several chapters later, he confesses to having backed himself into a corner by the question, which has no answer. Quoting the Nobel laureate neurophysiologist Sir Charles Sherrington, Schrödinger unmasks a disquieting paradox: "Thoughts and feelings are not amenable to the matter concept. They lie outside it. Mind goes, therefore, in our spatial world more ghostly than a ghost, invisible, intangible; it is a thing not even of outline; it is not a thing, but remains without sensual confirmation."[4]

The difficulty is as follows. All that science knows of the universe, it knows by "poking," by observation processed through the senses of sight, hearing, touch, taste, or smell. Our senses are organs of conscious awareness. But science, whose domain is exclusively the material world, allows no quarter for consciousness, which seems resolutely immaterial. The quandary for science is not just apparent. "The impasse is an impasse," states Schrödinger categorically.

> So we are faced with the following remarkable situation. While the stuff from which our world picture is built [comes] exclusively from the sense organs as organs of the mind, so that every [individual's] world picture is and always remains a construct of . . . mind, . . . yet the conscious mind itself remains a stranger within that construct, it has no living space in it, you can spot it nowhere in space.[5]

To Schrödinger, mind is a kind of double agent, like the painter who includes on his canvas a peripheral figure who is the artist himself (as, for example, in Diego Velázquez' enigmatic *Las Meninas* from 1656). "To me this seems to be the best simile of the bewildering double role of the mind," Schrödinger offers. "On the one hand, mind is the artist who has produced the whole; in the accomplished work, however, it is but an insignificant accessory that might be absent without detracting from the total effect."[6]

Paradoxes, although disquieting, mask opportunities. The resolution of a paradox almost always results in a quantum leap in human understanding. In his keynote address at the Second International Conference on Science and Consciousness in 2000, author Peter Russell suggested that the science of consciousness today is where the science of cosmology was before the Copernican revolution. We're still "epicycling," searching desperately for an elusive fix to an inadequate paradigm.[7] What is needed is an altogether new paradigm.

Schrödinger began *Mind and Matter* from the philosophical position of the scientific metaphysic, which presumes the primacy of matter. Through the eyes of the scientific paradigm, the fundamental question is:

How does consciousness emerge from a material universe?

In the domain of philosophy, the question is posed: "How does physical reality generate the mental representation?" Kant, whose love of astronomy and an interest in nebulae led to a lifelong preoccupation with issues of perception, argued conversely: "It is the representation that makes the object possible rather than the object that makes the representation possible."[8] By flipping the scientific paradigm, Kant believed that he had instigated a new "Copernican" revolution:

How does material reality emerge in a conscious universe?

Herein lies much of the tragedy of the human condition; two rigid metaphysics compete for human allegiance: *transcendental monism* (spirit first) and *materialistic monism* (matter first), the former the metaphysic of faith and the latter that of science. "Do we really need to make this tragic choice?"[9] pleads chemistry Nobel laureate Ilya Prigogine.

There is a third way. "If one can remain with this tension of opposites long enough – 'sustain it, be true to it – [one] can become a vessel within which the divine opposites come together and give birth to a new reality.' "[10] Such is the struggle common among mystics.

A God of Iron

World-renowned paleontologist, Jesuit priest, and intellectual: Pierre Teilhard de Chardin was many things. He was also a marvelous specimen of a human being, beloved for his humor, warmth, courage, and charisma.

Among Teilhard's devoted friends was the secular humanist Julian Huxley, whose observation on evolution and consciousness concludes the first paragraph of this chapter. Teilhard took special delight in Huxley's aphorism on the human and evolution because it so perfectly encapsulated his own thinking. To Teilhard, evolution was no threat to faith; it was *the* guiding principle of the cosmos.

> Is evolution a theory, a system, or a hypothesis? It is much more: it is a general condition to which all theories, all hypotheses, all systems must bow and which they must satisfy henceforward if they are to be thinkable and true. Evolution is a light illuminating all facts, a curve that all lines must follow.[11]

Uniquely endowed to unite disparate ideas that to others seem irreconcilable, Teilhard was also a mystic, a new shoot from the ancient root of the "perennial philosophy." Christian mysticism had struck a deep vein in the Middle Ages. In the gardens of the German monasteries along the Rhine River grew the "Rhineland mystics" Hildegard of Bingen, Mechthild of Magdeburg, and Meister Eckhart. A near-death experience transformed Julian of Norwich into the greatest of early English mystics. Meanwhile Spanish Catholicism nurtured the mysticism of Saint Teresa of Avila and Saint John of the Cross.

Teilhard's own mysticism was inborn. The fourth of 11 children, Teilhard was born in France on May 1, 1881, to a pious Catholic family. Upon his mother Berthe fell responsibility for his religious and moral upbringing. Pierre's father Emmanuel, on the other hand, instilled in his children a love of the outer world and of nature study.[12] Two loves – religion and natural science – formed the solid core of the child Pierre and the mystic he would become.

His family's eighteenth-century manor house in provincial Sarcenat commanded a view of the vast Clermont plain, peppered by rounded hills and extinct volcanoes. In the distance lay the Puy Mountains. The land abounded in rocks and minerals and in botanical and zoological diversity. From an early age, Pierre showed far stronger affinity for minerals than for the evanescence of flowers or the iridescence of butterflies that so delighted his father and siblings.

Not yet ten years old, Teilhard once stubbed his toe on a pebble. "I picked it up, and I loved it." The humble anecdote attests to one of Teilhard's unusual qualities: a lifelong "feeling for matter and the durable."[13] Throughout his life, Teilhard religiously carried a geology hammer and a magnifying glass on every outing, enamored by all aspects of the material world.

His affinity for matter bordered on the mystical attraction of alchemists for the elements. Awakened from a reverie by a teacher who demanded to know about what Teilhard had been thinking, the boy replied honestly, "Pebbles." A childhood memory of his first "idol" – an iron plow-spanner – speaks to the depths of Teilhard's attraction to matter. "You should have seen me as in profound secrecy and silence I withdrew into the contemplation of my God of Iron, delighting in its possession, gloating over its existence. A God, note, of iron; and why iron? Because in all my childish experience, there was nothing in the world harder, tougher, more durable than iron."[14] Childhood infatuations are often crushed, and so it was with Teilhard's idol.

> I can never forget the pathetic depths of a child's despair, when I realized one day that iron can be scratched and can rust. I had to look elsewhere for substitutes that would console me. Sometimes in the blue flame (at once so material, and yet so pure and intangible) flickering over the logs in the hearth, but more often in a more translucent and more delightfully colored stone: quartz or amethyst crystals, and most of all glittering fragments of chalcedony such as I could pick up in the neighborhood.[15]

One senses, mirrored in Teilhard, Plato's anguish over the death of his teacher Socrates and subsequent longing for the durable in the face of the impermanence of "that which flows." Similarly, Teilhard quested after his own Holy Grail: a substance as durable as he had once believed iron to be.

At eleven, Teilhard was packed off to a Jesuit boarding school near Villefrance-sur-Saone to begin his secondary education. "Disheartening well-behaved," his new teacher, Henri Bremond, described him. Teilhard excelled in scholastics except in the study of religion, where he chafed at the rote training in the catechism.

Despite this aversion to the dull surface of Catholic rituals, Teilhard sensed unfathomed depths beneath the surface. By the end of his secondary studies, he had settled upon a religious vocation. Barely eighteen, Teilhard entered the Jesuit Novitiate at Aix-en-Provence. Upon completing the novitiate, he transferred to Laval for the Jesuit "juniorate," two years of intense concentration on the classical languages. In 1901, when France's draconian Law of Associations forbade all religious orders,

Laval's Jesuits abruptly relocated to Jersey, where Teilhard completed his juniorate. Having chosen philosophy for further study, Teilhard earned a B.A. in 1902 at Caen.[16]

The volcanic Jersey terrain revived Teilhard's childhood interest in geology but precipitated an internal crisis. Naturally predisposed to science but religiously pious, Teilhard felt pulled between two worlds and anguished over whether to abandon secular pursuits to devote himself more fully to the life of the spirit. The sudden death of his eldest brother, Albéric, to tuberculosis only exacerbated his crisis of purpose. A wise rector in whom the distraught Teilhard confided encouraged the "development of natural gifts every bit as much as sanctification." From that day onward, a rare kind of integrity graced Teilhard's life, simultaneous faithfulness to two complementary though sometimes conflicting aspects of his persona and of the universe itself: matter and spirit.[17]

This tension between the material and the spiritual had dogged humanity for centuries, played out through the animosity of science and religion. For Teilhard, the tension was personal – and painful. On the one hand, the world advanced at breakneck pace, propelled by scientific discovery. On the other hand, the Church provided stability, a port in the storm of change. Psychologically, Teilhard felt "torn between two loyalties to which he wished at all costs to be true."[18] But like diamonds, mystics are forged under the harshest of conditions:

> For the mystic . . . the struggle for identity and integrity assumes epic, even cosmic proportions because it presents itself existentially as an intensely personal life-and-death struggle. He becomes a mystic precisely because he experiences a breakthrough in terms of a unifying principle, a coincidence of opposites, an overarching integration in which conflicting worlds become one world. He impresses his fellows as both strangely at home in the world and mysteriously at peace with himself. It is as if the mystic's integrity . . . is achieved because he has suffered, at least in imagination, a potential dichotomy that threatens to tear him to pieces.[19]

In the introduction to Teilhard's seminal work, *The Phenomenon of Man*,[20] Julian Huxley alluded to the extraordinary result of Teilhard's uncommon integrity: "[Teilhard] has effected a threefold synthesis – of the material and physical world with the world of mind and spirit; of the past with the future; and of variety with unity, the many with the one." It is curious that Einstein, Teilhard's contemporary and a mystic in his own way, also occasioned a threefold synthesis: matter and energy, space and time, gravitation and acceleration.

Having completed studies in scholastic philosophy in 1905, Teilhard was sent to Cairo to teach chemistry and physics at the College of the Holy Family. There, he recalled, "The East flowed over me in a first wave of exoticism: I gazed at it and drank it in eagerly – the country itself, not its peoples or its history (which as yet held no interest for me), but its light, its vegetation, its fauna, and its deserts."[21] Despite the demands of teaching, Teilhard found time to indulge his old passion of geology and to awaken a nascent interest in paleontology, which was to bloom into his scientific calling. Upon his return from Cairo in 1908, Teilhard began theological studies anew at Ore Place, near Hastings in Sussex, England, where Jesuit theologians had taken refuge following their expulsion from France in 1901. Sussex awakened Teilhard in new ways. Despite diligent preparation for his theology exams (which he passed with flying colors in August 1911) Teilhard carved out ample time for scientific study and practice. He devoured Henri Bergson's *Creative Evolution* and relished the paleontology holdings of the British Museum, to which he and a fellow student contributed some 260 fossil specimens.[22]

Bergson's influence on Teilhard was profound. A first-rate philosopher, Bergson reasoned persuasively that science and intuition are equal methodologies for apprehending reality. Each can be precise and trustworthy. But each is incomplete: inner and outer modes of knowing are complementary.[23] Bergson also exposed deep philosophical issues concerning the human perception of time, notions that insinuated their way into Teilhard's evolving theology.

The Sussex Weald exerted a tug on Teilhard that turned his allegiances topsy-turvy. Nature, whose attraction had heretofore been her material side, burst to life before Teilhard's eyes. "My sense of plenitude had become, as it were, reversed. And in the light of this new orientation, I have never ceased to observe and proceed ever since."[24]

Throughout his life Teilhard remained true to matter and to spirit, but, as he entered his thirties, spirit began to supplant matter as his first love. The boy whose idol was iron and who loved pebbles would later write:

> Throughout my whole life, during every moment I have lived, the world has gradually been taking on light and fire for me, until it has come to envelop me in one mass of luminosity, glowing from within. . . . The purple flush of matter fading imperceptibly into the gold of spirit, to be lost finally in the incandescence of a personal universe. . . . This is what I have learnt from my contact with the earth – the diaphany of the divine at the heart of a glowing universe, the divine radiating from the depths of matter aflame.[25]

It was as if, over the course of a lifetime, the matter that Teilhard so loved dissolved before his very eyes to reveal an inner essence that was yet more precious. Paradoxically, the durability he coveted in matter was to be found in the most insubstantial of all substances. "The felicity that I had sought in iron," Teilhard recalled, "I can now find only in Spirit."[26]

Serendipitously, in the hard-boiled world of science, matter's primacy was also being questioned. Rutherford's gold foil experiments revealed the atom, nature's building block, to be mostly empty space. Moreover, Einstein's most famous paper, which equated matter and energy, suggested that energy is the primary stuff of the cosmos. Adding insult to injury, de Broglie established that matter, like light, exhibits wave-particle duality. And in the Copenhagen interpretation of quantum mechanics, matter waves express the *probability* of the material object being found in a given place when observed. That is, matter manifests only a tendency to exist until observed. The dissolution of the material world at the hands of science provoked one well-respected physicist to quip, "Whatever matter is, it isn't made of matter."[27]

On a more serious note, Eddington (the revered British astronomer of Chapters 5 through 8) admitted, "The frank realization that physical science is concerned with a world of shadows is one of the most significant of recent advances." To illustrate, he spoke fondly of his favorite writing desk, as recounted by Schrödinger:

> Some may remember A. S. Eddington's "two writing desks"; one is the familiar old piece of furniture at which he is seated, resting his arms on it, the other is the scientific physical body which not only lacks all . . . sensual qualities but in addition is riddled with holes; by far the greatest part is empty space, just nothingness, interspersed with innumerable tiny specks of something, the electrons and nuclei whirling around.[28]

Even if the world is primarily material, matter is not what it has seemed.

The Within and the Without

Following theological studies, Teilhard joined the Museum of Natural History in Paris, taking up the study of paleontology under the direction of the eminent professor Marcellin Boule. At the museum, he struck up a friendship with prehistorian Henri Breuil, who afforded Teilhard a golden opportunity to visit paleolithic caves in Altamira, Spain. The trip into humanity's dim past was the first of many upon which Teilhard embarked with relish. But the later journeys would have to wait for Teilhard's baptism by fire.

"The shot heard round the world" plunged Europe into war. His studies interrupted by conscription, Teilhard was sent to the front as a stretcher bearer for a regiment of Moroccan Tirailleurs (sharpshooters).[29] The unit had no chaplain. Into that void stepped Teilhard, who assumed the solemn duty of consoling the wounded and dying, most of whom were Muslim. Extreme heroism and devotion to the soldiers and officers of the Eighth Regiment earned Teilhard a title of the highest esteem: *"Sidi Marabout"* – one who is "closely bound to God."[30] Although he saw action in many great battles of the First World War, Teilhard survived miraculously unscathed, without the slightest wound. Decorated for valor time and again, he declined all promotions in order to remain at the front, dedicated to those who most needed him and, by all indications, a complete stranger to fear.[31]

Despite its unspeakable horror, life at the front elicited in Teilhard a strange exhilaration. There, all conventional notions were swept away, and life was reduced to its raw essence. "The man at the front is no longer the same man," Teilhard understood. In the crucible of battle, his calling came into ever-sharper focus. In free moments amid chaos, sometimes under heavy artillery bombardment, Teilhard composed nearly twenty essays that collectively laid out a bold new view of Christianity. The essays would later serve as seed stock for his great works, among them *The Phenomenon of Man*.

Teilhard's soul remained unscathed by war as well. He returned from the front lines without cynicism, committed more deeply than ever to "the great work of creation and of sanctifying humanity."[32] In an essay entitled "Sketch of a Personalistic Universe," Teilhard set the stage for *The Phenomenon of Man* and, in the bargain, gave humanity a new metaphysic: "There is neither spirit nor matter in the world; the stuff of the universe is spirit-matter. No other substance but this could produce the human molecule."[33]

Where then does consciousness originate? At what level of complexity can we say an organism is conscious? Science had no answer, but Teilhard reasoned, "In a coherent perspective of the world: life inevitably assumes 'pre-life' for as far back as the eye can see. . . . Refracted rearwards along the course of evolution, consciousness displays itself qualitatively as a spectrum of shifting shades whose lower terms are lost in the night."[34]

In short, Teilhard recognized consciousness as an integral "aspect or dimension in the stuff of the universe."[35] How does consciousness arise? Why, it was there all along.

To Teilhard spirit and matter were complementary aspects of the same reality, which he termed the *Within of Things* and the *Without*. Like two sides of a coin, or the two faces of the moon, matter has two faces. For centuries, astronomers saw but one face of the moon. Similarly, scientists normally deal only with the exterior face of matter, its "Without." But matter, like the moon, has a hidden face. "Co-extensive with the Without, there is a Within to things. . . . We have recognized the existence of a conscious inner face that everywhere duplicates the 'material' external face, which alone is commonly considered by science."[36]

It delighted Teilhard to find sympathy for his ideas from unlikely sources. For example, several years after writing *Phenomenon*, he unearthed a gem of resonance by J. B. S. Haldane, Britain's iconoclastic mathematical biologist.

We do not find obvious evidence of life or mind in so-called inert matter, and we naturally study them most easily where they are most completely manifested; but if the scientific point of view is correct, we shall ultimately find them, at least in rudimentary forms, all through the universe. ... Now, if the co-operation of some thousands of millions of cells in our brains can produce our consciousness, the idea becomes vastly more plausible that the co-operation of humanity, or some sections of it, may determine what Comte calls a "Great Being."[37]

Teilhard had his detractors as well. Spirit-matter is a difficult sell for Westerners, steeped as we are in a tradition of metaphysic 1: materialism. Smacking of animism, Teilhard's position was anathema to most scientists, for "modern science was born out of the breakdown of the animistic alliance with nature."[38] As pantheistic and "pagan," Teilhard's hybrid metaphysic was also anathema to traditional religion, which had gone to extraordinary lengths, including genocide, to expunge the world of pagan notions.

But the ancient and ubiquitous notion of pervasive consciousness is tenacious. Consider, for example, the Sanskrit aphorism, "God sleeps in minerals, awakens in plants, walks in animals, and thinks in humans."[39] What distinguishes the artist, poet, or mystic from the run-of- the-mill human being? Each shares a common attribute: the rare ability to see the hidden "Within" of things, to recognize the "humanity" of a stone, a tree, or an enemy. "Everything that is dead quivers," expressed the Russian artist Wassily Kandinsky. "Not only the things of poetry, stars, moon, wood, flowers, but even a white trouser button glittering out of a puddle in the street. . . . Everything has a secret soul, which is silent more often than it speaks."[40] Disturbed in 1798 "with the joy of elevated thoughts" – lingering "oceanic" feelings of mystical experience, a sense of union with all-that-is – William Wordsworth wrote "Lines Composed a Few Miles above Tintern Abbey," which concludes wondrously with allusions to a cosmic consciousness:

> A motion and a spirit, that impels
> All thinking things, all objects of all thought,
> And rolls through all things.

Ubiquitous too among indigenous peoples the world over is the notion of the world soul: the collective spirit shared by all beings and objects of creation – humans, animals, plants, and stones.

It is perhaps to the "simpler and higher" view of nature characteristic of the aboriginal human that the great nineteenth-century psychologist and philosopher William James referred when he wrote, "The truth of things is after all their living fullness, and some day, from a more commanding point of view than was possible to any one in [a previous] generation, our descendants, enriched with the spoils of all analytic investigations, will get round to that higher and simpler way of looking at nature."[41]

For centuries, science contented itself with Descartes' *res extensa* as its exclusive domain, assiduously avoiding the *res cogitans*. But as James prophesied, the analytical investigations of science have turned science back to the neglected problem of consciousness. Science has no alternative, for consciousness is the culprit in the demise of all its sacred cows –

dualism, materialism, determinism, and reductionism. The physics of the future must bridge mind and matter if we are to heed the admonitions of the greatest modern physicists.

" . . . the stuff of the world is mind stuff."

– Arthur Eddington[42]

". . . the stream of knowledge is heading toward a non-mechanical reality; the Universe begins to look more like a great thought than like a great machine."

– Sir James Jeans, thermodynamicist, astronomer, and popularizer of scientific thought[43]

"Yet, I am arguing for some kind of active role [in physics] for consciousness, and indeed a powerful one . . ."

– mathematical physicist Roger Penrose in The Emperor's New Mind[44]

"It would be most satisfactory of all if physics and psyche could be seen as complementary aspects of the same reality."

– Wolfgang Pauli, recipient of the 1945 Nobel Prize in physics for the Pauli exclusion principle[45]

". . . consciousness must be introduced into the laws of physics!"

– Eugene Wigner, recipient of the 1963 Nobel Prize in physics for contributions to the theory of the atomic nucleus and elementary particles[46]

Cosmogenesis

"In science, truth always wins."[47] The admirable inclination of most scientists to seek out a higher level of understanding when faced with the "damned facts" lies in stark relief to the historical tendencies of ecclesiastical authorities. Too often, the inclination of religious authority at the first sign of "heresy" is to bring out the instruments of torture and light the stake. Teilhard soon ran afoul of the guardians of orthodoxy.

In 1919, his education having been delayed for five years by war, Teilhard returned to his scientific studies in Paris with renewed vigor and purpose. In 1922, at the age of forty-one, he completed his doctorate. Despite a preference for fieldwork, Teilhard accepted an academic post at Paris' Institut Catholique, where he quickly earned the admiration of colleagues and students. His influence spreading, Teilhard received numerous offers to lecture at prestigious institutions including the Ecole Normale, the Poly-technique, and the School of Mineralogy. His luminous lectures held wide appeal, but particularly to the young, those receptive to intellectual and spiritual adventure and willing to challenge orthodox beliefs.[48]

In 1921, through his association with the Museum of Natural History, Teilhard was placed in charge of an acquisition of mammalian fossils col-lected during France's initial paleontological expedition to China under Father Emile Licent. The two Jesuits struck up a correspondence, and on April 10, 1923, Teilhard set sail for China to join Licent in Tientsin. Lit-tle did Teilhard know that China would quarantine him for most of the remainder of his life.

Upon arrival in Tientsin, the middle-aged Teilhard was stricken by an agonizing and rare collapse of self-confidence. By sheer force of will, he fulfilled his commitment to accompany Licent's expedition to inner Mon-golia's Ordos Desert. Gradually, Teilhard regained his emotional sea legs. In the Ordos, the two Jesuits unearthed the first incontrovertible evidence of paleolithic humans in that remote corner of the earth. And there, in the healing stillness and solitude of the vast desert, Teilhard composed his mystical "Mass upon the World," which rivals the psalms of David in its lyrical beauty.

In 1924, Teilhard returned to Paris, by all fair measures a conquering hero. His first-rate anthropological work in China had garnered world-wide notoriety. But instead of praise, censure awaited.

Throughout his adult life, Teilhard struggled with the inner imperative to reconcile his bedrock Christian faith with the evolutionary theory that was his vocation and passion. Rather than dismissing the facts of natural philosophy contained in the fossil record, some of which he had collected with his own hands, Teilhard made a bold move. He shifted to a philo-sophical perspective so expansive that it embraced both science and faith in its panorama. Of Teilhard's sweeping view, his friend Julian Huxley wrote:

> Through this combination of wide scientific knowledge with deep religious feeling . . . he has forced theologians to view their ideas in the new per-spective of evolution, and scientists to see the spiritual implications of their knowledge. He has both clarified and unified our vision of reality. In the light of that new comprehension, it is no longer possible to maintain that science and religion must operate in thought-tight compartments or con-cern separate sectors of life; they are both relevant to the whole of human existence.[49]

One who heard Teilhard's lectures at Grandes Ecole noted: "He uncov-ered the timidity of our irresolute faith, acts of intellectual cowardice . . . and lastly the puerility of our beliefs . . . the malign duplicity of our spir-itual lives."[50] It is possible to choose religion over science or science over religion, or to artificially compartmentalize the two, but the result is a spiritually and/or intellectually stunted life. Needless to say, such pronounce-ments did not sit well with the timid keepers of orthodoxy. For offering a brazen new evolutionary view of Christianity, Teilhard was reprimanded and forbidden to teach. The wound ran deep for one whose every fiber embraced the message and hope of a "cosmic Christ." Still, mindful of the damage that controversy could inflict upon the Church, Teilhard submit-ted to his superiors and returned to China as ordered – banished, some would say.

In a "nice stroke of irony," Teilhard's banishment allowed him ulti-mately to hone and refine his "dangerous thoughts." In China, Teilhard completed drafts of *Le milieu divin* in 1927 and *Le phenomene humain* in 1930. Year after year, he requested permission to publish. Year after year, authorities denied permission, and Teilhard, the obedient servant, acqui-esced. Of his eighteen major works, only one saw print within Teilhard's lifetime.[51]

The Phenomenon of Man (1957) is widely considered Teilhard's greatest posthumous legacy. The panoramic sweep of his thought is breathtak-ing. The book's domain is the entire universe, and its temporal scale spans from the moment of the Big Bang until the distant horizons of the future. In content it embraces cosmology, evolution, and religious faith. From this huge perspective, Teilhard understood intuitively something that sci-ence's more restrictive view had failed to grasp: evolution is not just about life. Life is the conveyor of consciousness, and, in general, the more com-plex the organism, the higher its level of awareness. Evolution therefore is inherently about consciousness. And to Teilhard, consciousness was syn-onymous with spirit. In this regard, he was in perfect resonance with the German philosopher Hegel, for whom "nature's purpose is eventual self-realization of its spiritual component."[52]

The major innovation of *The Phenomenon of Man* is the outline of a gen-eralized theory of evolution – Evolution with a big "E" – for which Teil-hard coined the term *cosmogenesis*. The term evokes a restless cosmos, ever creative and continually unfolding. When was the moment of creation? Now. How different is the cosmos of Teilhard, recreated at each instant, from that of Aristotle, perfected by the creator for all eternity!

How could a man of such faith arrive at so bold a view of evolution? Keenly aware of twentieth-century developments in cosmology, among them that the universe is expanding from Lemaítre's "primeval atom," Teilhard grasped that evolution applies not only to biological processes but also to the universe as a whole. The Copernican revolution and its aftershocks in modern cosmology had freed us from the Aristotelian mis-perception that the physical universe is immutable. Similarly, Darwinism liberated us from the misapprehension that biology is static. Teilhard saw a trend: the nature of nature is to change.

Most significantly, however, Teilhard's deep spirituality allowed him to intuit that the cosmic evolutionary processes are not directionless but

converge to create beings of greater biological complexity and concomitantly higher consciousness, through a process he termed *complexification*
(or *complexity-consciousness*). And by following the arc of complexification
into the future, Teilhard envisioned the *Omega Point* on the far horizon of
evolution, that point toward which all creation advances, however haltingly.

The Alpha and the Omega

His superiors occasionally permitted Teilhard brief sojourns to Paris from
exile, but the injunctions against teaching and publishing remained in effect. Still, Teilhard visited India and then America, the latter twice. The
weeks aboard ship afforded welcome opportunity for uninterrupted devotion to writing and clarifying his thought.

In 1939, China fell under Japanese occupation. For their protection, the
Jesuits relocated to Peking and founded the new Institute for Geobiology.
In Peking, Teilhard endured six dispiriting years, essentially under house
arrest. By strength of will, he maintained an outward optimism, but the
dual stresses of occupation and renunciation were at times unbearable,
and during those times Teilhard teetered on the edge of despair.

Throughout the period of occupation, a community of friends gathered daily at around 5:00, their companionship a salve to Teilhard's beleaguered spirit. Pierre Leroy, a young Jesuit scientist who shared the confinement for its duration, found in Teilhard's presence a constant source
of "optimism and confidence."

> He had a fine sense of humor: his face would light up like a child's at a
> good joke; and if sometimes he could not resist an inviting target for his sly
> wit – after all, on his mother's side the blood of Voltaire flowed in his veins
> – it was done with such unaffected good humour that no-one could take
> it in bad part. It was one of his outstanding characteristics that he never
> gave way to bitterness, not even when decisions were taken that prevented
> the dissemination of his ideas. No wonder he was universally loved and
> admired.[53]

Following the end of World War II, Teilhard returned to France. There
he was struck down by a debilitating heart attack while preparing for
travel to South Africa to assist in excavations of *Australopithecus*, an early
hominid. Two full years elapsed before he recovered sufficiently for the
rigors of fieldwork.

In 1951, France's highest scientific body, the *Academie de Sciences*, honored Teilhard by inducting him into its revered ranks. Ever wary of Teilhard's popularity, Church authorities, seeking to limit Teilhard's influence,
shipped him off to America. There he lived in New York City, continuing his anthropological work through an affiliation with the Wenner Gren
Foundation. Teilhard returned to his homeland just once, in 1954. While

there he visited the caves of Lascaux, renowned for their magnificent paleolithic art dating back 18,000 years. Subjected to additional restrictions while in France, Teilhard prematurely terminated his visit home and returned to America, resigned to exile.

To the end, Teilhard remained faithful to the Church that had silenced and banished him. Ever devout in his faith, Teilhard prayed for a good end and expressed a wish to die at Easter. Both requests were answered on April 10, 1955, when he collapsed of a massive heart attack just before afternoon tea on Easter Sunday.

Pierre Teilhard de Chardin was buried two days later near Poughkeepsie, New York, at Saint Andrews-on-Hudson, in the graveyard of the Jesuit Fathers. The service, on a gray and rainy day, was attended by just ten of his friends.

Teilhard's death preceded Einstein's by eight days. Each man belonged to the age and to the world more than to a race, a country, or a creed. Beneath the surface differences of religious and secular worldviews, there was more than a little resonance. "The most beautiful emotion we can experience is the mystical," Einstein held. "It is the source of all true art and science."[54] To their dying days, each remained in rapt awe of the "Great Holy Mystery," a child to the last breath.

In Revelation 22:13, the following cryptic words are attributed to Jesus: "I am the Alpha and the Omega, the First and the Last, the Beginning and the End." Teilhard, perhaps more than anyone else, laid bare the meaning behind the cipher. By the first letter of the Greek alphabet – Alpha – Teilhard referred to the spark of the divine at the very heart of matter. Within each and every particle of the universe, he believed, resides the "Christ seed." And by the last letter of the alphabet – Omega – he pointed to the end game of creation, the direction of evolution, and the goal of complexification. From beginning to end, we are creatures of spirit, in every material atom of our being.

Epilogue

Immediately following Teilhard's death, friends and admirers, recognizing the world's desperate need for his holistic vision, set to the task of bringing his banned writings into the light of day. His major works were published in rapid succession in his native French, beginning with *Le phenomene humain* (1955) and *Le milieu divin* (1957), each translated shortly thereafter into English. And thus it was that, posthumously, Teilhard attracted followers the world over that he was denied in life.

Since Teilhard's death, numerous leading organizations have sponsored events to celebrate his life and legacy. Among these are UNESCO and the United Nations. More specifically, in 1983 an exhibition at the Chapter House of London's famed Westminster Abbey commemorated for two weeks the centenary of Teilhard's birth, and in 2005 conferences convened at the UN and around the world to commemorate the fiftieth anniversary of his death. One – the "Spirit of Fire" conference – gathered in November 2005 at Philadelphia's Chestnut Hill College, a Catholic school with a Jesuit history. Four plenary talks on Teilhard each filled to their 500-person capacity, populated by admirers from all walks of life, diverse faiths, and many countries. In her opening remarks, college president Carol Jean Vale referred to "Saint Teilhard." It was no slip of the tongue. In the view of Vale and most of the audience, Teilhard de Chardin deserves sainthood as surely as the Christian mystics in whose footsteps he followed, among them Saint Teresa of Avila and Saint John of the Cross.

Sixteen

The Winds of Change

As far as you are able, join faith to reason.

– **Boethius**

If the cosmos has an overarching trait, it is dynamism. Heraclitus was right: everything flows. We occupy a cosmos that unfolds continually. If All-That-Is exists in perpetual flux, what can be said of the universal winds of change?

In 1932, the year in which Heisenberg won the Nobel Prize in physics, the Nobel in physiology went to Sir Charles Sherrington and Edgar Adrian "for their discoveries regarding the functions of neurons." Sherrington observed two contrary winds blowing about the cosmos: "The universe of energy is, we are told, running down. It tends fatally towards an equilibrium which shall be final. An equilibrium in which life cannot exist. Yet life is evolved without pause. Our planet in its surround has evolved it and is evolving it. And with it evolves mind."[1]

In Sherrington's broad brush strokes, two opposing tendencies coexist under the mantle of cosmogenesis. On the one hand, the universe exists in inexorable decline. It slides irreversibly toward heat death, condemned to decay by the second law of thermodynamics. Iron rusts, bodies wither, aged stars collapse, their gravesites unmarked as black holes. "This is the way the world ends," penned T. S. Eliot, "not with a bang but a whimper."[2]

And yet decay is only half the story. Acting simultaneously – seemingly in defiance of the second law – is Teilhard's counterentropic trend of complexification by which life, ever more ordered and biologically complex, swims upstream against the tide of entropy. This begs a question: what is the relationship, if any, between the two megatrends?

Far from Equilibrium

In thermodynamic terms, the universe is comprised of three types of systems: isolated, closed, and open, each defined by its boundary as addressed in Chapter 14. In bondage to the second law, isolated systems evolve toward the equilibrium of disorder and heat death. In contrast, systems with permeable boundaries, at least some of them, give rise to complexity, the antithesis of disorder. Post-Einstein, the distinction between closed and open systems has dissolved, for matter and energy are equivalent. Thus, the earth is open, her boundary permeable to the influx or efflux of matter/energy. Open systems are the incubators of life. How life bucks the trend in a universe that is running down is a story worth telling.

Although the cosmos is "going to hell in a handbasket," it doesn't do so uniformly. Here and there, scattered about, are pockets of activity that defy the overall trend, remote outposts that haven't gotten the message that the war against entropy is lost.

Think of the energy content of the universe as a flowing stream, whose current seeks the lowest point on its downhill course: the equilibrium of the vast ocean. In coastal areas – near equilibrium – the stream is brackish and sluggish. Further from equilibrium, in the Piedmont, the stream is more varied; ripples and occasional rapids punctuate flat water. The Piedmont is a good place for timid canoeists who nevertheless relish the sporadic rush of adrenalin of class one or two whitewater. Far from equilibrium (FFE), high in the mountains, where the stream plummets, class four and five rapids abound. Headwaters are the domains of kayakers and avid canoeists. There, the adrenalin rush is continual.

Almost continual, but not quite – amid the holes, waves, and hydraulics of whitewater are blessed pockets of relative calm. Forming on the leeward side of boulders, eddies offer dependable oases of tranquility in otherwise turbulent surroundings. Within these eddies, fish take refuge and kayakers take breathers. Eddies don't form in placid water but only where the stream is FFE.

Open systems are eddies in the thermodynamic stream of entropy. As water flows into and out of eddies, matter/energy flows into and out of open systems. Paradoxically, both give rise to order out of chaos. What ingredients are necessary for order to arise in a thermodynamic stream? The necessary conditions are three.[3] First, the system must be open. Isolated systems are constrained by the second law; they cannot give rise to order greater than that of their initial state. Second, the system must exist in a state FFE. Put another way, there must be a strong flux of energy driving the system, as in the example of a mountain stream whose gradient is steep and whose flow rate is fast. Finally, the system itself must be *nonlinear*. The term characterizes the mathematics of the system.[4] Simply put, a nonlinear system is one that exhibits a strong feedback mechanism upon itself.

By and large, nature has a strong predilection for the nonlinear. Newton's inverse square law of gravitation is nonlinear. Maxwell's equations of electrodynamics are not, but they are the exceptions. General relativity is strongly nonlinear. So are the strong and weak nuclear forces. As a result, so is most chemistry. And because life involves chemistry, life's basic mechanisms are inherently nonlinear.

The emergence of life in a decaying universe is possible only because all three criteria are satisfied within the cauldron of life. Biochemistry takes place inside cells, which are open systems that depend upon the transport of energy and material across their boundaries. Isolated from their environment and deprived of sustenance, cells falter and die like a city under siege. The earth, where life abounds, is also an open system. And thanks to our proximity to the sun, conditions on Earth are FFE. We've got what it takes here in the "Goldilocks zone." But the right conditions are no guarantee of life. Or are they?

The Glorious Garbage Bag

Forty years after Erwin Schrödinger's seminal lecture series *What Is Life?*, another prescient thinker unafraid to venture beyond his primary areas of expertise confronted a related question. The year was 1985, the venue was Cambridge University's Trinity College, the forum was the Tarner lecture series, and the speaker was mathematical physicist and polymath Freeman Dyson.

After paying homage to Schrödinger, Dyson began an exploration of the *Origins of Life* (1985) with questions that his predecessor had not addressed four decades previously. Chief among them: how did life emerge from the matrix of matter?

Life on Earth is complex. It assumes a bewildering array of forms from myriad species of bacteria and phytoplankton to toadstools and redwoods; from Portuguese men-of-war to monarch butterflies and dung beetles; from hummingbirds to humpback whales to humans. Are there characteristics common among all denizens of life's profligate diversity? It turns out there are. Life, for all its forms and complexities, has but two fundamental and distinct attributes: the ability to replicate and the ability to metabolize.

A catchall phrase for the magical chemical processes that occur in a living organism, *metabolism* involves the synthesis (constructive metabolism) of complex organic compounds or the breakdown (destructive metabolism) of other complicated molecules such as sugars. Both, for example, occur during the metabolic reaction of *glycolysis* through which cells break down glucose and synthesize ATP (adenosine triphosphate). The concentrated fuel of living cells, ATP is a biological universal. To risk oversimplification, metabolism incorporates all chemical processes associated with eating, drinking, and respiring,[5] whether these occur in plants or animals.

Dyson mildly chastised Schrödinger for devoting four chapters of *What Is Life?* to replication but only one chapter to metabolism. Schrödinger's bias may have simply reflected the times. At Cal Tech, Schrödinger's friend Max Delbrück was engaged in groundbreaking experiments with *bacteriophages*.[6] Viruses that infect bacteria, phages are naked DNA with no cell structure and no ability to metabolize. One who studied life through the phages would necessarily concentrate on replication. Schrödinger "took it for granted that the replicative aspect of life was primary, the metabolic aspect secondary."[7] Schrödinger was not alone. Historically, most definitions of life have focused upon replication. NASA's more recent definition of life as "a self-sustained chemical reaction capable of undergoing Darwinian evolution"[8] strikes a better balance by recognizing the primacy of both metabolism and replication.

However, in the early euphoria over the discovery of the structure of DNA in 1953, life became synonymous with replication. Nobel laureates gushed that the gene was "the living material, the present-day representative of the first life."[9] Such enthusiasm glossed over difficult questions. For example, when the double helix of DNA unzips during mitosis, it does so with the aid of numerous proteins.[10] This raises a classic chicken-or-egg question: "which large molecule appeared first – proteins (the chicken), or DNA (the egg)?"[11]

Before attending the question of priority, let's first establish an essential fact about life – its inherent nonlinearity. At the nexus of the cell's replicative and metabolic aspects life's nonlinearity lies exposed. Nonlinear chemistry involves autocatalytic and/or cross-catalytic reactions. A *cross-catalytic* reaction involves at least two products, each of which catalyzes the reaction for the other. A product that catalyzes its own reaction is *autocatalytic*. In cellular biochemistry, the two products of primary interest are DNA and proteins. DNA carries the blueprint for protein synthesis. Proteins in turn produce nucleic acids. Each strongly influences the other, which renders their interaction cross-catalytic. Thus, at its most basic level – cellular chemistry – life harbors nonlinear mechanisms.

Until perhaps the last decade, the prevailing theory favored the egg; that is, nucleotides – DNA or RNA – came first. "Replicator-first" theory was given a temporary boost shortly after the discovery of the double helix in 1953, when American biochemist Stanley Miller and his Ph.D. advisor Harold Urey demonstrated that amino acids, building blocks of life, form naturally when a spark discharge is applied to a mixture of simple gases then thought to be representative of the earth's early atmosphere.[12] Natural processes, they concluded, could have built up nucleotides from simpler amino acids.

But the egg-first theory had two Achilles heels. First, the earth's early atmosphere, that prevalent when life first emerged, was vastly different chemically from the atmosphere presupposed by Miller and Urey.

Second, and more damning, no natural processes have since been found that can synthesize nucleotides. It now appears that the probabilities of enormously complex molecules such as DNA or RNA appearing spontaneously are astronomically low. Thus, for the replicator-first scenario, the chances of life's natural emergence are vanishingly small. In this model, life becomes miraculous. Either science must embrace the supernatural or it must seriously explore other, more probable, mechanisms for life's origins.

Replication and metabolism – the two faces of life – are independent. It is conceivable that archaic organisms possessed one capability but not the other. Bacteriophages, for example, replicate but do not metabolize. In general, most parasites also fit this paradigm. They rely on their hosts for metabolic function. Can it work the other way? That is, is it possible for an organism to metabolize but not replicate? Examples are not so easy to come by, but the question is intriguing.

Dyson was quick to point out in *Origins of Life* that replication is not synonymous with reproduction. Replication is the process of making exact copies of cells by means of genetic duplication. Conceivably, early cells reproduced by simple fission, but lacking DNA, they failed to replicate. Replication requires a genetic code; reproduction does not. Reproduction could occur by simple mechanical means, by pinching a cell in two like a water droplet, for example.

Such conjecture leads to an intriguing hypothesis regarding life's origins, which Dyson unglamorously terms the "garbage-bag" theory.[13] As its name suggests, the theory is not elegant. Neither has it been the orthodox view among biologists. Nevertheless, Dyson and a growing number of adherents find "metabolism-first" to be chemically more plausible than "replication-first," and for some damn good reasons.

DNA and RNA are extraordinarily complex molecules. It is difficult to conceive of ordinary natural processes that could spontaneously generate such complexity. Just suppose, on the other hand, that metabolism came first. Living cells are chemical factories. Once the factories are up and running, they might be able to synthesize RNA and DNA over time. Although primary evidence for rudimentary metabolizing organisms does not exist, circumstantial evidence does. For example, "strikingly life-like phenomena" have been observed in laboratory experiments "when hot water saturated with iron sulfides is discharged into a cold water environment." Dyson continues: "The sulfides precipitate from the water as membranes and form gelatinous bubbles. The bubbles look like possible precursors of living cells. The membrane surfaces absorb organic molecules from solution, and the metal sulfide complexes catalyze a variety of chemical reactions on the surfaces."[14]

In the garbage-bag theory, life on Earth originated most probably near geothermal vents on the ocean floor. Thermodynamically, such vents exist at FFE conditions. Two fundamentally different types of thermal vents have been identified: those with acidic chemistry and those with alkaline chemistry.[15] Both are biologically fecund. In otherwise barren depths, thermal vents give rise to an amazing variety of life forms. So rich in biodiversity are these vents that they have been dubbed the rainforests of the deep ocean. Indeed, many ancient lineages of bacteria are thermophiles, strong circumstantial evidence that life may have originated in the exotic, sulfurous, and hellish environment near an undersea vent.

For its first 1.5 billion years, life on Earth was predominantly unicellular and prokaryotic. The DNA of prokaryotes, undifferentiated into chromosomes, is particularly rudimentary. Bacteria, the most common life forms on Earth, are prokaryotic. Their inventiveness at metabolism is legendary, which again lends credence to metabolism-first theory. Collectively, bacteria can eat almost anything and live almost anywhere, "growing in soil, acidic hot springs, radioactive waste, water, and deep in the earth's crust,"[16] not to mention on the skin and in the digestive track of human beings, whose bodies contain nearly ten times as many bacterial cells as human cells.

How prokaryotic cells developed nuclei to become eukaryotic is a matter of some speculation, but fascinating speculation nonetheless. It may have been a fluke of parasitism. Parasites invade the bodies of their hosts, usually to the host's detriment. But over time, a host-parasite relationship can transform into something mutually beneficial, even *endosymbiotic* (as defined in Chapter 10). Humans, in fact, are the beneficiaries of at least one form of endosymbiosis. For example, bacterial florae in the stomach and intestinal tract aid in digestion. Anyone whose intestinal florae have been suppressed by a powerful antibiotic quickly becomes aware of how dependent he or she is upon those friendly little former parasites. The endosymbiotic transformation of foe into friend progresses by stages whereby "the erstwhile disease organism became by degrees a chronic parasite, a symbiotic partner and finally an indispensable part of the substance of the host."[17]

Could cellular DNA be the lingering contribution of transformed parasitism? Lynn Margulis thinks so. A University of Massachusetts biologist, Margulis is considered "one of the chief bridge-builders in modern biology."[18] To Margulis, parasitism and symbiosis are the driving forces behind biological complexification.

A compelling example supporting Margulis' thesis is the presence of *mitochondria*. Mitochondria may be reformed parasites that now reside within the cells of all oxygen-breathing organisms. The "power plants of cells," mitochondria produce most of the cell's ATP.[19] Curiously, mitochondria "look like the free-living symbiotic bacteria from which they came. They do their own thing: they have their own private DNA and they grow and divide on their own inside each cell. Fortunately . . . mitochondria cannot abandon us, as they can no longer live outside of our cells."[20]

Dyson speculates that all cellular organelles were originally free-roaming parasitical "immigrants," which were appropriated by their host cells to perform specialized functions. As in a good crime scene investigation, DNA fingers the culprits. "The difference in genetic apparatus between organelles and nuclei is the strongest evidence confirming Lynn Margulis' theory that organelles of the modern eukaryotic cell were originally independent free-living cells and only later became parasites of the eukaryotic host."[21] Or it could have happened another way: one organism swallows another whole, but unable to digest its food, retains the genetic code of the prey, which it adapts to a specific purpose.[22]

Either way, such considerations contribute to a startling and relatively new theory about the origins of life. Life on Earth, it now seems plausible, originated twice. By the "double-origin hypothesis" we mean not that life originated full-blown in two distinct places or at two different times. Rather the term implies that the two fundamental characteristics of life – metabolism and replication – emerged at different times in evolutionary history, with metabolism first. In the double-origin schema, the earliest life forms contained no genetic code. Lacking both nuclei and genetic material, early living cells reproduced by the simple mechanical process of fission. Such rudimentary organisms "might have continued to exist for millions of years, gradually diversifying and refining their metabolic pathways,"[23] and in turn new modes of metabolism could lead to the synthesis of ever more complex molecules, ultimately including nucleic acids.

The two-origin hypothesis simply makes the most sense, at least to Dyson and a few iconoclasts: "The main reason why I find the two origin hypothesis philosophically congenial is that it fits well into the general picture of evolution portrayed by Margulis. According to Margulis, most of the big steps in cellular evolution were caused by parasites."[24]

If the endosymbiotic process of cells making peace with invading parasites is as fundamental to our origins as Margulis and Dyson propose, the process gives new meaning to Jesus' injunction to love one's enemies. The ability to embrace one's apparent enemies may lie at the very core of the evolutionary process. If cells can make peace with and transform their enemies, cannot humans learn to do likewise?

Metabolism-first theory is slowly gaining traction. In 2007, *Scientific American* laid out a strong case for the primacy of metabolism. Essentially, the theory proposes a bootstrap process of incremental changes rather than giant leaps. That is, metabolism-first theory embodies a sequence of natural events compatible with the grand sweep of evolution.

Let's summarize the modern version of the model. In the metabolism-first scenario, life is defined primarily by its thermodynamic properties rather than its genetic properties. For example, the late astronomer and cultural icon Carl Sagan defined life as "a localized [open] region that increases in order (deceases in entropy) through cycles driven by energy flow."[25] More specifically, the metabolism-first model has five basic ingredients.[26] First, a cell membrane separates life from nonlife. Cellular membranes are permeable, from which it follows that cells are thermodynamically open systems. Second, energy flows through the cell membrane to drive the system, and the state of the system must be FFE. Third, the auto-

and cross-catalytic chemical processes within the cell guarantee that cellular processes are inherently nonlinear. Fourth, new and more complex molecules are catalyzed within the cellular factory because cellular mechanisms satisfy all three of the conditions necessary for order to emerge from chaos. Fifth, the cells reproduce initially by simple mechanical means such as fission. The bottom line is that within the chemical cauldrons of the hypothesized early cells, the probabilities of the spontaneous emergence of complex nucleotides are greatly enhanced.

There is as yet no "smoking gun" for metabolism-first theory. There is no biological equivalent, for example, of the telltale microwave radiation that substantiated Big Bang cosmology. Still, recent experiments at the Technical University of Munich (TUM) add tantalizing evidence to support the theory. At TUM, German chemist Günter Wächtershäuser and colleagues have conducted a number of experiments with metal sulfide catalysts, such as those present at thermal vents. Their remarkable research finds that amino acids combine naturally in the presence of sulfide catalysts,[27] which suggests that the earth's earliest thermophiles could have catalyzed complex nucleotides.

Specifically, at the core of the metabolic processes of all life on Earth resides a universal biochemical reaction known as the Krebs cycle. Normally the Krebs cycle functions to release cellular energy in the form of ATP. Curiously, however, in FFE conditions the Krebs cycle can run in reverse. Astoundingly, the reverse cycle naturally synthesizes organic compounds, perhaps even proteins and RNA. Generally rare, the reverse Krebs cycle is common among the thermophilic bacteria that populate vent sites.[28] It's no smoking gun for metabolism-first theories, but it's damn close.

Meanwhile, the replicator-first versus metabolism-first debate rages on. Who is right may decide whether or not we are alone in the universe. If the proponents of replication-first theory are correct, the chances of life emerging spontaneously even once in the 13.7-billion-year history of the universe are exceedingly small. The chances of life happening more than once are nil. On the other hand, if metabolism-first theory is correct, "life is vastly more probable than we have supposed. Not only are we at home in the universe, we are far more likely to share it with as yet unknown companions."[29]

As fascinating as it is to speculate on extraterrestrial life or whether the chicken or egg came first, that is not really the central issue here. What is central? Life's two most fundamental attributes – replication and metabolism – must be considered independently. And of the two, metabolism has a direct cosmological connection.

Life Sucks

Why do we say that plants and animals are living but that rocks are not? What makes a living thing alive?

A cell is "living" precisely because of its ability to avoid decay, to maintain itself temporarily free from the universe's tendencies toward disorder. So-called inanimate matter does not possess this remarkable quality. For all their grandeur, mountains are powerless against the forces of erosion: wind, water, and acids. Each grain of sand on each beach testifies to the powerlessness of rocks and mountains in the face of gradual but unrelenting natural forces.

Insects, fungi, and bacteria soon reclaim the once majestic oak, fallen in the woods. But while living, it was able, if but for a time, to stave off the forces of decay. Similarly for man, it is written in *Genesis* 3:19, "for dust thou art, and unto dust shalt thou return." We come from the substance of the earth, whose origin is itself the cosmic dust, and to the dust of the earth we shall return – but not while living. Life exists as the interval between periods of dust, for life is, above all, "an organism's astonishing gift of concentrating a 'stream of order' on itself and thus escaping the decay into atomic chaos – of 'drinking orderliness' from a suitable environment."[30]

If life can be defined thermodynamically, then so can death. When no longer able to swim against the tide of entropy, an organism dies. Thermodynamic equilibrium is thus equivalent to biological death.[31] To remain alive, the organism must reside FFE, bathing in the entropic stream.

How exactly does a living organism avoid decay? The question is among the most profound addressed by Schrödinger in *What Is Life?* In a word, it does so by metabolism, that characteristic of life that Dyson deems most fundamental.[32]

Metabolism originates in a Greek root meaning "exchange." Life therefore is associated first and foremost with some sort of exchange. But what is its currency? One may find it tempting to regard energy as the legal tender of life. But Schrödinger, ever the physicist, catches the common mistake. Life does not thrive on "energy" any more than a nation needs an "energy" policy *per se*. For the first law of thermodynamics guarantees that energy is conserved. For example, for an adult organism – one no longer growing and whose weight is constant – the net exchange of energy with the environment is zero. Energy consumed by the organism in the form of food is exactly balanced by the energy relinquished back to the environment in the form of low-grade heat. The forms of the input and output energies, however, differ vastly. Energy contained in food is highly concentrated, while that in waste heat is highly diffuse. Which brings us back to thermodynamics' second law. In the concise language of physics and the words of Schrödinger: "What an organism feeds on is negative entropy."[33]

To the ordinary person who is not a Nobel laureate, Schrödinger's statement seems at first inscrutable. It helps to recognize that, as a measure of the probability of the state of a thermodynamic system, entropy also quantifies the quality (availability) of the system's energy content. The more concentrated (available) the energy, the lower its entropy. Conversely, the more degraded or diffuse the energy, the higher its entropy. Because an organism consumes as food its energy at low entropy and expels that energy as heat at high entropy, the net difference – entropy in minus entropy out – is negative. Put in more down-to-earth terms, "the device by which an organism maintains itself stationary at a fairly high level of orderliness really consists in continually sucking orderliness from its environment."[34] Schrödinger's observation offers a 180-degree twist to the existentialist's lament, "Life sucks." If it didn't suck, it would not live.

With a quantum physicist's eye, Schrödinger saw something essential about metabolism that often goes unappreciated. Metabolism is not about energy; it's about entropy. The fuel for life is negative entropy. And for negative entropy to exist, the universe must be thermodynamically in decline.

If life is counterentropic, then so is evolution. In the introduction to *The Phenomenon of Man*, Sir Julian Huxley wrote: "[Evolution] is an antientropic process, running counter to the second law of thermodynamics with its degradation of energy and its tendency to uniformity. With the aid of the sun's energy, biological evolution marches uphill, producing increased variety and higher degrees of organization."[35]

Although evolution is counterentropic, it does not violate the second law. The second law of thermodynamics forbids the decrease of entropy associated with life for isolated systems. Fortunately, the earth is not isolated. Provided there exists a concentrated source of energy available elsewhere that the earth can tap, life, like the salmon, can continue to swim upstream. "Although elaborate events may occur in the world around us, such as the opening of a leaf, the growth of a tree, the formation of an opinion, and disorder thereby apparently recedes, such events never occur without somehow being driven. That driving results in an even greater proportion of disorder elsewhere."[36]

In other words, for life to evolve on Earth, entropy must increase in the whole of the universe. Life and evolution, therefore, occur FFE in the eddies of the entropic stream. Were it not for the stream and its eddies, the salmon would have nowhere to spawn. In *Order Out of Chaos*, Prigogine and Stengers summarize: "We know today that both the biosphere as a whole as well as its constituents, living or dead, exist in far-from-equilibrium conditions. In this context, life, far from being outside the natural order, appears as the supreme example of the self-organizing processes that occur."[37]

Crucial to life and evolution is a nearby well of low entropy. For life on Earth, that well is the sun. The sun's light fuels photosynthesis for the metabolic processes of plants. Plants further concentrate energy into sugars, which herbivores then appropriate for food. Carnivores in turn appropriate the hapless herbivores. At each level of the food chain – with warm-blooded carnivores at the top – energy is further concentrated. Warm-bloodedness is yet another marvelous adaptation of evolution, for as Schrödinger observes, "The higher temperature of a warm-blooded animal includes the advantage of enabling it to [expel heat and] get rid of its entropy at a quicker rate."[38]

Far from forbidding life, the second law of thermodynamics, through its production of entropy, provides the driving force for life. In *God, Mystery, and Diversity*, Harvard University divinity professor Gordon Kaufman makes explicit the profound connection: "It has recently begun to appear . . . likely . . . that the continuous increase in entropy over time in the universe may itself, in the natural course of events, give rise – through the development of so-called dissipative systems – to complex forms of organization, eventually including living systems."[39] Kaufman's theology is informed by "Thermodynamics and Life," a revelatory article by ordained scientist Arthur Peacocke.

We have arrived at the answer to our penultimate question. What is the relationship between the two universal countercurrents of cosmogenesis? The opposing currents are not just related; they are intimately related. "The ceaseless decline in the quality of energy expressed by the second law is a spring that has driven the emergence of all components of the biosphere. In a very direct way, all the kingdoms of creation have been hoisted out of organic matter as the universe has sunk ever more into chaos."[40]

To reiterate, the thermodynamic demise of the physical world and the running up of the biological world are the two faces of cosmogenesis. The former enables the latter. The very spinning down of the physical world propels the biological world upward.

Peacocke concludes "Thermodynamics and Life" with an awe-inspiring synthesis:

> The picture that is emerging in . . . recent thermodynamic analyses . . . [suggests that] the movement of the [entropic] stream itself inevitably generates, as it were, very large eddies *within* itself in which far from there being a decrease of order, there is an increase first in complexity and . . . functional organization. . . . There could be no self-consciousness and human creativity without living organization, and there could be no living dissipative systems unless the entropic stream followed its general, irreversible course in time. Thus does the apparently decaying, randomizing tendency of the universe provide the necessary and essential matrix (*mot juste!*) for the birth of new forms – new life through death and decay of the old.[41]

Epilogue

Here then is Teilhard's Omega Point revealed. Evolution is *not* directionless (as proclaimed by most evolutionary biologists, including the late Stephen Jay Gould).[42] It proceeds, however hesitantly and erratically, toward the Omega on the distant horizon of time. The motive power for biological complexification is built into the warp and weft of the universe. The running down of the physical universe drives the running up of the biological universe. Entropy, therefore, is the thermodynamic term for the self-sacrificial act of the physical universe for the sake of complexification. In human terms, we call such self-sacrificial tendencies by a familiar name: love.

For all their study, erudition, and knowledge, neither scientists nor ecclesiastics are best endowed to articulate wisdom. Such things are best left to prophets, poets, mystics, and minstrels, those capable of seeing with new eyes, hearing with new ears, and feeling with new hearts. Listen then to the "minstrel of the dawn":[43]

> Then all at once it came to me
> I saw the wherefore
> And you can see it if you try
> It's in the sun above
> It's in the one you love
> You'll never know the reason why.

Seventeen

Pale Images

The labor of seaweed as it concentrates in its tissues the substances scattered, in infinitesimal quantities, throughout the vast layers of the ocean; the industry of the bees as they make honey from the juices broadcast in so many flowers – these are but pale images of the ceaseless working-over that all the forces of the universe undergo in us in order to reach the level of the spirit.

– **Teilhard de Chardin**

In Chapter 15 we met Teilhard de Chardin: paleontologist, priest, and mystic. The passage above hails from Teilhard's *The Divine Milieu*. Reading Teilhard is difficult; one can get lost in the loftiness of his prose and the profusion of ecclesiastical terms. But buried in each of his works are precious gems, more than sufficient motivation to keep the reader digging. To the author, the passage above is the crown jewel of Teilhard's gems. Within its poetic metaphors lies one of the deepest secrets of the universe. Laying bare that secret, at least to some degree, is the subject of this, the final chapter.

The Ghost in the Machine

"With life evolves mind," noted Sherrington. The evolutionary arrow points toward greater biological complexity. But biological complexity has a corollary: greater self-awareness. In general, more complex organisms have higher degrees of consciousness. Thus, Teilhard's notion of "complexification" was alternately termed "complexity-consciousness."

Consider, for example, the human brain. "With its 10^{11} nerve cells each fed by and feeding into hundreds or thousands of neurons, and making contact through some 10^{14} synapses, [the human brain] is apparently the most complex piece of matter in our known universe."[1]

Hominids have not always sported such disproportionately large brains. Within a relatively short period – a mere three million years – the brains of our ancestors quadrupled in size.[2] It is now widely accepted that our brains grew of necessity so that humans could adapt to the one constant of existence – change. The period of bulging brains coincides conspicuously with the onset of a succession of ice ages. During the epoch of ice, on a cycle of roughly 100,000 years, glaciers periodically buried one quarter of the globe. Brain capacity was *the* adaptation most beneficial for human ancestors to cope with severe climate oscillations.

As a consequence of increased brainpower, today's adaptable human has perfected survival in the most inhospitable of climes – in the tropics, the desert, the arctic, and the inner city. For short durations, astronauts defy the deadly vacuum and blistering radiation of space to perform repairs on the International Space Station and the Hubble Telescope. And divers find their inner amphibian to bring up pearls from oysters and galleons from shipwrecks.

With brain mass and complexity comes unprecedented adaptability. But "brain" is not synonymous with "mind." Brains are easy to locate. The human brain is the soft, wet, convoluted organ of about three pounds that resides in the cranial cavity of the skull. The brain is a clearinghouse for sensory stimuli, the seat of the emotions, the control center for complex movement, the processor of language, and presumably, the originator of thought.

Mind, on the other hand, is the faculty of conscious, subjective experience: the "ghost in the machine." If free will exists, mind is the seat of free will. "Brains are automatic, but people are free," asserts at least one neuroscientist.[3] The chief attribute of the brain is its extraordinary complexity. The chief attribute of mind is its inexplicable unity. Certainly, mind is related to brain, but how remains an enigma.

The mind may be localized to the brain. Then again, it may not. Scientists, particularly neuroscientists, tend to view mind as an epiphenomenon of neural activity. In particular, this seems the stance of Eric Kandel, 2000 Nobel laureate in physiology, as articulated in *In Search of Memory: The Emergence of a New Science of Mind*. Study the brain long enough, neuroscientists presume, and we will eventually understand mind. But a lot of evidence – anecdotal and scientific – suggests otherwise.

Freud earned scientific immortality by naming and probing the subconscious layer of mind beneath conscious awareness. Carl Jung, Freud's protégé, parted company with his mentor over the nature of the subconscious. Freud saw the subconscious, a dark place of illicit desires, as the seat of mental ill health. Jung turned the tables by revealing that the deepest longings of the subconscious, when brought into the light of day, point in the direction of the subject's greatest authenticity. Repression was a perversion of the conscious mind, not of the subconscious. As a therapist, Jung's job was to liberate the aspirations of the subconscious.

Jung transcended Freud in another aspect as well: mind has both subconscious and supraconscious aspects. Individual consciousness taps into a shared realm of symbolic archetypes. Dreams, Jung found, are particularly susceptible to influences from the "collective unconscious," which he defined as "a second psychic system of a collective, universal, and impersonal nature which is identical in all individuals."[4] Jung, a Swiss with deep affinity for Native American culture and beliefs, spent considerable time in the American Southwest studying aboriginal ways. The aboriginal "world soul" and Jung's "collective unconscious" are perhaps different cultural expressions for the same phenomenon: shared consciousness. Experiments with group dreaming – for example, the Dream Helper Ceremony of former Princeton University psychologist Henry Reed and University of Virginia psychologist Robert Van de Castle – strongly suggests that Jung was right.[5] If consciousness has a shared aspect, then brain and mind are not synonymous. Brain is local; mind is not. Brain is particle. Mind is wave.

Consider also the relatively common phenomenon of near-death experience (NDE). A substantial body of anecdotal and scientific evidence has been amassed regarding NDEs by a number of reputable researchers, among them Elisabeth Kübler-Ross of the University of Colorado, Kenneth Ring of the University of Connecticut, and Raymond Moody of the University of Virginia. Despite differences in religion, nationality, race, and age of those who ventured to the edge of death and returned, NDE accounts are remarkably consistent.[6] Many NDE episodes involve so-called out-of-body experiences (OBEs). In the classical OBE, the victim describes – in vivid detail and from a perspective outside of the physical body – what is transpiring in attempts to revive the body, even though he or she may be medically unconscious or lifeless at the time. Hardcore neuroscientists dismiss such reports as hallucinations of an oxygen-deprived brain. Curiously, however, among those able to report visual details of near-death OBEs are persons born blind.

Beyond the world of ordinary experience lies a universe of the paranormal. Paranormal or "psi" experiences include telepathy, precognition, clairvoyance, distant healing, and a host of yet more exotic phenomena including the aforementioned OBEs. Psi experiences are ubiquitous. Perhaps half the human race, if honest, has had some memorable or moving experience that falls outside normal sensory awareness.

Despite the prevalence of paranormal experience, science has not been kind to the paranormal. Psi phenomena lie outside the bounds of known physical laws and of temporal causality. Thus mainstream science has dismissed such accounts with varying degrees of disdain: by ignoring the phenomena altogether, by labeling investigations of psi experience as pseudoscience, or by declaring psi experiments the work of charlatans.

Such was the attitude of newly minted experimental psychologist Lawrence LeShan when he emerged from graduate studies in the 1950s. As a young Turk, LeShan set out to debunk all paranormal "nonsense." But LeShan, by his own admission, made a "mistake." He examined the century of psi data that had accumulated; the data were compelling. LeShan converted, if reluctantly, from skeptic to believer. In the process, he found his life's work – the quest to develop a theory of the paranormal. The theory began to take shape in the groundbreaking *The Medium, the Mystic, and the Physicist* (1966) and continues in his latest book *A New Science of the Paranormal* (2009) published as he neared ninety years of age. LeShan is no wild-eyed mystic (if there is such a thing). LeShan has opened to mystical insights because of his scientific integrity, not in spite of it.

Similarly, in *The Conscious Universe* (1997), Dean Radin, a brilliant young experimental parapsychologist, confronts head-on the prejudices toward psi of the scientific establishment. He does so by meticulous statistical meta-analyses of more than a century of data obtained by the most reputable psi researchers. These include J. B. Rhine at Duke University and researchers of the Princeton Engineering Anomalies Research (PEAR) Laboratory. Radin's statistics offer seemingly iron-clad evidence for the legitimacy of many types of psi phenomena: ESP playing-card tests, telepathy experiments, and the highly refined Ganzfeld remote-viewing experiment, among others. The "hit rate" of each phenomenon significantly exceeds that of chance alone.

Psi challenges the metaphysical assumptions of science: dualism, materialism, determinism, causality, and reductionism. But then so does quantum mechanics. The response of the scientific community to these two challenges, however, remains markedly different. Science has reluctantly embraced quantum theory. In contrast, on the whole it consistently finds the world of the paranormal to be anathema.

Bergson, Whitehead, and Prigogine have all argued for a more tolerant science, each for his own reasons not related to the paranormal. Prigogine's motivation is to heal the "schizophrenia" that plagues humankind. He pleads, "Whitehead's case as well as Bergson's convince us that only an opening, a widening of science, can end the dichotomy between science and philosophy."[7] Suggesting a way through the impasse, LeShan invokes the wisdom of the French philosopher Ernest Renan, who asserted that there exist two legitimate scientific methodologies: *la science de la nature* and *la science de l'humanité*.[8] The former studies what can be quantified; its methodologies involve general laws and measurement. The latter studies what cannot be quantified; its methodologies involve deep contemplation of the subject: listening, empathy, and sympathy. The German language also affords dual words for science: *Naturwissenschaft* and *Geistewissenschaft*, the "science of nature" and "science of the spirit," respectively.[9] "For the study of psi, we need both methods," argues LeShan.

In toto, psi phenomena strongly suggest that consciousness is both nonlocal and collective. Were mainstream science more open to these extensively documented "damned facts," rigorous scientific investigations could help invaluably to confirm or refute this hypothesis, to shed light on the numinous qualities of the cosmos, and to probe the full range of capabilities of the human being. Indeed, there is the beginning of movement in this direction within the scientific community, as noted in the Introduction and as addressed eloquently by pioneering transpersonal psychologist Charles Tart in *The End of Materialism.*

But whether or not consciousness is confined to individual brains is, once again, not quite the central issue. The central issue is that the universe appears to be in the business of consciousness.[10] Evolution is a long and winding road, but its destination is clear: "There has been, by and large, a general increase in complexity during the broad sweep of the evolutionary process,"[11] and "with life evolves mind."

Through a Glass Darkly

The poetic quote with which this chapter began suggests that the cosmos, from the widest possible perspective, is a distillery of consciousness. Key to the process is entropy.

"The concept of entropy plays a central role in the description of evolution."[12] Far from precluding evolution, as creationists have sometimes tried to argue, the "entropic stream" of the second law of thermodynamics makes evolution possible. As discussed in the previous chapter, entropy plays a fundamental role at two different levels of scale. At the level of the organism, life can be defined by its ability to keep death temporarily at bay by "drinking orderliness" in the form of negative entropy from the environment. At the level of the universe as a whole, the entropic stream creates eddies – localized open systems far from equilibrium (FFE), like the earth itself – in which order can emerge from chaos and organisms can tend over time toward greater complexity.

In order for the entropic stream to flow – for the universe to manifest a tendency toward physical decay – the early universe had to exist in a FFE state. According to Peacocke, "A state of low entropy is necessary . . . for biological complexity and organization to occur."[13] Where then is the vast reservoir of low entropy needed for both life and evolution? As a city needs a water reservoir, a universe in flux needs a reservoir of low entropy.

At UC Berkeley in 2000, Freeman Dyson addressed the question above in an Oppenheimer lecture. The lecture was pitched to a general audience and simply titled "Gravity Is Cool." And like all of Dyson's writings and talks, it combined genius with down-to-earth accessibility. The universe's well of low entropy is the sum of all its gravitational potential energy, a highly available form of energy. In other words, gravity is where it's at. Moreover, gravity has a remarkable attribute essential to the processes of life: it can run cold or hot.

When a planet orbits a star, gravity is cool. By this Dyson meant that the exchange of kinetic and potential energies is completely reversible: entropy remains constant. Following a perfectly elliptical orbit as observed by Kepler, the planet cyclically speeds up when closer to the star and slows down when farther away, keeping the sum of its kinetic and potential energies constant. No energy is dissipated and no entropy created.

But in the cosmic bumper-car race of whirling particles and bodies, some are bound to collide, and when bodies collide, gravitational energy runs hot. Collisions are irreversible processes, which generate enormous heat. As a consequence, they produce entropy. More than four billion years ago, a cataclysmic collision took place between a nascent Earth and a wandering planetesimal somewhat larger than our moon. The impact savaged the earth. Kinetic energy, converted instantly to heat, warmed the earth to white hot and ejected massive quantities of debris into space. Over time, though, the debris settled into orbit around the stricken Earth and coalesced to form our moon.

Like the Hindu god Shiva, gravity has two faces: creator and destroyer. Following each destructive event come new opportunities. Both sides were right in the old debate between catastrophism and uniformitarianism. The aggregation of small changes shapes the earth, but catastrophic events, though rare, shape it too. Both, observes Dyson, are essential to life and evolution. "The existence of life depends crucially on the fact that time has these two faces, the quiet and the violent. The violent face created the stuff that we are made of. The quiet face sustains us and allows us to evolve."[14]

Follow cosmogenesis backward in time, and one finds the hand of gravity behind all evolutionary processes. Under the tug of gravity, clouds of hydrogen created at the Big Bang compress to form and ignite stars. When the fusion furnaces of the stars burn out, gravity crushes their wasted hulks. Sometimes gravity leaves neutron stars as remnants. Sometimes, in their death throes, the collapsing stars rebound in supernovae, creating a panoply of elements, the ingredients of life. And sometimes, all that remains from the crush of gravitational attraction is a black hole.

Black holes too perform vital roles in cosmogenesis. As early thermodynamicists learned, a heat engine cannot run without a cold sink – a collecting reservoir into which heat can flow. Black holes are the cold sinks of the universe. Nearly every galaxy has one. When all the energy of the universe is impounded in black holes, the world will grind to a halt. The universe will have reached its final equilibrium. But until then, as Sherrington marveled, "Life is evolved without pause."

The universe that gave birth to life and nurtured it for eons through evolutionary processes to beget beings capable of pondering those very processes began in a state of extraordinarily low entropy. Low entropy means "unlikely." The initial state of the universe was extremely improbable – astronomically improbable, it seems.

Cosmology has coined the term *anthropic principle* to describe the unlikely state of the early universe. The cosmos seems finely tuned to facilitate life and sentience. Too finely tuned, in fact. If any of a number of physical constants had even slightly different values – the gravitational constant, the mass of the proton, or the age of the universe, for example[15] – life as we know it could not and would not exist. To some it appears that the universe has been "designed" for life and consciousness. To others, the apparent coincidence is tautological. We perceive the universe as finely tuned because we would not exist to perceive it were it not exactly as it is. Still, it makes one wonder. How, and perhaps why, did the universe originate in a state of such low entropy?

Closely associated with entropy is thermodynamic irreversibility, the culprit behind "the arrow of time." Humans perceive time to flow because most processes in human experience are irreversible. Otherwise, we could magically "refill the petrol tanks of our cars by pushing them backwards!"[16] From whence comes irreversibility?

In *Order Out of Chaos*, Prigogine and Stengers probe the nature of time for more than 300 pages with only partial success. The Holy Grail of science and philosophy remains elusive. Why time flows is yet mysterious. Nevertheless, Prigogine and Stengers cast considerable light on the subject.[17] Their analysis appears to suggest that irreversibility stems from the combination of two phenomena: instability and randomness.[18] The first trait arises naturally in nonlinear dynamical systems when FFE. Small changes to the initial states of such systems can produce wildly different results. The more modern term for "instability" is *sensitive dependence to initial conditions*, known to the layperson as the "butterfly effect," as discussed briefly in Chapter 11. We have already established that the universe is governed predominantly by nonlinear mechanisms and that the chemistry of life, in particular, is both nonlinear and FFE. When such processes combine inordinate sensitivity to initial conditions with randomness in those conditions, then the evolution of those processes becomes both unpredictable and, as a consequence, irreversible.

Let's peel back one more layer of the onion: randomness. Whence randomness? With the discovery of radioactivity in 1895, the Trojan horse of randomness entered the encampment of physics, bringing with it quantum mechanics. Quantum indeterminacy gave Einstein heartburn. In a letter to fellow physicist Max Born, he grumbled, "I find the idea quite intolerable that an electron exposed to radiation should choose *of its own free will* [emphasis added] not only its moment to jump off, but also its direction. In that case, I would rather be a cobbler, or even an employee in a gaming house, than a physicist."[19]

Einstein was speaking euphemistically. A dyed-in-the-wool determinist, he did not for a moment believe that electrons possess some sort of "free will." On the other hand, maybe – just maybe – they do. If so, what is perceived as randomness is altogether something else.

In June 2009, the author had the rare opportunity for a face-to-face visit with Professor Freeman Dyson, whose ideas grace many of these pages. The venue was his natural habitat: the Institute for Advanced Studies

(IAS) in Princeton, New Jersey, where Dyson has thrice held residence since 1948.

Despite the drizzle, simply setting foot on the hallowed grounds of IAS was thrilling. After all, the ghosts of twenty-two Nobel laureates and thirty-four Fields Medalists roam the halls and walkways. With a little imagination, one could feel the lingering presence of Einstein commuting to and from Mercer Street in his slippers; of chain-smoking Robert Oppenheimer, director of the Manhattan Project; of Hungarian computing pioneer John von Neumann; or of logician Kurt Gödel, immortalized in the book *Gödel, Escher, and Bach*. The office of Edward Witten, generally regarded as the greatest living physicist, sat just across the narrow hallway from Professor Dyson's. IAS is a heady place.

Freeman Dyson, then eighty-six and having just returned from his native England, had, through his assistant, graciously consented to meet with me over an extended lunch. Sitting opposite Professor Dyson, I could not help but feel like Luke Skywalker in the presence of the Jedi master Yoda. Having read several of Dyson's articles and books, I wanted to verify that I understood his theses – some gloriously unorthodox. I was taken aback by his graciousness, self-deprecating humor, and kindness and also by an unexpected mannerism in addressing questions. While contemplating, Dyson had the habit of pausing for what seemed an eternity. Then, just when I feared the question had been lost or dismissed as banal, he would offer the most articulate and well-reasoned response, fully formed. Each wait produced a marvel of clarity.

The question I most wanted to put to him concerned the following passage from the Templeton Prize-winning *Disturbing the Universe*:

> I cannot help but think that the awareness of our brains has something to do with the process that we call "observation" in atomic physics. That is to say, I think our consciousness is not just a passive epiphenomenon carried along by the chemical elements in our brains, but is an active agent forcing the molecular complexes to make choices between one quantum state and another. In other words, mind is already inherent in every electron, and the processes of human consciousness differ only in degree but not in kind from the processes of choice between quantum states which we call "chance" when they are made by electrons.[20]

Upon first reading the passage, I had been struck by multiple flashes of déjà vu. Here again was a parallel between setting intention in human consciousness and the collapse of the wavefunction in quantum mechanics. And here again was Teilhard's belief that it is spirit (or consciousness) "all the way down," even to the level of the electron.

And so, I asked Dyson point-blank: Am I correct that you believe electrons to be conscious at some level? That elementary particles possess a "quantum" of consciousness? Moreover, that a quantum of consciousness grants a quantum of free will? Professor Dyson conceded that the viewpoint is animistic and that, as such, it is a highly unpopular among most scientists. Nevertheless, this is his view. The so-called randomness observed in nature is actually the manifestation of the consciousness of quantum particles, each of which really does "choose of its own free will" when to jump and in which direction.

If the aboriginals and the renegade scientists are correct, consciousness does not gush magically from a material spring. It is a primary attribute of the stuff of the universe, even at the most elemental level. The universe does not create consciousness. It distills the protoconsciousness scattered about the cosmos, collecting it into higher consciousness through counter-entropic evolutionary processes, as bees collect nectar for honey "from the juices broadcast in so many flowers."

An Infinite Storm of Beauty

To the early environmentalist John Muir, the universe's dynamism seemed as a great storm. Muir observed that, "when we contemplate the whole globe as one great dewdrop, striped and dotted with continents and islands, flying through space with other stars all singing and shining together as one, the whole universe appears as an infinite storm of beauty."[21]

The universe, the "storm of beauty" we call home, has evolved for some 13.7 billion years. It began with a bang, and from a state of extraordinarily low entropy it has been sliding toward equilibrium ever since. Within the eddies of its entropic stream, the downhill trend is temporarily reversed. In the eddies, the universe gives rise to an ever changing pageantry of complex forms. In the eddies, each moment is the moment of creation and the moment of Armageddon: birth and death; death and rebirth.

In one small and remote eddy, life emerged 3.5 billion years ago and has been experimenting ever since. On Earth life has given rise to a profusion of forms, from simple to exceedingly complex. One unique biological form, *Homo sapiens*, is a relative newcomer, just 100,000 years old. The scientific name *Homo sapiens* means "wise man" or "knowing man." The name is fitting because within us "the ceaseless working-over of all the forces of the universe" has coalesced into an unprecedented level of consciousness. Human beings are the first inhabitants of planet Earth to piece together their story.

We have reached the temporary end of that ongoing saga. Perhaps the drama is being played out elsewhere in the cosmos. Perhaps not, and we are alone. But that is unlikely, for there is no telling what wild impulses the universe entertains on its celestial Galapagos archipelagos.

Epilogue

A new mentality is needed, and this implies above all a recovery of ancient and original wisdom. And a real contact with what is right under our noses.

– Thomas Merton, in a letter to Thich Nhat Hanh

In 1875, Austrian geologist Edward Suess made an evolutionary leap. In his popular geology text *The Face of the Earth*, Suess introduced a new notion: the *biosphere*. As an academic geologist, Suess' domain had been limited to the physical earth, the geosphere: the earth's core, mantle, and crust. Within the geosphere, by definition, life has no quarter and plays no role. The geosphere is the material world, the *res extensa* of Descartes and the majestic clockwork of Newton.

By pushing outward by a few miles the boundaries of the geosphere, Suess also expanded his awareness. Now a universally acknowledged concept, the biosphere envelops the geosphere, incubating life. Within its air, topsoil, and water life breathes, blossoms, and flourishes. Outside it, life withers and dies. By radial extension, Suess' geosphere grew to encompass the material world and the biological world, the worlds of Newton and of Darwin.

While in the trenches during World War I, on night watch under a full moon, Teilhard de Chardin pushed the envelope of human awareness outward yet further. In his mind's eye, Teilhard witnessed something rise above the biosphere: an ephemeral substance, a whiff, a ghostly vapor. What Teilhard envisaged to waft upward from the narrow zone of life was consciousness itself.[1] In the most unlikely of circumstances, in a mystical state on a battlefield of unspeakable horror, Teilhard intuited the meaning of the cosmos:

**Consciousness (spirit) is not a byproduct of evolution;
it is the purpose of evolution.**

To express this epiphany in language, Teilhard, like Suess before him, coined a wholly new term: *noosphere*, from the Greek *nous*, meaning "mind." To Teilhard, enveloping the biosphere lay another concentric shell, the noosphere, within which all sentience is contained. The three

315

shells – the geosphere, the biosphere, and the noosphere – lie gently nestled, each inside the next, like Russian dolls. Conversely, each sphere
contains the previous and thus symbolizes a bigger and more expansive
worldview. Within the largest, the noosphere, is contained the collective
consciousness of the earth and all her inhabitants.[2]

On its long journey to maturity, a Chesapeake Bay crab molts seventeen times on average. Copernican revolutions are molting processes for
the human psyche. With each molt, an antiquated and restrictive worldview is cast off and a new "skin" formed. Each new notion – geosphere,
biosphere, noosphere – expands the sphere of awareness: awareness of
our place in the physical universe, in the biological universe, and in the
spiritual universe, respectively. During the molting process, humans feel
particularly vulnerable. Growth is often a delicate balance between the
terror of leaving behind the familiar and anticipation of the new.

In the four-and-three-quarter centuries since Copernicus, science – "the
engine of the Enlightenment"[3] – has driven the expansion of human self-
awareness, supplanting religion as the locus of authority and wisdom in
the Western world.

For all its achievement and all its promise, however, science has not
resolved the greatest problems of the human condition. Indeed, it has created as many as it has alleviated. Much of technological advancement –
fire, gunpowder, and atomic energy, for example – has served the cause
of warfare. Each innovation has a dark twin. The Wright Brothers' airplane has become the stealth bomber. The nuclear genie has unleashed the
mushroom cloud. Goddard's toy rocket has shape-shifted into the Nazis'
V2, the Taliban's rocket-propelled grenade, and the United States' cruise
missile. Science has helped humanity perfect the art of genocide, the human species being the only one to engage in the wholesale slaughter of
its own kind. The double-edged sword of science was well appreciated
by Winston Churchill. England barely survived World War II, thanks in
large measure to the inventions of radar and digital computing, the latter
of which deciphered the Nazis' cryptographic code. But following the destruction of Hiroshima and Nagasaki, Churchill fretted over the return of
the Stone Age "on the gleaming wings of Science."[4]

Beyond warfare, science has unleashed other dark powers. In the
spring of 2010, the wellhead of a British Petroleum rig blew out in the
Gulf of Mexico, daily spewing hundreds of thousands of barrels of crude
oil from a hole on the ocean's floor a mile beneath the surface. For six
months the oil gushed. Before the flow was staunched, the slick covered
an area the size of Pennsylvania. A year later, a Pacific tsunami unleashed
a Japanese nuclear disaster nearly on the scale of Chernobyl. And in the
Age of Information, Orwell's "Big Brother" seems all too real a possibility. In a slow but vile process, the automobile and the power plant pump
carbon into the atmosphere at an alarming rate, bringing unprecedented
change to the atmosphere, the oceans, and the climate. Ultimately, climate

destabilization may threaten life upon the earth more than nuclear holocaust. Reputable climatologists predict that a climate tipping point – a point of accelerated and possibly irreversible change – is near,[5] yet nearly half of Americans deny the evidence.

The religiously inclined may be tempted to harbor *Schadenfreude* at science's failure to live up to its utopian promise. But the history of religion is no less checkered than that of science. Religion unleashed the Crusades and the Inquisition. It is estimated that during the Middle Ages, some two million women were tortured and destroyed for suspicion of witchcraft.[6] Church authorities burned at the stake one of their own, Giordano Bruno, for espousing "heretical" beliefs, among them that the stars are suns like our own. Galileo, in turn, was threatened with torture and kept under house arrest for daring to promote Copernicanism. Hitler perverted Christianity as a blunt instrument to wield against Jews, as today's white supremacists do similarly against blacks. The roots of genocide all too often trace to religious intolerance. Fundamentalist terrorists now routinely kill innocents and set the world aflame in the name of Allah, the Judeo-Christian God, or righteousness. If these be the fruits of religion, then would not the world be better off if atheism and agnosticism prevailed? The godless at least do not hide behind God to justify their misdeeds.

And so, whether in ignorance, or in the name of science, or in the name of God, humans go about fouling their nest and perfecting the killing of their own kind. The sheer numbers of us on the planet mean that the stakes have never been higher for life on Earth. The impact of humans upon the planet is now so great that cultural evolution threatens to replace Darwinian evolution as the principal agent of change. Simply put, humans are changing the face of the earth at a rate far faster than biological processes.

Given human history and the nightly news, it is easy to regard the future with pessimism. At this crucial juncture in evolutionary history, the vision of Teilhard de Chardin offers both solace and cause for cautious optimism. The cosmology of Copernicus, Galileo, and Kepler; the physics of Newton and Einstein; the biology of Darwin and the genome; the quantum of Heisenberg and Bohr; and the thermodynamics of Boltzmann and Gibbs: all point toward a revolutionary new view of the cosmos that might best be termed the Teilhardian synthesis. Teilhard illuminates the future in three ways: by providing a new paradigm, by his long view of evolution, and by serving as a radiant example of a transformed human being.

First, enlightened awareness comes not by choosing sides between science and religion but, in the model of Teilhard, by embracing wholeheartedly their complementary aspects. "When we are aware of the universe through religious experience, nothing is quantitative," observes Freeman Dyson, "and when we are aware of the universe through scientific observation and analysis, nothing is sacred."[7] But where these two roads meet, there lies hope of salvation for the individual and the species.

Science and religion have failed not because either is wrong but because each is incomplete. Human beings remain partly amphibian, residents of two realities: the material world and the spiritual world, physical reality and mystical reality.

> The meaning and validity of our lives are given by that part of us that relates to the world of the One. The mechanics and techniques by means of which we live our lives are given by the part of us that relates to the world of the many. It is the One that gives meaning to the Many, and the Many that gives form to the One.[8]

To live solely or primarily in one world is to be less than whole. Only in the "conjunction of reason and mysticism," argued Teilhard, "will the human spirit attain its maximum vitality."[9] Similarly, the great German-American theologian Paul Tillich asserted, "There is no conflict between faith in its true nature and reason in its true nature."[10] There is conflict only when either or both leave the path of wisdom.

Second, Teilhard's long view of evolution elicits hope, if not for humans, then at least for nature. The greatest evolutionary leaps have always followed cataclysms, times of enormous stress and mass extinction, when the rule for survival was simple: evolve or die. Life on Earth is now in the midst of another major extinction spasm, its sixth, this one principally of humanity's doing. The lesson of evolutionary history is this: Whatever happens, nature will not only survive, she will flourish. Whether or not humans survive, however, remains an open question.

Were all of Teilhard's thought compressed into a single pithy injunction, it might be this:

**As humans, who occupy the pinnacle of the tree of life,
we have a sacred obligation to participate responsibly in evolution.**

How then does humanity participate responsibly in the evolutionary story? As a new species, there is so much we must learn. But by far, the greatest evolutionary leap human beings can take is by learning to love more fully.

Although love appears to be woven into the fabric of the universe, humans are not yet very adept at understanding love or the practice of loving. Most humans can find it within their hearts to love their children, their families, and perhaps their own kind – those of their religion or ethnicity. Where we fail is in widening the circle of compassion. "We are prisoners of our little loves," observed Teilhard.[11] Those who can love the birds, like Francis of Assisi, the outcasts, like Mother Teresa, or their enemies, like Jesus, are the saints. Individuals who can truly love are rare. Rarer still are nations capable of setting aside parochial self-interests for the good of the wider human community and the planet we call home. "The age of nations is past," wrote Teilhard. "The task before us now, if we would not perish, is to build the earth."[12]

The greatest need of human beings, as individuals and as a species, is therefore to extend the circle of compassion. Just how wide should this circle be? In a letter to a total stranger, Einstein set the goal:

> A human being is part of the whole, called by us "universe," a part limited in time and space. He experiences himself, his thoughts, and feelings, as something separate from the rest – a kind of optical delusion of his consciousness. This delusion is a kind of prison for us, restricting us to our personal desires and to affection for a few persons nearest to us. Our task must be to free ourselves from this prison by widening our circle of compassion, to embrace all living creatures and the whole of nature in its beauty. Nobody is able to achieve this completely, but the striving for such achievement is in itself a part of the liberation and a foundation for inner security.[13]

Einstein's vision is nothing other than that of mystical union as described in the "perennial philosophy"[14] of the ancients, in which *self* merges with *Self*. The highest revelation of all spiritual traditions is found in the three most seminal words of the Upanishads: That art Thou.[15] There is no separation between self and other. Separation is *maya*: illusion.

If our highest failure as human beings is to remain prisoners of our little loves, then the greatest among us are those who break free of their prisons. We are all too painfully aware that hearts can be broken. But ever so rarely, a human heart is broken *open*. And when a heart breaks open, it grows so large that nothing lies beyond its embrace. Here too, we find a shining example in Teilhard. Battered by the continual rejection of the Church he loved and served, Teilhard's heart ultimately grew to embrace the universe in its entirety, like an all-consuming flame. What remained to Teilhard after his fiery inner transformation was "a diaphany of the divine at the heart of a universe on fire."[16]

When one is loved by a heart so expansive, one cannot but be drawn upward and outward. Such was the case of Jeanne, who chanced upon Teilhard when he was an elderly priest in exile in New York City and she an awkward teenage girl.

In the habit of running with her dog Champ to Central Park, Jeanne twice bumped into a frail, gentlemanly stranger, the first time knocking the wind out of him. On the second occasion, he resolutely informed her, "I will go with you." Thereafter, for more than a year, Jeanne and the old gentleman met several times a week in Central Park. Jeanne called him "Mr. Tayer," as close as she could get to the first part of his long French name.

Walks with Mr. Tayer were full of magic and delight. His unpretentiousness and childlike wonder transformed the most ordinary experience

– like stumbling upon a caterpillar – into a moment of enchantment, as Jeanne recounts.

> "Jeanne, can you feel yourself to be a caterpillar?"
> "Oh yes," I replied with the baleful knowing of a gangly, pimply faced teenager.
> "Then think of your own metamorphosis," he suggested. "What will you be when you become a butterfly, *une papillon*, eh? What is the butterfly of Jeanne?" . . .
> Perhaps the most extraordinary thing about Mr. Tayer was the way that he would suddenly look at you. He looked at you with wonder and astonishment joined to unconditional love, joined to a whimsical regarding of you as the cluttered house that hides the holy one. I felt myself primed to the depths by such seeing. I felt evolutionary forces wake up in me by such seeing, every cell and thought and potential palpably changed. I was quickened, greened, awakened by such seeing and the defeats and denigrations of adolescence redeemed. I would . . . tell my mother, who was a little skeptical about my walking with an old man in the park so often, "Mother, I was with my old man again, and when I am with him, I leave my littleness behind." . . . You could not be stuck in littleness and be in the radiant field of Mr. Tayer.[17]

Jeanne saw Mr. Tayer for the last time on a Thursday before Easter Sunday, 1955. Strangely, when the time came for Jeanne to say *"Au revoir, Mr. Tayer,"* Champ refused to leave. The following Tuesday, Mr. Tayer did not return, nor did he ever.

Years thereafter a friend sent Jeanne a copy of *The Phenomenon of Man*. The writing seemed oddly familiar, and turning to the page bearing the author's photograph, Jeanne discovered the identity of "old Mr. Tayer" as none other than the great paleontologist and mystic Teilhard de Chardin. The Jeanne of the story is now Dr. Jean Houston, world-renowned protégé of Margaret Mead, prolific author, and a founder of the human potential movement.

Houston's chance encounter and subsequent meetings with "old Mr. Tayer" changed her life as no other event ever did. "I leave my littleness behind, because he saw God in me and I had to rise."[18] How desperately we humans need to leave our littleness behind.

If human beings are to survive, it will be by accepting Teilhard's sacred charge to participate responsibly in evolution, by shedding the tight egos that have bound us in littleness, and by rising above ourselves to a new and more expansive view of the universe and our place within it. Ever the optimist, Teilhard predicted: "Some day after mastering the winds, the waves, the tides, and gravity, we shall harness for God the energies of Love. And then, for the second time in the history of the world, man will have discovered fire."[19]

It remains to be seen whether Teilhard's faith in humanity is well founded or misplaced. Assuming such hopefulness to be justified, then life, as expressed in *Homo sapiens sapiens*, its most self-aware form, stands at a crossroad never before faced by another species on this beautiful and fragile planet. To survive, we must make the first conscious choice to transmute rather than to stagnate. In this event, however likely or unlikely, the branch labeled *Homo sapiens sapiens* on the tree of life shall purposefully sprout a higher branch.

With what scientific name should the new offshoot of *Homo sapiens sapiens* be christened? To those who recognize their true nature, that name is obvious:

Homo sapiens spiritus.[20]

Notes

Preface

1. A central insight of Matthew Fox's *Creation Spirituality*. Fox is America's leading mystical theologian.
2. For riveting interviews with *Wisdomkeepers*, see Wall and Arden's book by that name.
3. Because of an implied confidentiality, the details of this otherwise true event have been slightly altered.

Introduction

1. Goethe, p. 618.
2. Science and religion's "separation" and "divorce" adapted from Peck, p. 40.
3. Jacques Monod, (late) eminent biologist and Nobel laureate, as quoted by Prigogine and Stengers, p. 22.
4. Composite quote of Prigogine, drawn from Prigogine and Stengers, pages 7 and 96.
5. Michael Meade, "Mythic Visions for Uncertain Times," *Shift* (periodical of the Institute of Noetic Sciences), No. 23 (Summer 2009), pp. 18-23.
6. From Fox's plenary address at the 2013 Science and Nonduality Conference in San Jose, CA.
7. Einstein, *Ideas and Opinions*, p. 46.
8. Berry extends the dialogue further. In *The Great Work: Our Way into the Future* (2000), he speaks of the fourfold wisdom necessary "to remind us of our place in the world and our right relationship with the natural world and with each other." The components of the fourfold wisdom: the wisdom of science, the wisdom of the classical religious traditions, the wisdom of the indigenous, and the wisdom of women.
9. Kandel, preface, p. xi.
10. See www.metanexus.net/essay/ten-reasons-constructive-engagement-science-and-religion.
11. See for example Carl Sagan's *The Varieties of Scientific Experience* which "describes his personal search to understand the nature of the sacred in the vastness of the cosmos."
12. Pope John Paul II, p. 246.
13. Pope John Paul II, p. 247.

14. Freud, pp. 137-144, "A Difficulty in the Path of Psycho-Analysis." In the insightful article Freud defines the three successive blows to human narcissism as cosmological, biological, and psychological.
15. Teilhard de Chardin, *The Phenomenon of Man*, p. 221.
16. As quoted by Russell, p. 139.

Chapter 1 The Clockmaker and the Clockwork

1. Kraft, p. 322.
2. Newman, Vol. I, p. 275.
3. *To the Moon.*
4. *To the Moon.*
5. As quoted by J. S. Dorian in *A New Day: 365 Meditations for Personal and Spiritual Growth*, New York: Bantam, 1989.
6. *To the Moon.*
7. Bronowski, p. 222.
8. Newman, Vol. I, p. 277, attributed to Aldous Huxley.
9. Gleick, *Newton*, p. 20ff.
10. Simmons, p. 804, attributed to I. Bernard Cohen, Introduction to Newton's *Principia*, Harvard University Press, 1971, pp. 291-292.
11. Newman, Vol. I, p. 261.
12. Simmons, p. 806.
13. Simmons, p. 806, citing I. Bernard Cohen, Introduction to Newton's *Principia*, Harvard University Press, 1971, pp. 47-54.
14. Inertia and momentum are closely related but subtly different. By the definition of momentum, an object at rest carries no momentum by virtue of having no velocity. Nevertheless, an object at rest retains inertia in that it will resist attempts to set it into motion. Whereas momentum depends upon the frame of reference, inertia does not. In essence, mass alone quantifies inertia. In contrast, the product of mass and velocity quantifies momentum.
15. Bronowski, p. 223.
16. Cernan, Chapter 13, "The Space Walk from Hell," provides a firsthand account of the astronaut's life-threatening experience.
17. Gleick, *Chaos: Making a New Science*, pp. 11-13.
18. Newton, Preface to the First Edition of *Principia Mathematica*.
19. Newman, Vol. I, pp. 266-267.
20. Bronowski, p. 234.
21. Bronowski, p. 236.
22. Simmons, p. 808.
23. Gleick, *Newton*, p. 101.
24. Gleick, *Newton*, p. 99.
25. Gleick, *Newton*, p. 106.
26. Quoted in Freeman Dyson's Witherspoon Lecture, 2003, and attributed to Frank E. Manual, *The Religion of Isaac Newton* (London: Oxford University Press, 1974), pp. 135-136.
27. Simmons, p. 808.
28. Simmons, p. 808.

29. Gleick, *Chaos: Making a New Science*, p. 14.
30. Prigogine and Stengers, p. 50.
31. Lord Maynard Keynes' eulogy was read by his brother Geoffrey, the former having died shortly before the occasion.
32. Lord John Maynard Keynes' eulogy of Newton, 1946.

Chapter 2 The Music of the Spheres

1. Mitchell, p. 48.
2. *National Geographic*, March 2008, pp. 36-61.
3. Marc Bekoff, "We Second That Emotion," *Yes!*, pp. 38-40.
4. Ferris, *Coming of Age*, pp. 20-22.
5. Kuhn, pp. 8-11.
6. This approximation glosses over the fact that a *sidereal day*, the time required for the earth to rotate exactly once relative to the stars is shorter by 3 minutes, 56 seconds than a *solar day*, which measures the time between solar high noon on successive days. The measures differ because the earth also revolves about the sun as it is rotating. The mean solar day is exactly 24 hours long.
7. Ferris, *Coming of Age*, p. 21
8. Kuhn, pp. 20-22.
9. Kuhn, p. 24.
10. Kuhn, p. 24.
11. Kuhn, p. 28.
12. Kuhn, Chapter 1, "The Ancient Two-Sphere Universe," pp. 1-44.
13. Kuhn, p. 29, Fig. 10.
14. *Great Books*, Vol. 16, "Translator's Introduction," p. 481.
15. Ferris, *Coming of Age*, p. 28.
16. Kuhn, p. 72.
17. Kuhn, p. 109.
18. Kuhn, p. 111.
19. *Great Books*, Vol. 16, "Biographical Note," p. 499.
20. *Great Books*, Vol. 16, p. 510.
21. *Great Books*, "Biographical Note," p. 499.
22. *Great Books*, Vol. 16,"Biographical Note," p. 499.
23. *Great Books*, Vol. 16, *De Revolutionibus*, Book One, p. 511.
24. *Great Books*, Vol. 16, p. 508.
25. *Great Books*, Vol. 16, *De Revolutionibus*, Book One, p. 510.
26. *Great Books*, Vol. 16, *De Revolutionibus*, Book One, p. 510.
27. *Great Books*, Vol. 16, *De Revolutionibus*, Book One, p. 511.
28. *Great Books*, Vol. 16, *De Revolutionibus*, Book One, p. 514.
29. Kuhn, p. 42.
30. *Great Books*, Vol. 16, *De Revolutionibus*, Book One, p. 519.
31. *Great Books*, Vol. 16, *De Revolutionibus*, Book One, p. 557.
32. Shea, pp. 92-103.
33. *Great Books*, Vol. 16, "Biographical Note," p. 506.
34. Bourne, Chapter 7, pp. 105-116.
35. Goethe, p. 618.

36. *Great Books*, Vol. 16, "Translator's Note," p. 490.
37. *Great Books*, Vol. 16, "Translator's Note," p. 490.
38. Ferris, *Coming of Age*, pp. 137-141.
39. Dennis Danielson, "The Bones of Copernicus," *American Scientist*, Vol. 97 (January-February 2009), pp. 50-57.

Chapter 3 The Starry Messenger

1. Newman, Vol. 2, p. 734.
2. A common paraphrase of Galileo's most famous observation: "Philosophy is written in the vast book which stands forever open before our eyes, I mean the universe; but it cannot be read until we have learnt the language and become familiar with the characteristics in which it is written. It is written in mathematical language, and the letters are triangles, circles, and other geometrical figures, without which means it is humanly impossible to comprehend a single word." From his polemic *Il Saggiatore*.
3. Sobel, p. 365.
4. Sobel, p. 17.
5. Provided the amplitude of the oscillations is not too great.
6. Sobel, p. 18.
7. Sobel, p. 19.
8. Sobel, p. 20, and table of monetary measures on p. 375.
9. Newman, Vol. 2, p. 726.
10. Newman, Vol. 2, p. 731.
11. Newman, Vol. 2, p. 734.
12. Newman, Vol. 2, p. 731.
13. Newman, Vol. 2, p. 739.
14. Newman, Vol. 2, p. 747.
15. Newman, Vol. 2, p. 727.
16. Newman, Vol. 2, p. 728.
17. Newman, Vol. 2, p. 755, Theorem I, Proposition I, of the fourth day of *Dialogue Concerning Two New Sciences*.
18. Hey and Walters, *Einstein's Mirror*, p. 169.
19. Sobel, p. 31.
20. Bronowski, p. 204.
21. Sobel, p. 36.
22. Sobel, p. 38.
23. Koestler, p. 440.
24. Koestler, p. 440.
25. Geymonat, p. 77.
26. Geymonat, p. 76.
27. Langford, p. 70.
28. Langford, p. 71.
29. Langford, p. 66.
30. Langford, p. 71.
31. Sobel, p. 66.
32. Historians disagree as to whether Clavius' denial of the Medicean stars was genuine or simply in jest due to his close friendship with Galileo.

33. Sobel, p. 70.
34. Langford, p. 62.
35. Koestler, p. 443.
36. De Santillano, p. 142.
37. Geymonat, p. 75.
38. Geymonat, p. 75.
39. Langford, p. 61.
40. Koestler, p. 455.
41. Sobel, p. 73.
42. Koestler, p. 459.
43. Sobel, p. 48
44. Sobel, p. 66.
45. Koestler, p. 472.
46. Pedersen, p. 13.
47. Sobel. p. 78.
48. Sobel, p. 79
49. Sobel, p. 80.
50. Sobel, pp. 86-88.
51. Sobel, p. 93.
52. Langford, p. 58.
53. Sobel, p. 138.
54. Sobel, p. 143.
55. Newman, Vol. 2, p. 732.
56. Sobel, p. 222.
57. Geymonat, p. 136.
58. Geymonat, p. 137.
59. Sobel, p. 241.
60. Geymonat, p. 136.
61. Bronowski, p. 208.
62. Geymonat, p. 138.
63. Sobel, p. 247.
64. Sobel, p. 249.
65. Brownowski, p. 214.
66. Sobel, p. 256.
67. Sobel, p. 272.
68. All quotations from Sobel, p. 274.
69. Bronowski, p. 216.
70. Hey and Walters, *Einstein's Mirror*, figure caption, p. 49.
71. Sobel, p. 258.
72. Sobel, p. 317.
73. Sobel, p. 314.
74. Sobel, p. 344.
75. Sobel, pp. 5-6.
76. Sobel, p. 4.
77. Sobel, p. 361.
78. Sobel, p. 30.
79. Sobel, p. 326.
80. Newman, Vol. 2, p. 734.
81. Newman, Vol. 2, p. 738.

82. Newman, Vol. 2, p. 755.
83. Sobel, p. 348.
84. Sobel, p. 349.
85. Sobel, pp. 350-351.
86. Sobel, pp. 355-356.
87. Sobel, p. 326.
88. Sobel, p. 302.
89. As quoted by Gleick, *Newton*, p. 98.
90. Sobel, p. 368.
91. Pope John Paul II, p. 245.

Chapter 4 The Star-Crossed Prophet

1. Conner, p. 24. The narrative of the life of Johannes Kepler presented herein draws heavily from James Conner's excellent biography *Kepler's Witch*.
2. Conner, p. 26.
3. Conner, p. 13.
4. Conner, p. 45.
5. Kuhn, p. 106.
6. Conner, p. 47.
7. Conner, p. 61.
8. Conner, p. 60.
9. Conner, p. 44.
10. Newman, Vol. I, p. 218.
11. Conner, p. 68.
12. Conner, p. 73.
13. Conner, p. 79.
14. Newman, Vol. I, p. 221.
15. Newman, Vol. I, p. 223.
16. Conner, p. 113.
17. Conner, p. 138.
18. Newman, Vol. I, p. 224.
19. http://www.hps.cam.ac.uk/starry/keplertables.html.
20. Newman, Vol. I, p. 224.
21. Conner, p. 153.
22. Conner, p. 155.
23. Newman, Vol. I, p. 227.
24. Newman, Vol. I, p. 230.
25. Conner, p. 224.
26. Conner, p. 229.
27. Newman, Vol. I, p. 232.
28. Conner, p. 281ff.
29. Conner, p. 302ff.
30. Conner, p. 354.
31. Conner, p. 357.
32. Conner, p. 363.
33. Conner, p. 364.

34. Carl Sagan, *Cosmos: A Personal Voyage*, 13-part PBS television show, Episode III, 1980.

Chapter 5 The Jewish Saint

1. Einstein to Queen Elizabeth of Belgium, November 20, 1933, in Michelmore's *Einstein*, from end notes of Brian, p. 464.
2. Brian, p. 283.
3. Bronowski, p. 256.
4. Rhodes, p. 171.
5. Paraphrase of Albert Einstein, "Autobiographical Notes," in *The World Treasury of Physics, Astronomy, and Mathematics*, ed. Timothy Ferris, p. 580.
6. Rhodes, p. 196.
7. Isaacson, p. 8.
8. Brian, p. 3.
9. Albert Einstein, "Autobiographical Notes," in *The World Treasury of Physics, Astronomy, and Mathematics*, ed. Timothy Ferris, p. 578.
10. Albert Einstein, "Autobiographical Notes," in *The World Treasury of Physics, Astronomy, and Mathematics*, ed. Timothy Ferris, pp. 582-583.
11. Brian, p. 20.
12. Brian, p. 31.
13. Brian, p. 45.
14. John Archibald Wheeler, "Albert Einstein," in *The World Treasury of Physics, Astronomy, and Mathematics*, ed. Timothy Ferris, p. 568.
15. Brian, p. 52.
16. Brian, p. 54.
17. Brian, p. 55.
18. Brian, p. 83.
19. John Archibald Wheeler, "Albert Einstein," *The World Treasury of Physics, Astronomy, and Mathematics*, ed. Timothy Ferris, p. 575.
20. Brian, p. 258.
21. Isaacson, p. 358.
22. Isaacson, p. 311.
23. Isaacson, p. 345.
24. Isaacson, p. 346.
25. Isaacson, p. 349.
26. Isaacson, p. 347.
27. John Archibald Wheeler, as quoted by Isaacson, p. 325.
28. Brian, p. 165.
29. Isaacson, p. 363
30. Isaacson, p. 363.
31. Isaacson, p. 468.
32. Isaacson, p. 411.
33. Rhodes, p. 186.
34. All previous short quotes from Isaacson, pp. 373-376.
35. Isaacson, p. 374.
36. Rhodes, p. 186.
37. Isaacson, p. 360.

38. Brian, p. 232.
39. Brian, p. 231.
40. Isaacson, pp. 407-408.
41. Isaacson, p. 447
42. Isaacson, p. 401.
43. Isaacson, p. 426.
44. John Archibald Wheeler, "Albert Einstein," in *The World Treasury of Physics, Astronomy, and Mathematics*, ed. Timothy Ferris, p. 576.
45. From a review of Isaacson's biography of Einstein by John Updike, *The New Yorker*, April 2, 2007.
46. Isaacson, p. 7.
47. Isaacson, p. 67.
48. *Great Books*, Vol. 34, p. 8.
49. Isaacson, p. 37.
50. Isaacson, p. 350.
51. Isaacson, p. 352.
52. Isaacson, p. 461.
53. Isaacson, p. 101.
54. Isaacson, p. 317.
55. Rhodes, p. 168.
56. Isaacson, p. 183.
57. Isaacson, p. 88.
58. Isaacson, p. 246.
59. Isaacson, p. 274.
60. Depiction of Einstein's relationship with his sister and stepdaughter and the quotations therein come from Isaacson, pp. 517-518.
61. Brian, p. 229.
62. Brian, p. 185, during an interview with German-born George Sylvester Viereck, the Barbara Walters of his day, who specialized in interviewing "big game."
63. *Time*, December 11, 1999, p. 65
64. Einstein, *Ideas and Opinions*, p. 118, from *Atlantic Monthly*, November, 1945.
65. Isaacson, p. 533.
66. Isaacson, p. 494.
67. Isaacson, p. 550.
68. Isaacson, p. 242.
69. Isaacson, p. 445.
70. This amusing anecdote, including the quotation, is described by Isaacson, pp. 369-370.
71. John Archibald Wheeler, "Albert Einstein,", in *The World Treasury of Physics, Astronomy, and Mathematics*, ed. Timothy Ferris, p. 575.
72. Isaacson, p. 30.
73. Brian, p. 161.
74. Brian, p. 186.
75. John Archibald Wheeler, "Albert Einstein," in *The World Treasury of Physics, Astronomy, and Mathematics*, ed. Timothy Ferris, p. 573.
76. John Archibald Wheeler, "Albert Einstein," in *The World Treasury of Physics, Astronomy, and Mathematics*, ed. Timothy Ferris, p. 573.
77. Brian, p. 186.

78. John Archibald Wheeler, "Albert Einstein," in *The World Treasury of Physics, Astronomy, and Mathematics*, ed. Timothy Ferris, pp. 573-574.
79. Einstein, *Ideas and Opinions*, p. 342.
80. Einstein, *Ideas and Opinions*, p. 39.
81. Einstein, *Ideas and Opinions*, p. 39.
82. Einstein, *Ideas and Opinions*, p. 45.
83. Brian, p. 226.
84. Brian, pp. 186-187.
85. Isaacson, p. 536.
86. Isaacson, p. 389.
87. Isaacson, p. 390.
88. Einstein, *Ideas and Opinions*, p. 11.
89. Rhodes, p. 169.
90. Rhodes, p. 169.
91. Brian, p. 282.
92. Brian, p. 230.
93. Brian, p. 351.
94. Einstein, *Ideas and Opinions*, p. 9.
95. Albert Einstein, "Autobiographical Notes," in *The World Treasury of Physics, Astronomy, and Mathematics*, ed. Timothy Ferris, p. 573.
96. Einstein, *Ideas and Opinions*, p. 12
97. Brian, p. 264.
98. Brian, p. 159.
99. Brian, p. 433.
100. Brian, p. 257, and last plates, between p. 240 and p. 241.
101. *Time*, December 31, 1999, p. 62.
102. Isaacson, p. 508.

Chapter 6 A Wrinkle in Time

1. Brian, p. 363.
2. The title of Madeline L'Engle's fantasy classic.
3. Bronowski, p. 247.
4. Brian, pp. 160-161.
5. Hey and Walters, *Einstein's Mirror* pp. 36-37, and Ferris, *Coming of Age*, p. 179.
6. http://www.youtube.com/watch?v=MnqP81BUwgs.
7. See, for example, the Appendix of *Einstein's Mirror*, pp. 258-263.
8. Hey and Walters, *Einstein's Mirror*, pp. 18-19.
9. Einstein, *Relativity*, p. 150.
10. Hey and Walters, *Einstein's Mirror*, p. 99.
11. Einstein, *Relativity*, pp. 45-46.
12. Einstein, *In His Own Words*, p. 237-238.

Chapter 7 Einstein's Happiest Thought

1. Ohanian, p. 75.
2. Brian, p. 91.
3. Hey and Walters, *Einstein's Mirror*, p. 181.

4. Brian, p. 85.
5. Hey and Walters, *Einstein's Mirror*, p. 167.
6. Einstein, *Ideas and Opinions*, p. 339.
7. Hey and Walters, *Einstein's Mirror*, pp. 172-173.
8. Galileo, as quoted by Ohanian, p. 18.
9. Hey and Walters, *Einstein's Mirror*, p. 170.
10. Einstein, *Relativity*, pp. 78-82.
11. Davis and Hersh, pp. 217-222.
12. Hey and Walters, *Einstein's Mirror*, p. 184.
13. Hey and Walters, *Einstein's Mirror*, p. 186.
14. Simmons, p. 845.
15. Simmons, p. 846.
16. Hey and Walters, *Einstein's Mirror*, p. 188, attributed to Einstein's *Notes of the Origins of General Relativity*, 1934.
17. Brian, p. 92.
18. Simmons, p. 846.
19. Simmons, p. 847.
20. Attributed to John Archibald Wheeler, in Brian, p. 91.

Chapter 8 Lights All Askew

1. Hey and Walters, *Einstein's Mirror*, p. 210.
2. Narrative with quotations adapted from Ferris, *Coming of Age*, pp. 147-148.
3. Ferris, *Coming of Age*, p. 161.
4. Ferris, *Coming of Age*, p. 169.
5. Hey and Walters, *Einstein's Mirror*, p. 213, and Ferris, *Coming of Age*, pp. 169-170.
6. Circumstances of the debate adapted from Ferris, *Coming of Age*, pp. 170ff.
7. Ferris, *Coming of Age*, p. 172.
8. Ferris, *Coming of Age*, p. 172.
9. Hey and Walters, *Einstein's Mirror*, p. 211.
10. Hey and Walters, *Einstein's Mirror*, pp. 211-212.
11. Hey and Walters, *Einstein's Mirror*, p. 216.
12. Ferris, *Coming of Age*, p. 210.
13. Hey and Walters, *Einstein's Mirror*, p. 214.
14. Hey and Walters, *Einstein's Mirror*, p. 214.
15. Based on a dramatization of the meeting between Einstein, Hubble, and Lemaître presented in *Stephen Hawking's Universe: The Big Bang*, Vol. 1, Part II.
16. Isaacson, p. 255.
17. Georges Lemaître, "The Primeval Atom," in *The World Treasury of Physics, Astronomy, and Mathematics*, ed. Timothy Ferris, pp. 360-364.
18. http://www.aip.org/history/cosmology/ideas/expanding.htm. Historical notes on Big Bang cosmology from American Institute of Physics website.
19. Ferris, *Coming of Age*, p. 274.
20. Ferris, *Coming of Age*, p. 275.
21. David M. Schramm, "The Age of the Elements," in *The World Treasury of Physics, Astronomy, and Mathematics*, ed. Timothy Ferris, p. 174.

22. Hey and Walters, *Einstein's Mirror*, p. 217.
23. The *New York Times*, August 18, 2007. At the end of this life, Ralph Alpher, who died in August 2007, remained embittered about the lack of credit he received for Big Bang cosmology. When the cosmic background radiation was discovered by Penzias and Wilson in 1964, "no one involved in the discovery so much as tipped a hat in [Alpher's and Herman's] direction."
24. Thorne, p. 122.
25. Thorne, p. 122.
26. Hawking, pp. 104-105, 123.
27. Hawking, p. 105.
28. Thorne, p. 123.
29. Thorne, p. 124.
30. Thorne, p. 134.
31. Historical note and quotation from Thorne, pp. 140-144.
32. Kameshwar C. Wali, "Chandra: A Biographical Portrait," *Physics Today*, December 2010, pp. 38-43.
33. Thorne, p. 151.
34. Thorne, p. 197.
35. Hawking, p. 104.
36. Thorne, p. 460.
37. Thorne, p. 462.
38. Thorne, p. 463.
39. Thorne, p. 420.
40. Thorne, p. 421.
41. *Stephen Hawking's Universe: The Big Bang*, Vol. 1, Part II.
42. *Stephen Hawking's Universe: The Big Bang*, Vol. 1, Part II.
43. http://lambda.gsfc.nasa.gov/product/cobe.
44. Freeman Dyson, "Chandrasekhar's Role in 20th-Century Science," *Physics Today*, December 2010, p. 46.
45. Thorne, p. 299.

Chapter 9 The Entangled Bank

1. Darwin, *Voyage*, p. 323.
2. From the introduction by David Quammen to the commemorative *Origin Illustrated*, p. v.
3. From the biographical prologue to the *Voyage of the Beagle* by Recorded Books, Camp Hill, PA: History Book Club, 1994.
4. Fletcher, p. 6.
5. *The Autobiography of Charles Darwin* as excerpted in *Origin Illustrated*, p. 75.
6. Fletcher, p. 6.
7. *The Autobiography of Charles Darwin* as excerpted in *Origin Illustrated*, p. 93.
8. Fletcher, p. 7.
9. *The Autobiography of Charles Darwin* as excerpted in *Origin Illustrate*, p. 98
10. Fletcher, p. 7.
11. Fletcher, p. 9.
12. Darwin, *Voyage*, p. 1.
13. Gould, pp. 30-31.

14. Darwin, *Voyage*, p. 213.
15. Darwin, *Voyage*, p. 11.
16. Darwin, *Voyage*, as excerpted in *Origin Illustrated*, p. 50.
17. Darwin, *Voyage*, p. 36.
18. Darwin, *Voyage*, p. 40.
19. Ferris, *Coming of Age*, p. 220.
20. Ferris, *Coming of Age*, p. 225.
21. Ferris, *Coming of Age*, p. 228.
22. Darwin, *Voyage*, p. 12.
23. Darwin, *Voyage*, p. 87.
24. Lyell, p. 140.
25. Darwin, *Voyage*, pp. 181-182.
26. Darwin, *Voyage*, p. 182.
27. Darwin, *Voyage*, p. 86.
28. Thanks to the hospitality of Andy Currant, curator of mammalia, British Natural History Museum, London.
29. Carey, p. 214.
30. Darwin, *Voyage*, pp. 182-183.
31. Darwin, *Voyage*, pp. 183-184.
32. Darwin, *Voyage*, pp. 183-184.
33. Darwin, *Voyage*, pp. 183-184.
34. Darwin, *Voyage*, pp. 184-185.
35. Darwin, *Voyage*, pp. 329-330.
36. Darwin, *Voyage*, p. 331.
37. Darwin, *Voyage*, p. 331.
38. Andy Currant, curator of mammalia, British Natural History Museum, London, personal communication.
39. Darwin, *Voyage*, pp. 338-339.
40. Darwin, *Voyage*, p. 339.
41. Ferris, *Coming of Age*, p. 225.
42. Darwin, *Voyage*, p. 355.
43. Darwin, *Voyage*, pp. 354-355.
44. Darwin, *Voyage*, p. 355.
45. Fletcher, p. 24.
46. Darwin, *Voyage*, p. 403.
47. Darwin, *Voyage*, p. 400.
48. Darwin, *Voyage*, p. 410.
49. Darwin, *Voyage*, pp. 404-405.
50. Darwin, *Voyage*, pp. 404-405.
51. Weiner, p. 25.
52. Keynes, Richard Darwin, ed., *Charles Darwin's Beagle Diary*, Cambridge: Cambridge University Press, 2001, p. 389.
53. Darwin, *Voyage*, p. 225.
54. Darwin, *Voyage*, p. 226.
55. Darwin, *Voyage*, p. 430.
56. Darwin, *Voyage*, p. 444.
57. Darwin, *Voyage*, p. 456.
58. Darwin, *Voyage*, p. 482.
59. Darwin, *Voyage*, p. 482.

60. Ferris, *Coming of Age*, p. 233.
61. Ferris, *Coming of Age*, p. 233.
62. Darwin, *Voyage*, p. 532.
63. Darwin, *Voyage*, pp. 535-536.
64. Darwin, *Voyage*, pp. 535-536.
65. Darwin, *Voyage*, p. 535.
66. Darwin, *Voyage*, pp. 535-536.
67. Darwin, *Voyage*, p. 532.
68. Darwin, *Voyage*, p. 536.
69. Weiner, p. 29.
70. Keynes, p. 3.
71. Keynes, p. 3.
72. Keynes, p. 7.
73. Keynes, p. 10.
74. Francis Darwin, ed., *The LIfe and Letters of Charles Darwin*, Vol. 1, 1887, p. 70.
75. *Autobiography of Charles Darwin*, as excerpted in *Origin Illustrated*, p. 250.
76. *The Life and Letters of Charles Darwin*, as excerpted in *Origin Illustrated*, p. 258.
77. American Museum of Natural History website
      (http://www.amnh.org/exhibitions/darwin/work/day.php
78. *The Autobiography of Charles Darwin*, as excerpted in *Origin Illustrated*, p. 213.
79. Keynes, p. 4.
80. Fletcher, p. 26.
81. Malthus, T. R., *An Essay on the Principle of Population*, 1798.
82. Darwin Correspondence Project, Letter 729 – Darwin, C. R., to Hooker, J. D.,
      January 11, 1844.
83. Weiner, p. 141.
84. Weiner, p. 55.
85. Weiner, p. 144.
86. Keynes, p. 135.
87. Bronowski, p. 306.
88. Darwin Correspondence Project website (http://www.darwinproject.ac.uk).
89. Keynes, p. 133.
90. Keynes, pp. 214-217.
91. Keynes, p. 232.
92. Weiner, p. 49.
93. *Origin Illustrated*, introduction by David Quammen, p. xiii.
94. Bronowski, p. 306.
95. Bronowski, p. 308.
96. Keynes, p. 247.
97. *The Autobiography of Charles Darwin*, as excerpted in *Origin Illustrated*, p. 282.
98. Fletcher, p. 33.
99. *The Autobiography of Charles Darwin*, as excerpted in *Origin Illustrated*, p. 282.
100. *Origin Illustrated*, introduction by David Quammen, p. x.
101. *The Life and Letters of Charles Darwin*, as excerpted in *Origin Illustrated*, p. 350.
102. *The Life and Letters of Charles Darwin*, as excerpted in *Origin Illustrated*, pp.
      340-341.
103. *The Life and Letters of Charles Darwin*, as excerpted in *Origin Illustrated*, p. 208.
104. Darwin, *Origin*, p. 133.
105. *The Life and Letters of Charles Darwin*, as excerpted in *Origin Illustrated*, p. 316.

106. Darwin, *Origin*, p. 134.
107. Darwin, *Origin*, p. 408.
108. Darwin, *Origin*, p. 133.
109. *Great Books*, Vol. 49, *Darwin*, pp. 108-109.
110. Darwin, *Origin*, p. 409.
111. *The Autobiography of Charles Darwin*, as excerpted in *Origin Illustrated*, p. 443.
112. Fletcher, p. 44.
113. Fletcher, p. 44.
114. Darwin, *Origin*, p. 429.
115. *The Autobiography of Charles Darwin*, as excerpted in *Origin Illustrated*, p. 486.
116. Gallup Poll News Service release, March 5, 2001.
117. The Clergy Letter Project (http://www.theclergyletterproject.org).

Chapter 10 The Time Bomb

1. This quote is a composite extracted from a letter from Darwin to Wallace, dated July 12, 1871, in the aftermath of a review of *Descent* by St. George Jackson Mivart in the *Quarterly Review*, July 1871. Information provided by Tori Reeves, Collections Curator, English Heritage, London.
2. Weiner, p. 7.
3. Ferris, *Coming of Age*, p. 244.
4. Ferris, *Coming of Age*, p. 245.
5. Bronowski, p. 308.
6. One may download this work for free from Project Gutenberg (www.gutenberg.org).
7. Website: www.victorianweb.org/science, entry on J.-B. Lamarck by David Clifford, Ph.D., Cambridge University.
8. Bronowski, p. 308.
9. Bronowski, p. 302.
10. Darwin, *Origin*, p. 3.
11. Weiner, p. 49.
12. Weiner, p. 64.
13. From Darwin's paper of 1858 presented before the Linnean Society of London, before which Wallace's paper was also presented.
14. Darwin, *Origin*, p. 266.
15. Ferris, *Coming of Age*, p. 245.
16. Ferris, *Coming of Age*, p. 247.
17. Ferris, *Coming of Age*, p. 247.
18. Both quotations from Ferris, *Coming of Age*, p. 247.
19. Bronowski, p. 301.
20. *National Geographic*, "Was Darwin Wrong?" November 2004, p. 12.
21. *National Geographic*, "Was Darwin Wrong?" November 2004, p. 13.
22. Attributed to Richard Owen (1804-1892), the British anatomist who established London's Natural History Museum.
23. Weiner, pp. 230-231.
24. Weiner, p. 182.
25. Darwin, *Origin*, p. 66.
26. Weiner, p. 109.

27. Weiner, p. 110.
28. Weiner, p. 7.
29. Weiner, pp. 71-78.
30. Weiner, p. 78.
31. Weiner, p. 36.
32. *National Geographic*, "Was Darwin Wrong?" November 2004, p. 21.
33. Weiner, p. 265.
34. Population genetics is the modern incarnation of biogeography, whereby population demographics of species are overlaid with their genetic markers to map global distributions of allied species.
35. The advent of the digital computer in the mid-twentieth century opened up an entirely new field: computational biology, whose golden egg is genomics, the study of an organism's entire genetic code.
36. Darwin, The Complete Works of Charles Darwin Online (http://darwin-online.org.uk), Evolution Notebook C, pp. 196-197.
37. This chronology of the early universe is liberally adapted from Steven Weinberg, "The First Three Minutes," in *The World Treasury of Physics, Astronomy, and Mathematics*, ed. Timothy Ferris, pp. 395-409.
38. The chronology of life is adapted from *A Walk Through Time: From Stardust to Us*, by Liebes et al., online version at www.globalcommunity.org/wtt.
39. Wilson, p. 183.
40. *A Walk Through Time Online*, 395 MYA.
41. Wilson, pp. 25-26.
42. *National Geographic*, March 2006, pp. 64-65. Recent evidence now suggests that humans may have occupied North America for 35,000 years.
43. Wilson, p. 94.
44. In the *punctuated equilibrium* paradigm of Stephen Jay Gould and Niles Eldridge.
45. Wilson, p. 31.
46. Wilson, p. 31.
47. Attributed to the late Russian Geneticist Theodore Dobzhansky, as quoted in Weiner, p. 267.

Chapter 11 Through the Looking Glass

1. As quoted in Bryson, p. 17.
2. Terminology and scenarios adapted from Wolfson, 1997.
3. Isaacson, p. 96.
4. Hey and Walters, *The Quantum Universe*, p. 38.
5. Lindley, p. 51.
6. Heisenberg, p. 34.
7. Lindley, p. 53., not directly attributed to Rutherford.
8. Lindley, p. 138.
9. Hey and Walters, *The Quantum Universe*, p. 27.
10. Hey and Walters, *The Quantum Universe*, p. 27.
11. More precisely, the probability density illustrated in Fig. 24 is the square of the modulus of the object's wavefunction, namely $|\Psi(x)|^2$. The probability

of observing the object between locations $x = a$ and $x = b$, $P(a, b)$, is $\int_a^b |\Psi(x)|^2 dx$.

12. Heisenberg, p. 2.

13. Lindley, p. 85.

14. The reasons for the irreconcilable breakdown of the relationship between Bohr and Heisenberg during the Second World War remains itself shrouded in uncertainty and is the subject of Michael Frayn's captivating play *Copenhagen*.

15. Lindley, pp. 1-2.

16. Shortly before he succumbed to cancer in 1988, Feynman was thrust into the public eye by serving prominently on the Challenger Commission, which investigated the January 28, 1986, disaster that destroyed the space shuttle *Challenger* and killed its seven crew members, including America's teacher in space, Christa Corrigan McAuliffe.

17. In truth, photographic paper is not sufficiently sensitive to respond to a single photon. It takes an accumulation of several photons in the same region to instigate chemical change. That said, the charge-coupled devices (CCDs) of digital photography do respond to individual photons, so the argument above essentially holds.

18. A composition of quotes from *The Feynman Lectures of Physics*, Vol. 3 (Quantum Mechanics), Lectures 1-6.

19. Heisenberg, p. 38.

20. Heisenberg, pp. 40-41.

21. Analogy by Charles Peskin of Courant Institute, conveyed during personal conversation.

22. Heisenberg, p. 35.

23. Heisenberg, p. 35.

24. Hey and Walters, *The Quantum Universe*, p. 27.

25. Widely attributed to Schrödinger; see for example www.brainyquotes.com.

26. Hey and Walters, *The Quantum Universe*, p. 1.

27. Prigogine and Stengers, p. 300.

28. Heisenberg, p. 24.

29. Lindley, p. 137.

30. Zukav, p. 316.

31. Zukav, p. 321.

32. S. J. Freedman and J. F. Clauser, *Phys. Rev. Lett.*, Vol. 28, 1972, pp. 938-941.

33. Penrose, p. 286.

34. Zukav, p. 326. The author had the pleasure of meeting Stapp at a conference in October 2015. Stapp has quite a pedigree. Following his Ph.D in physics from UC Berkeley, Stapp did postdoctoral work with both Wolfgang Pauli and Werner Heisenberg before a distinguished career at Lawrence Berkeley Laboratory.

35. S. Groeblacher et al., "An Experimental Test of Non-Local Realism," *Nature* 446 (April 19, 2007), p. 871.

36. S. Groeblacher et al., "An Experimental Test of Non-Local Realism," *Nature* 446 (April 19, 2007), p. 871.

37. J. D. Jost et al., "Entangled Mechanical Oscillators," *Nature* 459 (June 4, 2009), p. 683.

Chapter 12 Parallel Universes

1. Campbell, *The Power of Myth*, No. 3: *The First Storytellers*.
2. Lao Tzu, p. 2.
3. Christians might argue that God manifests through the incarnation of Jesus.
4. James, p. 305.
5. Campbell, *The Power of Myth*, No. 3: *The First Storytellers*.
6. LeShan, *How to Meditate*, p. 18.
7. LeShan, *How to Meditate*, p. 18.
8. Smith, p. 82.
9. No other quote of Goethe's is more popular. However, it appears to be a composite, with contributions from Goethe and others. For clarification, follow the quotations link at the website of the Goethe Society of North America, www.goethesociety.org.
10. Attributed to Heisenberg and quoted in LeShan, *How to Meditate*, p. 49.
11. Eco-theologian Thomas Berry's most oft-quoted aphorism (www.thomasberry.org).
12. Rowling, p. 333.

Chapter 13 The Arrows of Time

1. Prigogine and Stengers, p. 291.
2. Adapted from Prigogine and Stengers, p. 17.
3. Prigogine and Stengers, p. 310.
4. Prigogine and Stengers, p. 94.
5. Hey and Walters, *Einstein's Mirror*, p. 56.
6. John Greenleaf Whittier, *Maud Miller* (1856), in *One Hundred Choice Selections*, ed. Phineas Garrett, Philadelphia: Penn Publishing Company, 1897.
7. Penrose, p. 302.
8. Hawking, pp. 184-185.
9. Charles Dickens, opening line of *A Tale of Two Cities* (1859), in *World's Classics*, ed. Andrew Sanders, Oxford: Oxford University Press, 1980.
10. Atkins, p. 111.
11. Atkins, p. 113.
12. Atkins, p. 115.
13. Prigogine and Stengers, p. 8.
14. Prigogine and Stengers, p. 233.
15. This classification pre-dates Einstein's equivalence of mass and energy, following which thermodynamic systems are essentially either isolated or open.
16. A detailed consideration of the possibilities of dynamical systems in far-from-equilibrium conditions is the central focus of *Order Out of Chaos*. The interested reader is encouraged to read this thought-provoking best-seller by Prigogine and Stengers.
17. See for example the description of the "Brusselator" in Prigogine and Stengers, pp. 140ff.
18. From Strogatz' February 2004 TED lecture. See www.ted.com/talks/steven_strogatz_on_sync.html.

19. Prigogine and Stengers, p. 300.
20. Several of these light-hearted alternatives of the second law of thermodynamics are taken or adapted from Chapter 4 of Trefil and Hazen, especially pp. 75-79.
21. In truth, regenerative braking long predates the hybrid automobile, having emerged earlier in electric locomotives for rail transport.
22. Hawking, p. 189.
23. Prigogine and Stengers, p. 303.
24. Paraphrase of Prigogine and Stengers, p. 129.

Chapter 14 What Is Life?

1. Annexation of Austria.
2. Dyson, *Origins of Life*, p. 1.
3. Schrödinger, *What is Life? with Mind and Matter*, from the biographical notes.
4. Schrödinger, p. 3.
5. Schrödinger, 1. 1.
6. Schrödinger, p. 1.
7. Roger Penrose, from the foreword to the 1991 Canto edition of *What is Life? with Mind and Matter*.
8. Roger Penrose, from the foreword to the 1991 Canto edition of *What is Life? with Mind and Matter*.
9. Dyson, *Origins of Life*, pp. 2-3.
10. This account of the discovery of the structure of DNA is adapted from James Watson's riveting but infamous book, *The Double Helix*. Watson's account is unabashedly one-sided, and his portrayal of Rosalind Franklin is unforgivingly misogynistic. When *The Double Helix* was first proposed under the title *Honest Jim*, it caused considerable consternation among many of those closest to the story. Lawrence Bragg, the director of the Cavendish Laboratory, initially refused to write the foreword, doing so only after Watson toned down some of the excesses. Still, the book chronicles high drama, and Watson's storytelling is captivating.
11. The interested reader is referred to Georgina Ferry's delightful biography *Max Perutz and the Secret of Life*, Cold Spring Harbor Laboratories Press, 2007. In addition to being a fascinating read in its own right, by focusing on Max Perutz' catalytic role at the Cavendish's Laboratory for Molecular Biography, it sets the stage for the story of DNA.
12. Watson, Chapter 2.
13. Watson, p. 15.
14. Brenda Maddow, "The Double Helix and the 'Wronged Heroine,' " *Nature* 421 (January 23, 2003), pp. 407-408.
15. Watson, p. 44.
16. Ferry, p. 147.
17. Ferry, p. 147.
18. Watson, p. 49.
19. Watson, p. 76.
20. Whether or not they had Peter Pauling's permission is unclear from Watson's account.

21. Watson, p. 209.
22. Ferry, p. 268.
23. Ferry, p. 268.
24. Human Genome Project (www.genome.gov).
25. Human Genome Project (www.genome.gov).
26. From Collins' presentation "Art Imitates Life: The Dance of DNA, Decoding, & Doctoring," James Madison University, January 22, 2011.
27. Watson, p. 226.
28. Ferry, p. 153.

Chapter 15 A World Aflame

1. Adapted from www.ucmp.berkeley.edu/history/thuxley.html.
2. Attributed to Johns Hopkins University neurologist Vernon Mountcastle by Kandel, p. 302.
3. Vol. 297, No. 4, p. 76.
4. Schrödinger references Sherrington's quote on p. 121. Here, however, I have used a slightly different variant of Sherrington's quote as given on pages 201-202 of the afterword by Edgar N. Jackson of LeShan's *How to Meditate*. The two versions are substantially the same.
5. Schrödinger, p. 122.
6. Schrödinger, p. 137.
7. From Peter Russell's keynote address at the Second International Conference on Science and Consciousness, 2000, Albuquerque, New Mexico.
8. Kant, p. 118, a liberal translation of Kant's thought as expressed in his letter of 1798 to Marcus Herz.
9. Prigogine and Stengers, p. 7.
10. Tarnas, p. 24ff.
11. Teilhard de Chardin, *The Phenomenon of Man*, p. 219.
12. Teilhard de Chardin Centenary Exhibition publication, p. 27.
13. Teilhard de Chardin Centenary Exhibition publication, p. 111111111111111111111111, as recounted by Pierre Emmanuel.
14. Teilhard de Chardin, *The Divine Milieu*, biographical preface, p. 18.
15. Teilhard de Chardin, *The Divine Milieu*, biographical preface, p. 18.
16. Teilhard de Chardin Centenary Exhibition publication, p. 28.
17. King, pp. 18-23.
18. "On My Attitude to the Official Church," in Teilhard's *The Heart of Matter*, p. 115.
19. Yungblut, p. 178.
20. Since this work was begun, a new and more accurate translation of Teilhard's *Le phenomene humain* by Sarah Appleton-Weber has appeared under the gender-neutral English title *The Human Phenomenon*.
21. Teilhard de Chardin, *The Divine Milieu*, biographical preface, p. 20.
22. King, pp. 36-40.
23. Prigogine and Stengers, p. 91.
24. Teilhard de Chardin, *The Heart of Matter*, p. 26.
25. Teilhard de Chardin, *The Divine Milieu*, foreword by Pierre Leroy, p. 13.
26. Teilhard de Chardin, *The Divine Milieu*, foreword by Pierre Leroy, p. 22.

27. Attributed to Hans-Peter Duerr, former director of Munich's Max Planck Institute.
28. Schrödinger, p. 121.
29. New research suggests that the unit may have been part of an Algerian regiment. (Ursula King, personal communication.)
30. King, pp. 51-52.
31. Teilhard de Chardin Centenary Exhibition publication, p. 30.
32. Teilhard de Chardin, *The Divine Milieu*, foreword by Pierre Leroy, p. 23.
33. Teilhard de Chardin, *Sketch of a Personalistic Universe*, 1936.
34. Teilhard de Chardin, *The Phenomenon of Man*, pp. 57-60.
35. Teilhard de Chardin, *The Phenomenon of Man*, p. 55.
36. Teilhard de Chardin, *The Phenomenon of Man*, pp. 57-60.
37. Teilhard de Chardin, *The Phenomenon of Man*, p. 57.
38. Prigogine and Stengers, p. 75.
39. This ubiquitous quote is claimed by many, ancient and modern. Its very ubiquity suggests that the source is indeed ancient.
40. Jung, p. 292.
41. William James, as quoted by Oliver Sachs in *A Leg to Stand On*, New York: Touchstone, 1984, p. 166.
42. From Arthur Eddinton's *The Nature of the Physical World*, New York: The Macmillan Company, 1928, p. 276.
43. Sir James Jeans (1877-1946), *The Mysterious Universe*, Cambridge: Cambridge University Press, 2009, p. 139 (first published in 1930).
44. Penrose, p. 446.
45. As quoted in Friedman, p. 79.
46. Friedman, p. 34.
47. Attributed to Max Perutz, 1962 Nobel laureate in chemistry. Quote from the dust jacket of Ferry's *Max Perutz and the Secret of Life*.
48. Teilhard de Chardin, *The Divine Milieu*, biographical preface, p. 23.
49. Teilhard de Chardin, *The Phenomenon of Man*, p. 26.
50. King, p. 36.
51. Dillard, p. 103.
52. Adapted from Prigogine and Stengers, p. 89.
53. Teilhard de Chardin, *The Divine Milieu*, biographical preface, p. 31.
54. As quoted in Ferris, *The Mind's Sky*, p. 92.

Chapter 16 The Winds of Change

1. Schrödinger, p. 136.
2. *The Hollow Men*, 1925.
3. The necessary three ingredients for order to emerge from chaos have been well established on mathematical and physical grounds by the so-called Brussels school of thermodynamics, of which Prigogine is a principal player (Peacocke, p. 428). See also *Order Out of Chaos* by Prigogine and Stengers.
4. The interested reader is referred to James Gleick's best-seller *Chaos: Making a New Science* (1988), and excellent introduction for the lay reader to the fascinating world of nonlinear dynamics.

5. Schrödinger, p. 70.
6. Dyson, *Origins of Life*, p. 3. Max Delbrück, by the way, was the one who apprised his Cal Tech colleague Linus Pauling of Crick and Watson's success in inferring the structure of DNA. Despite having been sworn to secrecy by Watson, upon receiving Watson's letter of triumph, Delbrück immediately informed Pauling that he had been scooped.
7. Dyson, *Origins of LIfe*, p. 5.
8. Shapiro, p. 47.
9. Shapiro, quoting Nobel laureate, H. J. Muller, p. 47.
10. Shapiro, p. 47.
11. Shapiro, p. 48.
12. Shapiro, p. 48.
13. Freeman Dyson, "Gravity is Cool, or Why Our Universe is Hospitable to Life," Oppenheimer Lecture, University of California Berkeley, March 2000.
14. Freeman Dyson, "Gravity is Cool, or Why Our Universe is Hospitable to Life," Oppenheimer Lecture, University of California Berkeley, March 2000.
15. Lane, Chapter 1.
16. Although this quotation is ubiqitous, it is also routinely unattributed. A recent source, but certainly not the original, is the following: entry "bacteria" of Appendix C: Glossary of *Protecting the Frontline in Biodefense Research*, the National Academies Press, 2011. (See www.nap.edu/read/13112/chapter/14.)
17. Dyson, *Origins of Life*, p. 12.
18. Dyson, *Origins of Life*, p. 12.
19. From Wikipedia entry of *mitochondrion* (http://en.wikipedia.org/wiki/Mitochondrion), January 30, 2012.
20. *A Walk Through Time*, walking exhibit, 1600 MYA.
21. Dyson, *Origins of Life*, p. 25.
22. Suggested by biologist Jon Monroe, James Madison University, personal communication.
23. Dyson, *Origins of Life*, p. 13.
24. Dyson, *Origins of Life*, p. 13.
25. Shapiro, p. 50.
26. The five ingredients of the metabolism-first model are adapted from Shapiro, pp. 47-53.
27. Shapiro, p. 52.
28. Lane, Chapter 1.
29. Shapiro, quoting eminent biologist Stuart Kauffman, p. 53.
30. Schrödinger, p. 77.
31. Peacocke, p. 401.
32. Schrödinger, p. 70.
33. Schrödinger, p. 71. And by analogy, what nations desperately need today are entropy policies rather than energy policies. An entropy policy would set terms for the management of the quality of the energy so that sufficient high-quality energy is available now and into the future.
34. Schrödinger, p. 73.
35. Teilhard de Chardin, *The Phenomenon of Man*, p. 27.
36. Oxford University chemist Peter Atkins, *Galileo's Finger*, p. 124.
37. Prigogine and Stengers, p. 175.

38. Schrödinger, p. 73.
39. Kaufman, endnote 8 to Chapter 6, p. 219.
40. Atkins, pp. 125-126.
41. Peacocke, p. 430.
42. In the prologue of *Ever Since Darwin*, Gould boldly asserts, "Evolution is purposeless, nonprogressive, and materialistic." Gould, p. 14.
43. With kind permission of Moose Music, Toronto, from "The Wherefore and the Why" (*Did She Mention My Name*, 1968) by Canadian singer-songwriter Gordon Lightfoot, among whose more than 400 other songs is "The Minstrel of the Dawn." © Moose Music 1968, 1996 (renewed). Used by permission.

Chapter 17 Pale Images

1. Quoted in Peacocke, p. 408, and attributed to D. H. Hubel, "The Brain," *Scientific American Book*, San Francisco: Freeman, 1979.
2. Ferris, *The Mind's Sky*, p. 150.
3. Attributed to Michael Gazzaniga, cognitive neuroscience pioneer and member of the American Council of Bioethics, in Kandel, p. 390.
4. C. G. Jung, "The Concept of the Collective Unconscious," Chapter 4 of *The Portable Jung*, ed. Joseph Campbell, New York: Penguin Books, 1971, p. 60.
5. Reed and English, p. 16ff. Also, an excellent summary of the Dream Helper Ceremony (DHC), entitled "When Hearts Are Joined: My Story of Exploring Consciousness Thought Intuition," can be found at Reed's website: www.creativespirit.net/henryreed. The author has participated six times in the DHC.
6. Adapted from the chapter "The Death Trip," of *The Mind's Sky* by Ferris.
7. Prigogine and Stengers, p. 96.
8. LeShan, *A New Science of the Paranormal*, p. 50.
9. LeShan, *A New Science of the Paranormal*, p. 50.
10. The spiritual implications of flowering consciousness are the themes of Eckhart Tolle's best-sellers *The Power of Now* and *A New Earth*.
11. Peacocke, p. 408.
12. Prigogine and Stengers, p. 131.
13. Peacocke, p. 401.
14. Dyson, "Gravity is Cool," Oppenheimer lecture, UC Berkeley, March 2000.
15. Penrose, p. 433.
16. Peacocke, p. 399.
17. Prigogine's notions on the origins of irreversibility are not universally accepted. One finds both support (e.g., *Physical Review Letters 59*, No. 1, 1987) and critical (e.g. *Philosophy of Science 58*, No. 4, 1991) analyses in the scientific literature.
18. Prigogine and Stengers, Chapter IX in general and p. 276 in particular.
19. Fred R. Shapiro and Joseph Epstein, *The Yale Book of Quotations*, Yale University Press, 2006, p. 228.
20. Dyson, *Disturbing the Universe*, p. 249.
21. John Muir, *Travels in Alaska*, Seattle: CreateSpace, 2011, p. 2.

Epilogue

1. Teilhard's vision, dated January 15, 1918, but almost certainly from February 15 of the same year, is described in his early essay "The Great Monad: A Manuscript Found in a Trench," which is believed by some to be his literary masterpiece. "The Great Monad" was his initial term for the later and more fitting term "noosphere." The essay is bundled with *The Heart of Matter*, 1979, pp. 182-195.
2. For an overview of the impact on environmental science of Teilhard's notion of the noosphere, see *The Biosphere and Noosphere Reader: Global Environment, Society, and Change*, ed. David Pitt and Paul R. Samson, London: Routledge, 1999.
3. E. O. Wilson, "Back from Chaos," *The Atlantic Monthly*, March 1998, p. 44.
4. E. O. Wilson, "Back from Chaos," *The Atlantic Monthly*, March 1998, p. 54.
5. "Alarmingly, we may have already passed tipping points that are irreversible within the time span of our current civilization," from the introduction to the section Climate Change in the 2009 Year Book of the UN Environmental Program.
6. Tolle, audio version of *The New Earth*.
7. Dyson, Witherspoon lecture, 2003.
8. LeShan, *How to Meditate*, p. 185.
9. Teilhard de Chardin, *The Phenomenon of Man*, p. 285.
10. View of the twentieth-century theologian Paul Tillich, as quoted by Edgar N. Jackson in the afterword of *How to Meditate*, p. 285.
11. Ursula King, plenary address, "Spirit of Fire Conference," Chestnut Hill College, 2005. The address is contained (pp. 9-31) with many others in *Rediscovering Teilhard's Fire*, ed. Kathleen Duffy, S. S. J., Saint Joseph's University Press, 2010.
12. Attributed to Teilhard de Chardin by Yungblut, p. 105.
13. Brian, pp. 388-389.
14. See Aldous Huxley's timeless classic *The Perennial Philosophy*.
15. James, p. 329, and Huxley, Chapter 2.
16. Teilhard de Chardin, *The Heart of Matter*, p. 16.
17. Houston, pp. 89-95.
18. Attributed to Jean Houston by Ursula King, plenary address, "Spirit of Fire Conference," Chestnut Hill College, 2005. See Duffy, 2010.
19. Peking, February 1934, "The Evolution of Chastity," in *Toward the Future*, London: Collins, 1975, pp. 86-87. The version quoted is a more popular translation than that given in Teilhard's essay.
20. John Yungblut, personal communication, 1994.

Bibliography

- Atkins, Peter. *Galileo's Finger: The Ten Great Ideas of Science*. Oxford: Oxford University Press, 2003.

- Augustine, Saint. *Augustine Catechism: Enchiridion on Faith Hope and Charity*. Translated by Bruce Harbert. Preface by Boniface Ramsey. Hyde Park: New City Press, 2008.

- Beauregard, Mario, Gary E. Schwartz, Lisa Miller, Larry Dossey, Alexander Moreira-Almeida, Marilyn Schlitz, Rupert Sheldrake, and Charles Tart. "Manifesto for a Post-Materialistic Science." *Explore*. Vol. 10, No. 5, 2014.

- Berry, Thomas. *Dream of the Earth*. New York: Random House, Inc., 1988.

- Bourne, Edmund J. *Global Shift*. Oakland: New Harbinger Publications, 2008.

- Brian, Denis. *Einstein: A Life*. New York: John Wiley & Sons, Inc., 1996.

- Bronowski, Jacob. *The Ascent of Man*. Boston: Little, Brown, & Company, 1973.

- Bryson, Bill. *A Short History of Nearly Everything*. New York: Broadway Books, 2003.

- Campbell, Joseph, and Bill Moyers. *Joseph Campbell and the Power of Myth*. New York: Mystic Fire Video, 1988.

- Cardeña, Etzel. "A Call for an Open, Informed Study of All Aspects of Consciousness." *Frontiers in Human Neuroscience*. Vol. 8, Article 17, 2014.

- Carey, John. *Eyewitness to Science*. Cambridge: Harvard University Press, 1995.

- Carson, Rachel. *Silent Spring*. Boston: Houghton Mifflin, 1962.

- Cernan, Gene, and Donald A. Davis. *Last Man on the Moon: Astronaut Eugene Cernan and America's Race in Space*. 10th Anniversary Ed. New York: St. Martin's Griffin, 1999.

- Conner, James A. *Kepler's Witch*. New York: HarperCollins, 2004.

- Darwin, Charles Robert. *The Descent of Man, and Selection in Relation to Sex*. 1st Ed. London: John Murray, 1871.

- Darwin, Charles Robert. *The Voyage of the H.M.S. Beagle: Journal of Researches into the Natural History and Geology of the Countries Visited during the Voyage of H.M.S. Beagle Round the World, under the Command of Capt. Fitz Roy R.N.* Illustrated Ed. London: John Murray, 1890.

- Darwin, Charles Robert. *On the Origin of Species by Natural Selection*. 6th Ed. London: John Murray, 1876.

- Darwin, Charles Robert. *On the Origin of Species by Natural Selection: The Illustrated Edition*, with general editor David Quammen. New York: Sterling, 2008.

- Davis, Phillip J., and Reuben Hersh. *The Mathematical Experience*. Boston: Houghton Mifflin, 1981.

- De Santillano, Giorgio. *The Crime of Galileo*. Chicago: University of Chicago Press, 1955.

- Dillard, Annie. *For the Time Being*. New York: Alfred A. Knopf, 1999.

- Duffy, Kathleen, ed. *Rediscovering Teilhard's Fire*. Philadelphia: Saint Joseph's University Press, 2010.

- Dyson, Freeman. *Disturbing the Universe*. New York: Basic Books, 1979.

- Dyson, Freeman. *Origins of Life*. Cambridge: Cambridge University Press, 1985.

- Dyson, Freeman. "Gravity Is Cool, or, Why Our Universe Is Hospitable to Life." Oppenheimer Lecture. University of California, Berkeley, March 9, 2000.

- Dyson, Freeman. "The Varieties of Human Experience." Witherspoon Lecture. Center of Theological Inquiry, Princeton, New Jersey, November 6, 2003.

- Einstein, Albert. *Ideas and Opinions*. New York: Wing Books, 1954.

- Einstein, Albert. *Relativity: The Special and the General Theory*. New York: Crown Publishers, Inc., 1961.

- Einstein, Albert. *In His Own Words: Relativity and Out of My Later Years*. New York: Portland House (Random House), 2000.

- Fagg, Lawrence W. *The Becoming of Time: Integrating Physical and Religious Time*. Durham: Duke University Press, 2003.

- Ferris, Timothy. *Coming of Age in the Milky Way*. New York: Harper Perennial, 2003.

- Ferris, Timothy. *The Mind's Sky: Human Intelligence in a Cosmic Context*. New York: Bantam Books, 1992.

- Ferry, Georgina. *Max Perutz and the Secret of Life*. Cold Spring Harbor, NY: Cold Spring Harbor Laboratory Press, 2007.

- Feynman, Richard, Robert Leighton, and Matthew Sands. *Feynman Lectures on Physics (boxed set): The Millennium Edition*. New York: Basic Books (Perseus Books Group), 2011.

- Fletcher, F. D. *Darwin: An Illustrated Life of Charles Darwin 1809-1882*. Lifelines 34. Haverfordwest: Shire Publications, Ltd., 1980.

- Fox, Matthew. *Creation Spirituality*. San Francisco: HarperSanFrancisco, 1991.

- Friedman, Norman. *Bridging Science and Spirit*. St. Louis: Living Lake Books, 1990.

• Freud, Sigmund. *Standard Edition of the Complete Psychological Works of Sigmund Freud*. Vol. 17: 1917-1919. Translated by James Strachey. London: Hogarth, 1955.

• Galilei, Galileo. *Dialogue Concering Two Chief World Systems: Ptolemaic and Copernican*, 1st Ed. Translated by Stillman Drake and Sonja Bargmann. Foreword by Albert Einstein. Berkeley: University of California Press, 1953.

• Galilei, Galileo. *Dialogues Concerning Two New Sciences*. Translated by Henry Crew and Alfonso De Salvio. New York: Cosimo Classics, 2010.

• Geymonat, Ludovico. *Galileo Galilei: A Biography and Inquiry into His Philosophy of Science*. Translated by G. de Scantillana. London: McGraw-Hill, 1965.

• Gleick, James. *Chaos: Making a New Science*. London: Penguin (Viking Adult), 1987.

• Gleick, James. *Isaac Newton*. New York: Vintage, 2003.

• Goethe, Johann Wolfgang. *Zur Farbenlehre*. Edited by Peter Schmidt. Münchner Ausgabe, München, Wien: Carl Hanser Verlag, 2006.

• Gould, Stephen Jay. *Ever since Darwin*. New York: W. W. Norton & Company, Inc., 1977.

• *Great Books of the Western World*. Robert Maynard Hutchins, editor in chief. Vol. 16: *Ptolemy, Copernicus, Kepler*. Chicago: Encyclopaedia Britannica, Inc., 1952.

• *Great Books of the Western World*. Robert Maynard Hutchins, editor in chief. Vol. 34: *Newton, Huygens*. Chicago: Encyclopaedia Britannica, Inc., 1952.

• *Great Books of the Western World*. Robert Maynard Hutchins, editor in chief. Vol. 49: *Darwin*. Chicago: Encyclopaedia Britannica, Inc., 1952.

• Hawking, Stephen. *The Illustrated A Brief History of Time*. New York: Bantam Books, 1996.

• Heisenberg, Werner. *Physics and Philosophy: The Revolution in Modern Science*. New York: Harper & Row, 1958.

• Hey, Tony, and Patrick Walters. *Einstein's Mirror*. Cambridge: Cambridge University Press, 1997.

• Hey, Tony, and Patrick Walters. *The Quantum Universe*. Cambridge: Cambridge University Press, 1987.

• Hofstadter, Douglas R. *Gödel, Escher, Bach: An Eternal Golden Braid*. New York: Basic Books, Inc., 1979.

• Houston, Jean. *Godseed: The Journey of Christ*. Wheaton, Illinois: Quest Books, 1992.

• Huxley, Aldous. *The Perennial Philosophy*. New York: HarperPerennial, 2009.

• Isaacson, Walter. *Einstein: His Life and Universe*. New York: Simon & Schuster, 2007.

• James, William. *The Varieties of Religious Experience*. New York: Touchstone by Simon & Schuster, 1997.

- Jung, Carl G. *Man and His Symbols*. New York: Dell Publishing Company (Mass Market Paperback), 1968.

- Kandel, Eric R. *In Search of Memory: The Emergence of a New Science of Mind*. New York: W. W. Norton & Company, 2006.

- Kant, Immanuel. *Prolegomena to Any Future Metaphysics*. 2nd Ed., and the Letter to Marcus Herz, February 1772. Indianapolis: Hackett Publishing Company, Inc., 1977.

- Kaufman, Gordon D. *God, Mystery, Diversity: Christian Theology in a Pluralistic World*. Minneapolis: Augsburg Fortress Publishers, 1996.

- Keynes, Randall. *Darwin, His Daughter, & Human Evolution*. New York: Riverhead Books, 2001.

- King, Ursula. *Spirit of Fire: The Life and Vision of Teilhard de Chardin*. New York: Orbis, 1996.

- Koestler, Arthur. *The Sleepwalkers: A History of Man's Changing Vision of the Universe*. Harmondsworth: Penguin, 1964.

- Kraft, Christopher. *Flight: My Life in Mission Control*. New York: Penguin Putnam, Inc., 2002.

- Kuhn, Thomas S. *The Copernican Revolution: Planetary Astronomy in the Development of Western Thought*. Cambridge: Harvard University Press, 1957.

- Lane, Nick. *Life Ascending*. New York: W. W. Norton & Company, Inc., 2009.

- Langford, J. J. *Galileo, Science and the Church*. Ann Arbor: University of Michigan Press, 1966.

- Lao Tzu. *Tao Te Ching: The Definitive Edition*. Translation and commentary by Jonathan Star. New York: Tarcher/Putnam/Penguin, 2003.

- LeShan, Lawrence. *How to Meditate: A Guide to Self-Discovery*. New York: Little, Brown and Company, 1974.

- LeShan, Lawrence. *The Medium, the Mystic, and the Physicist*. New York: Helios Press, 2003.

- LeShan, Lawrence. *A New Science of the Paranormal: The Promise of Psychical Research*. Wheaton, Illinois: Quest Books, 2009.

- Liebes, Sidney, Elizabet Sahtouris, and Brian Swimme. *A Walk through Time: From Stardust to Us – The Evolution of Life on Earth*. Hoboken, NJ: John Wiley & Sons, 1998.

- Lindley, David. *Uncertainty: Einstein, Heisenberg, Bohr, and the Struggle for the Soul of Science*. New York: Doubleday, 2007.

- Lyell, Charles. *Principles of Geology*. 1st Ed. Vol. III. London: John Murray, 1833.

- Malthus, T. R. *Essay on the Principle of Population (Oxford World's Classics)*. Edited by Geoffrey Gilbert. Oxford: Oxford University Press, 2008.

- Mitchell, Edgar. *The Way of the Explorer*. Rev. Ed. Franklin Lakes, NJ: New Page Books, 2008.

- Neihardt, John G. *Black Elk Speaks*. Lincoln: University of Nebraska Press (Bison Books), 1988.

- Newman, James R., ed. *The World of Mathematics*. Vols. 1-4. New York: Simon & Schuster, 1956.

- Nietzsche, Friedrich. *The Gay Science*. Translation by Walter Kaufmann. New York: Vintage (Mass Market Paperback), 1974.

- Ohanian, Hans. *Gravitation and Spacetime*. New York: Norton, 1976.

- Paley, William. *Evidences of Christianity*. CreateSpace, 2013.

- Paley, William. *Natural Theology (Oxford World's Classics)*. Oxford: Oxford University Press, 2008.

- Peacocke, Arthur. "Thermodynamics and Life." *Zygon* 19, (1984).

- Peck, M. Scott. *People of the Lie*. New York: Simon & Schuster, 1983.

- Pedersen, O. "Galileo and the Council of Trent: The Galileo Affair Revisited." *Journal of the History of Astronomy* 1, (1983), pp. 1-29.

- Penrose, Roger. *The Emperor's New Mind*. Oxford: Oxford University Press, 1989.

- Pope Francis. *Laudato Sí (On Care for Our Common Home)*. Encyclical Letter. Vatican City: Libreria Editrice Vaticana, 2015.

- Pope John Paul II. Address to an International Symposium for the 350th anniversary of the publication of Galileo's *Dialoghi sui due massimi sistemi del mondo* (1633), May 9, 1983.

- Prigogine, Ilya, and Isabelle Stengers. *Order Out of Chaos: Man's New Dialogue with Nature*. New York: Bantam Books, 1984.

- Radin, Dean. *The Conscious Universe: The Scientific Truth of Psychic Phenomena*. New York: HarperSanFrancisco, 1997.

- Reed, Henry, and Brenda English. *The Intuitive Heart: How to Trust Your Intuition for Guidance and Healing*. Virginia Beach: A.R.E. Press, 2000.

- Rhodes, Richard. *The Making of the Atomic Bomb*. New York: Simon and Schuster, 1988.

- Rowling, J. K. *Harry Potter and the Chamber of Secrets*. New York: Scholastic Press, Arthur R. Levine Books, 1999.

- Russell, Peter. *Waking Up in Time*. Novata, CA: Origin Press, 1998.

- Sagan, Carl. *The Varieties of Scientific Experience*. Edited by Ann Druyan. New York: The Penguin Press, 2006.

- Schrödinger, Erwin. *What Is Life? with Mind and Matter*. Cambridge: Cambridge University Press, 1992.

- Shapiro, Robert. "A Simpler Origin for Life." *Scientific American*, June 2007, pp. 46-53.

- Shea, William. *Biografie: Nikolaus Kopernikus. Spektrum der Wissenschaft*, 2003.

- Simmons, G. F. *Calculus with Analytic Geometry*. New York: McGraw Hill, 1985.

- Smith, Huston. *The Religions of Man*. New York: Harper & Row (Perennial Library), 1965.

- Sobel, Dava. *Galileo's Daughter*. New York: Walker & Company, 1999.

- *Stephen Hawking's Universe: The Big Bang*. Vol. 1, Part II, Thirteen, WNET, Uden Associates, David Filkin Enterprises coproduction in association with BBC-TV; series producer David Filkin. Philip Martin Publisher, PBS Video, 1997 and 2005.

- Tarnas, Richard. "The Great Initiation." *Noetic Sciences Review*, 47 (Winter), 1998.

- Tart, Charles T. *The End of Materialism: How Evidence of the Paranormal is Bringing Science and Spirit Together*. Foreword by Huston Smith and Kendra Smith. Oakland: New Harbinger Publications, Inc., 2009.

- Teilhard de Chardin. *The Human Phenomenon*. Translation by Sarah Appleton-Weber. East Sussex: Sussex Academic Press, 2003.

- Teilhard de Chardin. *The Phenomenon of Man*. New York: Harper & Row (Perennial Library), 1959.

- Teilhard de Chardin. *The Divine Milieu*. New York: Harper & Row (Perennial Library), 1960.

- Teilhard de Chardin. *The Heart of Matter*. New York: Harcourt, Inc., 1978.

- *The Teilhard de Chardin Centenary Exhibition Catalogue*. London, 1983.

- *The World Treasury of Physics, Astronomy, and Mathematics*. Edited by Timothy Ferris. Boston: Little, Brown, and Company, 1991.

- Thorne, Kip S. *Black Holes & Time Warps: Einstein's Outrageous Legacy*. New York: W. W. Norton & Company, 1994.

- Tolle, Eckhart. *The Power of Now: A Guide to Spiritual Enlightenment*. Novata, CA: New World Library, 1999.

- Tolle, Eckhart. *A New Earth: Awakening to Your Life's Purpose*. New York: Penguin Group, 2008.

- Trefil, James, and Robert M. Hazen. *The Sciences: An Integrated Approach*. Preliminary Ed. New York: John Wiley & Sons, Inc., 1994.

- *To the Moon*. NOVA Production, South Burlington, VT: WGBH Educational Foundation, 2000.

- Wall, Steve, and Harvey Arden. *Wisdomkeepers: Meetings with Native American Spiritual Elders*. Hillsboro, OR: Beyond Words Publishing, Inc., 1990.

- Watson, James D. *The Double Helix: A Personal Account of the Discovery of the Structure of DNA*. New York: Atheneum, 1968.

- Wolfson, Richard. *Einstein's Relativity and the Quantum Revolution: Modern Physics for Non-Scientists* (Videorecording). The Great Courses. Course No. 153. Chantilly, Virginia: The Teaching Company, 1997.

Index

About the Author

DAVE PRUETT, a former NASA researcher, is an award-winning computational scientist and emeritus professor of mathematics at James Madison University. He has received honors both for research in computational fluid dynamics and for excellence in teaching. *Reason and Wonder* is his first book for the general public. Dr. Pruett lives in Harrisonburg, Virginia, with his wife, Suzanne Fiederlein, and daughter, Elena.